Kirtley Library
Columbia College
Columbia, Missouri 65216

D1521531

TABLES FOR MICROSCOPIC IDENTIFICATION OF ORE MINERALS

TABLES FOR MICROSCOPIC IDENTIFICATION OF ORE MINERALS

W. UYTENBOGAARDT

Emeritus Professor of Economic Geology,
University of Technology, Delft
Former Professor of Mineralogy and Petrology,
Free University, Amsterdam

E.A.J. BURKE

Senior Lecturer of Mineralogy,
Institute of Earth Sciences,
Free University, Amsterdam

Second revised edition

DOVER PUBLICATIONS, INC.
NEW YORK

KIRTLEY LIBRARY
COLUMBIA COLLEGE
COLUMBIA, MO 65216

Copyright © 1971 by Elsevier Publishing Company, Amsterdam.
All rights reserved under Pan American and International Copyright Conventions.

Published in Canada by General Publishing Company, Ltd., 30 Lesmill Road, Don Mills, Toronto, Ontario.
Published in the United Kingdom by Constable and Company, Ltd., 10 Orange Street, London WC2H 7EG.

This Dover edition, first published in 1985, is an unabridged and slightly corrected republication of the second revised edition (1971) of the work first published in 1951 by the Elsevier Publishing Company, Amsterdam.

Manufactured in the United States of America
Dover Publications, Inc., 31 East 2nd Street, Mineola, N.Y. 11501

Library of Congress Cataloging in Publication Data

Uytenbogaardt, W.
  Tables for microscopic identification of ore minerals.

  Bibliography: p.
  Includes index.
  1. Ores—Identification—Tables.   2. Optical mineralogy—Tables.
  I. Burke, E. A. J.   II. Title.
QE390.U97   1985   549'.1   85-1473
ISBN 0-486-64839-7 (pbk.)

PREFACE

Since the first edition of these tables was published, in 1951, much has changed in the world of mineralogy, not least in the field of the opaque minerals. The International Mineralogical Association was founded and several of its commissions, especially the Commission on Mineral Data, the Commission on New Minerals and Mineral Names and the Commission on Ore Microscopy have largely contributed to our present knowledge about the opaque minerals.

Of the utmost importance is the introduction on an increasingly large scale, during the last two decades, of electron-probe micro-analysis. Several problems hitherto unsolvable, especially those concerning chemical composition of minerals, were unravelled. Many disqualified minerals were redefined, several familiar mineral names disappeared from the list of valid ore minerals, whereas a very large number of new minerals were discovered. It seems as if we are only at the beginning of a new development in this respect.

Besides, the use of quantitative methods in ore microscopy has become more and more standard practice for workers in the field. This was stimulated by several international and regional summer schools on quantitative ore microscopy and greatly facilitated by the manufacturers of equipment for high-precision measurement of reflectance and micro-indentation hardness. As a consequence, the stream of publications with new quantitative data has increased rapidly, and this makes it highly desirable to make a new inventory of all this information, although the authors are the first to admit that there are still serious gaps in the accurate quantitative data available for several groups of ore minerals.

The question may arise if tables like these still are useful in a world of microprobes and computers. The answer must be as affirmative for reflected-light microscopy as it is for transmitted-light microscopy; the reflected-light microscope will always be the first instrument to be used by ore mineralogists, ore geologists and mining engineers for the study of ore samples or products of mineral dressing. Even if microscopic identification of an unknown mineral will not be possible in all cases, determinative tables will always form an indispensable tool in the study of ore minerals.

It is impossible to mention here the names of all those who have contributed to the completion of the present work by sending us mineral specimens, useful quantitative information or reprints of their publications, and to whom we are greatly indebted.

The authors will be grateful for any corrections or additions that are brought to their attention.

Department of Mineralogy and Petrology
Institute of Earth Sciences
Free University, Amsterdam   W. Uytenbogaardt
February 1970   E.A.J. Burke

INTRODUCTION

In compiling the new edition of these tables the aims were the same as for the first edition:(1) to provide as much useful information as possible on all ore minerals that have been sufficiently described; (2) to arrange this information as conveniently as possible. However, the number of valid ore mineral species has increased enormously in twenty years. About 500 minerals are listed in these tables against about 250 in the first edition. For the greater part of these minerals sufficient information is available to permit listing them on the basis of one of their main physical properties. They are tabulated in Part I, Tables I–XII. For a fairly large group of minerals, however, the available data are insufficient. They could, therefore, not be listed in one of the tables of Part I, and the descriptions of these minerals are added as "Notes" to the descriptions of the most closely related minerals, in Part II. Carbon compounds are not included.

The general framework of this new edition has been kept the same as before in order to facilitate its use by those familiar with the first edition. The main properties on which the determination was based in the 1951-edition were "hardness" and reflectance[1]. Both properties still form the main basis for identification.

HARDNESS

Whereas in 1951 the polishing hardness was used because of the lack of quantitative data, nowadays the micro-indentation hardness (*Vickers Hardness Number, VHN*) has been determined for the greater part of the ore minerals. This provides a means of listing these minerals in order of increasing (*VHN*), in spite of the overlapping (*VHN*)-ranges (Part I, Table I). The Talmage hardness is abandoned completely in this edition. For many workers, however, the polishing hardness will still be of great help in the identification of a mineral. The more so since the (*VHN*) for many minerals is not yet known and the light line is usually visible, even in well-polished sections. Therefore, the minerals are also listed (in some cases tentatively) in order of increasing polishing hardness (Part I, Table III).

For a discussion of theoretical and practical problems of micro-indentation hardness measurements reference may be made here to Bückle (1960), Gahm (1969), Pärnamaa (1963) and Uytenbogaardt (1967b). Some practical hints may follow here:

(1) As the (*VHN*) is load-dependent, the applied load should be indicated for all published (*VHN*) data.

(2) If possible, the load should be chosen in such a way that the

---

[1] Following the nomenclature of the International Union of Pure and Applied Physics, and a decision on this subject of the IMA-Commission on Ore Microscopy, "reflectance" is used here instead of "reflectivity", and "bireflectance" instead of "reflection pleochroism."

diagonals of the resulting indentation have a length of 20–25 microns. Too small indentations do not penetrate the harder surface layer. A standard load of 100 g should be tried first.

(3) At least 10 indentations should be made for every determination.

(4) The mean length of the smallest diagonals and the mean length of the largest diagonals should be used for evaluating the micro-indentation hardness range.

(5) The lowering time of the indenter and the time of indentation should be 15 seconds each.

The values obtained from polished sections cannot be compared with values measured on crystal faces.

REFLECTANCE

The reflectance is one of the most important properties of an ore mineral, and its determination should form a part of normal routine work. Although the available data have been obtained with different types of instruments, filters, standards etc., and the results are not always comparable, an attempt was made to list most of the ore minerals in order of increasing minimum reflectance in air (Part I, Table II). The values given in this table are rounded-off, and most of them were determined for green light with some type of photomultiplier equipment. More details are given in the mineral descriptions of Part II. An appeal is made here to all workers in the field to publish, if possible, the (R) values for the four standard wave lengths recommended by the IMA-Commission on Ore Microscopy, viz: 470, 546, 589 and 650 nm (nanometres). If (R) is given for only one wave length, 546 nm should be used. Of course, it would be ideal if complete spectral curves for (R), one for isotropic minerals, two for uniaxial minerals and three for biaxial minerals, were available for all minerals, but this seems to be far away at present. However, a joint effort in this direction should be made by all workers in the field.

For a theoretical discussion of reflectance measurements, measuring errors etc. see Piller and Von Gehlen (1964), Bowie (1967) and Demirsoy (1968).

PARAGENESIS

Apart from the tables already mentioned, which are intended to provide general information quickly, Part I contains nine other tables in which are grouped together those minerals that are closely related or that commonly occur together: the selenides (Table IV), the tellurides (Table V), the Ag-sulphosalts and Ag–Fe-sulphides (Table VI), the Pb–Sb-sulphosalts (Table VII), the Bi-sulphosalts (Table VIII), the Pb–As-sulphosalts (Table IX), the Sn-sulphosalts and Sn-sulphides (Table X), the platinoid minerals (Table XI) and the oxidic manganese minerals (Table XII). The use of these tables is clear: usually the paragenesis of the ore minerals to be studied will point in a certain direction, and the appropriate table can be consulted. The groups of minerals mentioned above are described in detail in the same order in Part II, after the descriptions of the minerals listed in Table III.

DESCRIPTIONS OF ORE MINERALS

In Part II the descriptions of the ore minerals are given in the same order as in Tables III–XII. Their properties are listed as follows.

In the first column the name, chemical composition and crystal system are given. For mineral names the standard English orthography is used. Mineral names not accepted by the IMA-Commission on New Minerals and Mineral Names appear between quotation marks.

In the second column non-quantitative optical information is given: (*1*) colour; (*2*) bireflectance; (*3*) anisotropy; and (*4*) internal reflections. As far as known, the colour or tint of colour is given in comparison with the associated ore minerals. Thus "⟶galena, greyish" means that the tint is greyish as compared with galena when the mineral in question is in contact with galena. If not specified, *all colours are given as for oil immersion,* as in most cases the bireflectance or difference in tint of colour appears more distinct in oil than in air. The same applies to anisotropy. Therefore, oil immersion is recommended here as a standard way for studying polished sections.

In the third column quantitative data are found on reflectance values in air, as far as known. Values given to one decimal place have been determined with some type of photomultiplier equipment at a certain, specified wave length. Values obtained with other methods are indicated in rounded off values as e.g., "about 45%".

In the fourth column information is given on polishing hardness (H) and micro-indentation hardness (VHN). If not specified, the (VHN)-values are determined with a load of 100 g. VHN_{50} means that the applied load was 50 g. $VHN_?$ indicates that the applied load was not indicated by the author in question. If not otherwise indicated, the range in (VHN) was obtained by combining the minimum and maximum values measured by Bowie and Taylor (1958), Burke (in Uytenbogaardt 1967a), Lebedeva (1963), Pärnamaa (1963) and Young and Millman (1964).

In the last column some typical properties or characteristic features are enumerated, such as: crystal shape, twinning, cleavage, intergrowths, occurrence, paragenesis, etc. Completeness was not pursued, since this was considered to be far beyond the present scheme; it has been adequately performed by others, most recently by Ramdohr (1960).

The reference numbers indicate the numbers of the papers cited in the bibliography.

ETCH TESTS

In this edition the etch tests are left out. Since the introduction, on an increasingly large scale, of quantitative methods, there is no more need for the highly inaccurate method of identification with chemical etch tests. Besides, for most of the minerals described during the last two decades no etch tests are given.

CONTACT PRINT METHODS

Contact print methods may give useful additional information about the elements present in a polished section. However, chemical analysis is beyond the scope of these tables. The reader is referred to publications on this subject, such as Štemprok (1953a,b) where further references may be found.

ROTATION PROPERTIES

Rotation properties are not included in these tables since all available information of this kind is given by Cameron (1961), Cameron et al. (1961), and Carpenter and Cameron (1963).

SUPERFLUOUS NAMES

In the first edition a chapter was devoted to superfluous ore mineral names. The number of such names has increased to such an extent that no attempt is made to give a new complete review on this subject. In the alphabetical index the reader will find those names between quotation marks and a reference to the real name of the mineral in question. For complete information on disqualified minerals reference may be made to Chudoba (1967) and preceeding volumes of that standard work.

SUGGESTED LITERATURE

For problems concerning genetic and age relations of ore minerals, their intergrowths, exsolution or replacement phenomena, paragenesis etc. the reader is referred to Bastin et al. (1931), Schwartz (1951), Edwards (1954), Ramdohr (1960), Oelsner (1961b) and Freund (1966).

LIST OF ABBREVIATIONS

Anisotr.	anisotropy, anisotropic	Min.	minimum
Ass.	associations, associated	Mono.	monoclinic
Birefl.	bireflectance	Ortho.	orthorhombic
Chem. comp.	chemical composition	Parag.	paragenesis
Compl.	completely	Ref.	references
D	distinct	Refl.	reflectance
Diff.	differences	S	strong
Esp.	especially	Spec.	specimen(s)
H	polishing hardness	Temp.	temperature
Hex.	hexagonal	Tetr.	tetragonal
Int. refl.	internal reflections	Tricl.	triclinic
I, Isotr.	isotropic	VHN	Vickers Hardness Number
M	moderate	W	weak
Max.	maximum		

CONTENTS

Preface	V
Introduction	VI
List of abbreviations	X
Part I. Determinative Tables	1
Table I. Ore minerals listed in order of increasing minimum micro-indentation hardness (VHN)	3
Table II. Ore minerals listed in order of increasing minimum reflectance in air (R)	12
Table III. Ore minerals listed in order of increasing polishing hardness (H)	17
Table IV. Selenides	22
Table V. Tellurides	24
Table VI. Ag-sulphosalts and Ag–Fe-sulphides	26
Table VII. Pb–Sb-sulphosalts	27
Table VIII. Bi-sulphosalts	28
Table IX. Pb–As-sulphosalts	29
Table X. Sn-sulphosalts and Sn-sulphides	29
Table XI. Platinoid minerals	30
Table XII. Oxidic manganese minerals	31
Part II. Mineral Descriptions	33
Description of minerals from Table III	34
Selenides	215
Tellurides	233
Ag-sulphosalts and Ag–Fe-sulphides	255
Pb–Sb-sulphosalts	271
Bi-sulphosalts	281
Pb–As-sulphosalts	297
Sn-sulphosalts and Sn-sulphides	307
Platinoid minerals	321
Oxidic manganese minerals	337
References	361
Index	425

PART I. DETERMINATIVE TABLES

TABLE I
ORE MINERALS LISTED IN ORDER OF INCREASING MINIMUM MICRO-INDENTATION HARDNESS (VHN)

In this table about 350 ore minerals are listed in order of increasing minimum micro-indentation hardness (Vickers Hardness Number, VHN). The ranges are also given. As the range in (VHN) for many minerals is rather large, the minerals are divided up into groups for which the (VHN) is in between 1 and 100, 100 and 200, and so on. The last group has a (VHN) greater than 1000. Consequently, many minerals appear in more than one group.

At the end of this table is an alphabetical list of 167 ore minerals for which a micro-indentation hardness has not yet been determined.

Mineral	VHN	Mineral	VHN
VHN between 1 and 100		Krennerite	36–130
		Imhofite	38
Lead	4–6	Lorandite	39–57
Sternbergite	5–74	Nagyagite	39–129
Graphite	7–12	Joseite-A	40–87
Molybdenite	8–101	Silver	40–118
Bismuth	9–26	Gold	41–94
Tin	10–15	Stibnite	42–153
Franckeite	13–108	Guanajuatite	42–210
Tungstenite	15–16	Clausthalite	43–74
Acanthite-argentite	20–61	Antimony	45–135
Orpiment	22–58	Realgar	47–60
Berzelianite	22–99	Kongsbergite	48–81
Jalpaite	23–30	Herzenbergite	48–114
Coloradoite	23–35	Copper	48–143
Eucairite	23–94	Proustite-pyrargyrite	50–156
Hessite	24–44	Cinnabar	51–98
Sulphur	24–66	Galena	51–116
Aguilarite	25–35	Greenockite	52–91
Tetradymite	25–75	Mackinawite	52–181
Tellurium	25–87	Volynskite	55–103
Tiemannite	26–39	Klockmannite	57–86
Stephanite	26–124	Arsenic	57–167
Naumannite	27–56	Whitneyite	58
Stromeyerite	27–62	Chalcocite	58–99
Lengenbachite	29–40	Schapbachite	59–91
Joseite-B	29–67	Covellite	59–129
Owyheeite	29–214	Mckinstryite	60
Valleriite	30	Lithiophorite	60–100
Getchellite	30–50	Sylvanite	60–250
Digenite	30–83	Chalcothallite	61–90
Hedleyite	30–89	Melonite	63–156
Paraguanajuatite	30–160	Giessenite	65
Teallite	31–125	Guadalcazarite	66–89
Cylindrite	31–131	Jamesonite	67–126
Tellurobismuthite	32–93	Bismuthinite	67–216
Altaite	34–60	Bornite	68–124
Electrum	34–82	Kobellite	69–173
Kostovite	35–43	Chalcophanite	71–246
Petzite	35–74	Metacinnabarite	73–86
Laitakarite	36–50	Djurleite	74–83
Kermesite	36–99	Livingstonite	74–131

TABLE I (continued)

Mineral	VHN	Mineral	VHN
Cosalite	74–161	Empressite	108–133
Stuetzite	75–90	Polybasite	108–141
Pyrolusite	76–1500	Kitkaite	109–119
Umangite	77–112	Semseyite	109–173
Wehrlite	81	Parkerite	111–142
"Teremkovite"	83–155	Wallisite	113–165
Freieslebenite	85–140	Meneghinite	113–183
Miargyrite	88–130	Koutekite	114–147
Galenobismutite	88–150	Platinum	114–274
"Castaingite"	90–160	Iron	116–288
Boulangerite	90–183	Robinsonite	118–123
Ramsdellite	93–1200	Plagionite	120–165
Geocronite	95–206	Lillianite	120–195
		Gratonite	123–156
VHN between 100 and 200		Zinkenite	123–207
		Baumhauerite	128–182
Molybdenite	8–101	Sphalerite	128–276
Franckeite	13–108	Bonchevite	129–205
Stephanite	26–124	Alabandite	129–266
Owyheeite	29–214	Zinc	130
Paraguanajuatite	30–160	Indium	130–159
Teallite	31–125	Twinnite	131–152
Cylindrite	31–131	Berryite	131–171
Krennerite	36–130	Bournonite	132–213
Nagyagite	39–129	Rickardite	133–167
Silver	40–118	Argyrodite-canfieldite	133–172
Stibnite	42–153	Enargite	133–383
Guanajuatite	42–210	Heteromorphite	137–187
Antimony	45–135	Andorite	140–206
Herzenbergite	48–114	Stannite	140–326
Copper	48–143	Madocite	141–171
Proustite-pyrargyrite	50–156	Eskebornite	141–202
Mackinawite	52–181	Lautite	142–147
Volynskite	55–103	Pearceite	142–164
Galena	56–116	Allargentum	143–157
Arsenic	57–167	Dufrenoysite	145–156
Covellite	59–129	Wurtzite	146–264
Sylvanite	60–250	Lepidocrocite	147–782
Melonite	63–156	Tintinaite	149–157
Jamesonite	67–126	Cd-metacinnabarite	149–161
Bismuthinite	67–216	Seligmannite	149–167
Bornite	68–124	Nuffieldite	149–178
Kobellite	69–173	Jordanite	149–204
Chalcophanite	71–246	Playfairite	150–171
Livingstonite	74–131	Cubanite	150–264
Cosalite	74–161	Zincite	150–318
Pyrolusite	76–1500	Sulvanite	152–165
Umangite	77–112	Dyscrasite	152–178
Freieslebenite	85–140	Quenselite	153–186
Miargyrite	88–130	Veenite	156–172
Galenobismutite	88–150	Emplectite	158–249
"Castaingite"	90–160	Rathite-I	159–163
Boulangerite	90–183	Marrite	161–171
Ramsdellite	93–1200	Benjaminite	161–194
Geocronite	95–206	Wittichenite	161–216
Crookesite	101–141	Aikinite	165–246
Berthierite	102–213	Mawsonite	166–210
Palladium bismuthide	105–125		

TABLE I (continued)

Mineral	VHN	Mineral	VHN
Hutchinsonite	170–171	Betafite	173–815
Stibarsenic	170–202	Chalcopyrite	174–219
Launayite	171–197	Idaite	176–260
Sorbyite	172–186	Cuprite	179–218
Liveingite	173–183	Chalcostibite	183–287
Betafite	173–815	Wairauite	185–329
Chalcopyrite	174–219	Nickel	186–210
Idaite	176–260	Millerite	192–376
Cuprite	179–218	Pentlandite	195–303
Guettardite	180–197	Manganite	195–803
Manjiroite	181	Diaphorite	197–242
Chalcostibite	183–287	Montbrayite	198–228
Wairauite	185–329	Calaverite	198–237
Nickel	186–210	Crednerite	200–357
Millerite	192–376	Cadmoselite	203–222
Sartorite	194–197	Tenorite	203–254
Pentlandite	195–303	Psilomelane	203–813
Manganite	195–803	Luzonite–stibioluzonite	205–397
Diaphorite	197–242	Domeykite	206–250
Montbrayite	198–228	Awaruite	209–420
Calaverite	198–237	"Imgreite"	210–220
Crednerite	200–357	Wulfenite	211–333
		Algodonite	217–255
VHN between 200 and 300		Heazlewoodite	221–321
		Pyrochroite	224–245
Owyheeite	29–214	"Dzhezkazganite"	230
Guanajuatite	42–210	Pyrrhotite	230–390
Sylvanite	60–250	Stannoidite	232–271
Bismuthinite	67–216	Kotulskite	236
Chalcophanite	71–246	Coffinite	236–333
Pyrolusite	76–1500	Columbite-tantalite	240–1021
Ramsdellite	93–1200	Roquesite	241
Geocronite	95–206	Zvyagintsevite	241–318
Berthierite	102–213	Violarite	241–458
Platinum	114–274	Rhodostannite	243–266
Iron	116–288	Cobaltpentlandite	245–363
Zinkenite	123–207	Aurostibite	248–262
Sphalerite	128–276	Argentopyrite	250–252
Bonchevite	129–205	Frohbergite	250–297
Alabandite	129–266	Tetrahedrite-tennantite	251–425
Bournonite	132–213	Freibergite	252–375
Enargite	133–383	Pyrochlore	255–826
Andorite	140–206	Wolframite-series	258–657
Stannite	140–326	Schwazite	262–373
Eskebornite	141–202	Montroseite	266–300
Wurtzite	146–264	Skutterudite-series	268–974
Lepidocrocite	147–782	Hollandite	272–1048
Jordanite	149–204	Braunite	280–1187
Cubanite	150–264	Scheelite	285–464
Zincite	150–318	Indite	293–325
Emplectite	158–249	Renierite	295–425
Wittichenite	161–216	Colusite	296–376
Aikinite	165–246	Osmiridium	297–645
Mawsonite	166–210		
Stibarsenic	170–202		

TABLE I (continued)

Mineral	VHN	Mineral	VHN
VHN between 300 and 400		Sinnerite	357–390
		Maghemite	357–988
Pyrolusite	76–1500	Coronadite	359–840
Ramsdellite	93–1200	Polydymite	362–449
Enargite	133–383	Loellingite	368–1048
Stannite	140–326	Germanite	372–450
Lepidocrocite	147–782	Novakite	387–398
Zincite	150–318	Ferberite	387–418
Betafite	173–815	Brannerite	387–907
Wairauite	185–329		
Millerite	192–376	VHN between 400 and 500	
Pentlandite	195–303		
Manganite	195–803	Pyrolusite	76–1500
Crednerite	200–357	Ramsdellite	93–1200
Psilomelane	203–813	Lepidocrocite	147–782
Luzonite-stibioluzonite	205–397	Betafite	173–815
Awaruite	209–420	Manganite	195–803
Wulfenite	211–333	Psilomelane	203–813
Heazlewoodite	221–321	Awaruite	209–420
Pyrrhotite	230–390	Columbite-tantalite	240–1021
Coffinite	236–333	Violarite	241–458
Columbite-tantalite	240–1021	Tetrahedrite-tennantite	251–425
Zvyagintsevite	241–318	Pyrochlore	255–826
Violarite	241–458	Wolframite-series	258–657
Cobaltpentlandite	245–363	Skutterudite-series	268–974
Tetrahedrite-tennantite	251–425	Hollandite	272–1048
Freibergite	252–375	Braunite	280–1187
Pyrochlore	255–826	Scheelite	285–464
Wolframite-series	258–657	Renierite	295–425
Schwazite	262–373	Osmiridium	297–645
Montroseite	266–300	Niggliite	306–537
Skutterudite-series	268–974	Niccolite	308–533
Hollandite	272–1048	Descloizite	310–491
Braunite	280–1187	Pitchblende	314–803
Scheelite	285–464	Obruchevite	317–412
Indite	293–325	Tyrrellite	336–469
Renierite	295–425	Siegenite	336–580
Colusite	296–376	Microlite	341–916
Osmiridium	297–645	Nsutite	350–1288
Niggliite	306–537	Carrollite	351–566
Niccolite	308–533	Linnaeite	351–566
Descloizite	310–491	Pandaite	353–550
Greigite	312	Maghemite	357–988
Manganosite	314–325	Coronadite	359–840
Pitchblende	314–803	Polydymite	362–449
Obruchevite	317–412	Loellingite	368–1048
Kësterite	320–322	Germanite	372–450
Polyxen	329–397	Ferberite	387–418
Tyrrellite	336–469	Brannerite	387–907
Siegenite	336–580	Gudmundite	402–1221
Microlite	341–916	Penroseite	407–550
Nsutite	350–1288	Breithauptite	412–584
Carrollite	351–566	Safflorite	430–988
Linnaeite	351–566	Idiomorphic villamaninite	440–520
Pandaite	353–550		

TABLE I (continued)

Mineral	VHN	Mineral	VHN
Magnetite	440–1100	Cooperite	505–588
Stibiotantalite	441–607	Murdochite	519–592
Gallite	446–471	Gersdorffite	520–907
Rammelsbergite	459–830	Goethite	525–1010
Ullmannite	460–560	Cryptomelane	525–1048
Hausmannite	466–724	Nodular villamaninite	535–710
Thoreaulite	473–797	Ludwigite	537–1486
Nowackiite	480–500	Jacobsite	575–875
Rijkeboerite	485–498	Anatase	576–623
Hauerite	485–508	Hetaerolite	585–813
Plattnerite	490–642		
Lueshite	490–760	VHN between 600 and 700	
		Pyrolusite	76–1500
VHN between 500 and 600		Ramsdellite	93–1200
Pyrolusite	76–1500	Lepidocrocite	147–782
Ramsdellite	93–1200	Betafite	173–815
Lepidocrocite	147–782	Manganite	195–803
Betafite	173–815	Psilomelane	203–813
Manganite	195–803	Columbite-tantalite	240–1021
Psilomelane	203–813	Pyrochlore	255–826
Columbite-tantalite	240–1021	Wolframite-series	258–657
Pyrochlore	255–826	Skutterudite-series	268–974
Wolframite-series	258–657	Hollandite	272–1048
Skutterudite-series	268–974	Braunite	280–1187
Hollandite	272–1048	Osmiridium	297–645
Braunite	280–1187	Pitchblende	314–803
Osmiridium	297–645	Microlite	341–916
Niggliite	306–537	Nsutite	350–1288
Niccolite	308–533	Maghemite	357–988
Pitchblende	314–803	Coronadite	359–840
Siegenite	336–580	Loellingite	368–1048
Microlite	341–916	Brannerite	387–907
Nsutite	350–1288	Gudmundite	402–1221
Carrollite	351–566	Safflorite	430–988
Linnaeite	352–566	Magnetite	440–1100
Pandaite	353–550	Stibiotantalite	441–607
Maghemite	357–988	Rammelsbergite	459–830
Coronadite	359–840	Hausmannite	466–724
Loellingite	368–1048	Thoreaulite	473–797
Brannerite	387–907	Lueshite	490–760
Gudmundite	402–1221	Plattnerite	490–642
Penroseite	407–550	Ilmenite	501–752
Breithauptite	412–584	Gersdorffite	520–907
Safflorite	430–988	Goethite	525–1010
Idiomorphic villamaninite	440–520	Cryptomelane	525–1048
Magnetite	440–1100	Nodular villamaninite	535–710
Stibiotantalite	441–607	Ludwigite	537–1486
Rammelsbergite	459–830	Jacobsite	575–875
Ullmannite	460–560	Anatase	576–623
Hausmannite	466–724	Hetaerolite	585–813
Thoreaulite	473–797	Maucherite	602–724
Hauerite	485–508	Oregonite	605–635
Plattnerite	490–642	Groutite	613–813
Lueshite	490–760	Uraninite	625–929
Ilmenite	501–752	Calzirtite	626–1035

TABLE I (continued)

Mineral	VHN	Mineral	VHN
Magnesioferrite	627–925	Fergusonite	683–897
Loparite	648–895	Davidite	693–890
Hollingworthite	657–848	Ferroselite	700–933
Tin-tantalite	660	Ilvaite	703–1055
Franklinite	667–847	Vonsenite	707–1003
Bravoite	668–1535	Titanomagnetite	715–734
Pararammelsbergite	673–824	Arsenopyrite	715–1354
Fergusonite	683–897	Nolanite	717–766
Davidite	693–890	Hematite	739–1114
Ferroselite	700–933	Braggite	742–1030
		Woodruffite	744
VHN between 700 and 800		Geikielite	750–930
		Marcasite	762–1561
Pyrolusite	76–1500	Bismutotantalite	764–824
Ramsdellite	93–1200	Wodginite	766–1080
Lepidocrocite	147–782	Formanite	772–870
Betafite	173–815	Trevorite	773
Manganite	195–803	Vaesite	773–856
Psilomelane	203–813	Langisite	780–857
Columbite-tantalite	240–1021	Tapiolite	796–1132
Pyrochlore	255–826	Marokite	800
Skutterudite-series	268–974		
Hollandite	272–1048	VHN between 800 and 900	
Braunite	280–1187		
Pitchblende	314–803	Pyrolusite	76–1500
Microlite	341–916	Ramsdellite	93–1200
Nsutite	350–1288	Betafite	173–815
Maghemite	357–988	Manganite	195–803
Coronadite	359–840	Psilomelane	203–813
Loellingite	368–1048	Columbite–tantalite	240–1021
Brannerite	387–907	Pyrochlore	255–826
Gudmundite	402–1221	Skutterudite-series	268–974
Safflorite	430–988	Hollandite	272–1048
Magnetite	440–1100	Braunite	280–1187
Rammelsbergite	459–830	Pitchblende	314–803
Hausmannite	466–724	Microlite	341–916
Thoreaulite	473–797	Nsutite	350–1288
Lueshite	490–760	Maghemite	357–988
Ilmenite	501–752	Coronadite	359–840
Gersdorffite	520–907	Loellingite	368–1048
Goethite	525–1010	Brannerite	387–907
Cryptomelane	525–1048	Gudmundite	402–1221
Nodular villamaninite	535–710	Safflorite	430–988
Ludwigite	537–1486	Magnetite	440–1100
Jacobsite	575–875	Rammelsbergite	459–830
Hetaerolite	585–813	Gersdorffite	520–907
Maucherite	602–724	Goethite	525–1010
Groutite	613–813	Cryptomelane	525–1048
Uraninite	625–929	Ludwigite	537–1486
Calzirtite	626–1035	Jacobsite	575–875
Magnesioferrite	627–925	Hetaerolite	585–813
Loparite	648–895	Groutite	613–813
Hollingworthite	657–848	Uraninite	625–929
Franklinite	667–847	Calzirtite	626–1035
Bravoite	668–1535	Magnesioferrite	627–925
Pararammelsbergite	673–824	Loparite	648–895

TABLE I (continued)

Mineral	VHN	Mineral	VHN
Hollingworthite	657–848	Vonsenite	707–1003
Franklinite	667–847	Arsenopyrite	715–1354
Bravoite	668–1535	Hematite	739–1114
Pararammelsbergite	673–824	Braggite	742–1030
Fergusonite	683–897	Geikielite	750–930
Davidite	693–890	Marcasite	762–1561
Ferroselite	700–933	Wodginite	766–1080
Ilvaite	703–1055	Tapiolite	796–1132
Vonsenite	707–1003	Nb-rutile	803–1180
Arsenopyrite	715–1354	Cassiterite	811–1532
Hematite	739–1114	Glaucodot	841–1277
Braggite	742–1030	Ixiolite	860–947
Geikielite	750–930	Spinel	861–1650
Marcasite	762–1561	Bixbyite	882–1168
Bismutotantalite	764–824	Ta-rutile	911–1239
Wodginite	766–1080	Pyrite	913–2056
Formanite	772–870	Thorianite	920–1235
Vaesite	773–856	Perovskite	925–1131
Langisite	780–857	Rutile	933–1280
Tapiolite	796–1132	Cobaltite	948–1367
Marokite	800	Cattierite	953–1113
Nb-rutile	803–1180	Sperrylite	960–1277
Cassiterite	811–1532	Irarsite	976
Gaudefroyite	840		
Magnetoplumbite	841–868	VHN higher than 1000	
Glaucodot	841–1277	Pyrolusite	76–1500
Ixiolite	860–947	Ramsdellite	93–1200
Spinel	861–1650	Columbite-tantalite	240–1021
Bixbyite	882–1168	Hollandite	272–1048
		Braunite	280–1187
VHN between 900 and 1000		Nsutite	350–1288
Pyrolusite	76–1500	Loellingite	368–1048
Ramsdellite	93–1200	Gudmundite	402–1221
Columbite-tantalite	240–1021	Magnetite	440–1100
Skutterudite-series	268–974	Goethite	525–1010
Hollandite	272–1048	Cryptomelane	525–1048
Braunite	280–1187	Ludwigite	537–1486
Microlite	341–916	Calzirtite	626–1035
Nsutite	350–1288	Bravoite	668–1535
Maghemite	357–988	Ilvaite	703–1055
Loellingite	368–1048	Vonsenite	707–1003
Brannerite	387–907	Arsenopyrite	715–1354
Gudmundite	402–1221	Hematite	739–1114
Safflorite	430–988	Braggite	742–1030
Magnetite	440–1100	Marcasite	762–1561
Gersdorffite	520–907	Wodginite	766–1080
Goethite	525–1010	Tapiolite	796–1132
Cryptomelane	525–1048	Nb-rutile	803–1180
Ludwigite	537–1486	Cassiterite	811–1532
Uraninite	625–929	Glaucodot	841–1277
Calzirtite	626–1035	Spinel	861–1650
Magnesioferrite	627–925	Bixbyite	882–1168
Bravoite	668–1535	Ta-rutile	911–1239
Ferroselite	700–933	Pyrite	913–2056
Ilvaite	703–1055	Thorianite	920–1235

TABLE I (continued)

Mineral	VHN	Mineral	VHN
Perovskite	925–1131	Magnesiochromite	1200–1385
Rutile	933–1280	Nigerite	1206–1561
Cobaltite	948–1367	Laurite	1393–2167
Cattierite	953–1113	Hercynite	1402–1561
Sperrylite	960–1277	Karelianite	1790
Staringite	1033–1187	Gahnite	1910–2420
Chromite	1036–2000	Eskolaite	2077–3200
Högbomite	1048–1214		

Ore minerals for which no VHN data are available

Achavalite	Feitknechtite	Minium
Akaganeite	Ferroplatinum	Modderite
Allopalladium	Freboldite	Mohsite
Anilite	Freudenbergite	Moncheite
Annivite	Frieseite	Monteponite
Antimonial silver	Froodite	Moschellandsbergite
Aramayoite	Fülöppite	Neyite
Argyropyrite	Galaxite	Onofrite
Arsenargentite	Geversite	Orcelite
Arsenolamprite	Goldfieldite	Osmium
Arsenopalladinite	Grimaldiite	Ottemannite
Arsensulvanite	Guyanaite	Palladium
Auricuprid	Häggite	Parajamesonite
Aurorite	Hastite	Paramontroseite
Aurosmiridium	Hatchite	Paratenorite
Avicennite	Hauchecornite	Partridgeite
Berndtite	Hawleyite	Patronite
Betekhtinite	Heterogenite	Pavonite
Billingsleyite	Hocartite	Paxite
Birnessite	Horobetsuite	Platiniridium
Blaubleibender covellite	Horsfordite	Platynite
Bornhardtite	Huntilite	Plumboferrite
Bracewellite	Hydrohetaerolite	Porpezite
Briartite	Ikunolite	Potarite
Bursaite	Iridium	Priderite
Cafetite	Iridosmium	Pseudobrookite
Cannizzarite	Jaipurite	Pyrophanite
Cesarolite	Kallilite	Pyrostilpnite
Chaoite	Kennedyite	Raguinite
Chilenite	"Klaprothite"	Rancieite
Cohenite	Koppite	Redledgeite
Corynite	Kullerudite	Rhodite
Coulsonite	Landauite	Rutheniridosmium
Crichtonite	Latrappite	Sakuraiite
Csiklovaite	Lithargite	Samsonite
Cuprobismutite	Magnocolumbite	Sanmartinite
Dadsonite	Mäkinenite	Schirmerite
Delafossite	Maldonite	Schreibersite
Derbylite	Massicotite	Sederholmite
Dienerite	Mcconnellite	Selenium
Dimorphite	Melanostibite	Selentellurium
Doloresite	Merenskyite	Senaite
Donathite	Metastibnite	Shandite
Dzhalindite	Michenerite	Smithite

Smythite	Tantalobetafite	Ustarasite
Stannite jaune	Tantalo-obruchevite	Vrbaite
Stannoenargite	Titanobetafite	α-vredenburgite
Stannopalladinite	Titanomaghemite	Vulcanite
Sterryite	Titano-obruchevite	Vysotskite
Stibiocolumbite	Todorokite	Weissite
Stibiopalladinite	Trechmannite	Westgrenite
Stilleite	Trogtalite	Wilkmanite
Sukulaite	Trüstedtite	Willyamite
Talnakhite	Ulvöspinel	Wuestite
Tantalcarbid	Uranothorite	Xanthoconite
		Zirconolite

TABLE II
ORE MINERALS LISTED IN ORDER OF INCREASING MINIMUM REFLECTANCE IN AIR (R)

In this table about 400 ore minerals are listed in order of increasing minimum reflectance in air (R). The ranges, partly due to bireflectance, partly to varying composition are also given. The minerals are divided up into groups for which (R) is in between 1 and 10, 10 and 20 and so on. If necessary, a mineral is placed in more than one group. All values are rounded-off in order to stress that slight deviations may occur, depending on differences in chemical composition or on apparatus, standards or the polishing methods used. For more complete information the reader is referred to the descriptions in Part II.

At the end of this table is an alphabetical list of 80 ore minerals for which reflectance data are not yet available.

Mineral	R	Mineral	R
Reflectance between 1 and 10%		Manganosite	14
		Wodginite	14-15
Graphite	5-20	Valleriite	14-20
Spinel-series	7-8	Blaubleibender covellite	14-25
Coffinite	7-10	Calzirtite	15
Covellite	7-24	Donathite	15
Nigerite	7.5	Formanite	15
Dzhalindite	8	Perovskite-group	15
Ilvaite	8-9	Pseudobrookite	15
Ludwigite	8-10	Sukulaite	15
Ramsdellite	9-41	Thorianite	15
		Tin-tantalite	15-16
Reflectance between 10 and 20%		Columbite-tantalite	15-18
Graphite	5-20	Descloizite	15-18
Covellite	7-24	Pyrochroite	15-18
Ramsdellite	9-41	Tapiolite	15-18
Högbomite	10	Davidite	15-20
Scheelite	10	Freudenbergite	15-20
Zincite	10	Goethite	15-20
Gaudefroyite	10-13	Hydrohetaerolite	15-20
Pitchblende	10-15	Plumboferrite	15-20
Sulphur	10-15	Wolframite-series	15-20
Vonsenite	10-15	Manganite	15-21
Lithiophorite	10-20	Lepidocrocite	15-25
Heterogenite	10-25	"Dzhezkazganite"	15-30
Aurorite	10-27	Psilomelane	15-30
Chalcophanite	10-27	Thoreaulite	16-18
Cassiterite	11-13	Wulfenite	16-18
Chromite	11-13	Marokite	16-19
Fergusonite	11-14	Ilmenite	16-20
Staringite	12-14	Brannerite	17
Geikielite	12-15	Murdochite	17
Rancieite	12-15	Titanomagnetite	17
Pyrochlore-series	12-16	Uraninite	17
Umangite	12-16	Plattnerite	17-19
Groutite	12-20	Sphalerite	17-20
Klockmannite	12-36	Wurtzite	17-20
Ixiolite	13-14	Hausmannite	17-21
Hetaerolite	13-18	"Rhombomagnojacobsite"	17-21

TABLE II (continued)

Mineral	R	Mineral	R
Magnesioferrite	17.5	Stannite jaune	23–28
Franklinite	18	Mawsonite	24–25
Karelianite	18	Stibioluzonite	24–26
Ta-rutile	18–20	Stannoidite	24–27
α-vredenburgite	18–20	Crednerite	24–35
Quenselite	18–21	Eskebornite	24–35
Bornite	18.5	Alabandite	25
Jacobsite	19–20	Berndtite	25
Nb-rutile	19–20	Birnessite	25
Eskolaite	19–21	Coulsonite	25
		Hauerite	25
Reflectance between 20 and 30%		Kermesite	25
Covellite	7–24	Metacinnabarite	25
Ramsdellite	9–41	Orpiment	25
Heterogenite	10–25	Rickardite	25
Aurorite	10–27	Këserite	25–26
Chalcophanite	10–27	Luzonite	25–28
Klockmannite	12–36	Proustite	25–28
Blaubleibender covellite	14–25	Cinnabar	25–29
Manganite	15–21	Enargite	25–29
Lepidocrocite	15–25	Cuprite	25–30
"Dzhezkazganite"	15–30	Stephanite	25–30
Psilomelane	15–30	Villamaninite	25–30
Hausmannite	17–21	Xanthoconite-pyrostilpnite	25–31
"Rhombomagnojacobsite"	17–21	Raguinite	25–32
Quenselite	18–21	Selenium	25–35
Eskolaite	19–21	Niggliite	25–65
Anatase	20	Maghemite	26
Bismutotantalite	20	Woodruffite	26
Cadmoselite	20	Getchellite	26–27
Greenockite	20	Stannite	26–29
Lithargite	20	Hematite	26–30
Massicotite	20	Stromeyerite	26–30
Minium	20	Hollandite	26–32
Stibiotantalite	20	Cryptomelane	27
Realgar	20–21	Tennantite	27–31
Braunite	20–23	Coronadite	27–32
Todorokite	20–23	Franckeite	27–34
Rutile	20–24	Pyrolusite	27–47
Delafossite	20–25	Briartite	28
Tenorite	20–25	Cesarolite	28
Germanite	20–33	Rhodostannite	28
Magnetite	21	Samsonite	28
Trevorite	21	Colusite	28–30
Digenite	21–24	Cylindrite	28–31
Idaite	21–31	Imhofite	28–31
Argyrodite-canfieldite	22	Pyrargyrite	28–31
Gallite	22	Berzelianite	28.5
Sakuraiite	22	Indite	29
Roquesite	22–23	Sulvanite	29–31
Magnetoplumbite	22–24	Sinnerite	29–32
Molybdenite	22–42		
Mackinawite	22–45	Reflectance between 30 and 40%	
Bixbyite	23	Ramsdellite	9–41
Hocartite	23–24	Klockmannite	12–36
Renierite	23–26	Germanite	20–33

TABLE II (continued)

Mineral	R	Mineral	R
Idaite	21–31	Liveingite	34–36
Molybdenite	22–42	Lengenbachite	34–37
Mackinawite	22–45	Plagionite	34–38
Crednerite	24–35	Rathite-I	34–38
Eskebornite	24–35	Baumhauerite	34–39
Xanthoconite-pyrostilpnite	25–31	Pyrrhotite	34–39
		Empressite	34–50
Raguinite	25–32	Smythite	34.5
Selenium	25–35	Aguilarite	35
Niggliite	25–65	Crookesite	35
Hollandite	26–32	Cubic cubanite	35
Tennantite	27–31	Freieslebenite	35
Coronadite	27–32	Giessenite	35
Franckeite	27–34	Naumannite	35
Pyrolusite	27–47	Penroseite	35
Cylindrite	28–31	Schirmerite	35
Imhofite	28–31	Bournonite	35–39
Pyrargyrite	28–31	Sartorite	35–39
Sulvanite	29–31	Dadsonite	35–40
Sinnerite	29–32	Livingstonite	35–40
Andorite	30	Wittichenite	35–40
Chalcocite	30	Guettardite	35–42
Djurleite	30	Tyrrellite	35–45
Eucairite	30	Horobetsuite	35–50
Freibergite	30	Vulcanite	35–60
Jalpaite	30	"Castaingite"	36–39
Nowackiite	30	Sterryite	36–39
Ottemannite	30	Playfairite	36–40
Pearceite	30	Dufrenoysite	36–41
Schwazite	30	Jamesonite	36–41
Stilleite	30	Emplectite	36–42
Tiemannite	30	Seligmannite	36–42
Weissite	30	Tintinaite	36–43
Acanthite-argentite	30–31	Chalcostibite	36–44
Hutchinsonite	30–31	Petzite	37
Vrbaite	30–33	Stuetzite	37–39
Aramayoite	30–35	Heteromorphite	37–41
Fülöppite	30–35	Boulangerite	37–42
Mckinstryite	30–35	Sorbyite	37–43
Miargyrite	30–35	Twinnite	37–43
Berthierite	30–40	Launayite	37–44
Nsutite	30–40	Kobellite	37–45
Chalcothallite	30.5	Breithauptite	37–48
Vaesite	31	Diaphorite	38
Lorandite	31–33	Wallisite	38
Tetrahedrite	31–33	Jordanite	38–40
Marrite	31–34	Geocronite	38–42
Stibnite	31–47	Veenite	38–43
Bravoite	31–52	Zinkenite	38–43
Goldfieldite	32	Bismuthinite	38–50
Lautite	32	Cooperite	39
Cattierite	33	Nagyagite	39
Gratonite	33–34	Talnakhite	39
"Klaprothite"	33–40	Cubanite	39–40
Coloradoite	34	Hessite	39–41
Braggite	34–36	Nuffieldite	39–45
		Aikinite	39–46

TABLE II (continued)

Mineral	R
Reflectance between 40 and 50%	
Ramsdellite	9–41
Molybdenite	22–42
Mackinawite	22–45
Niggliite	25–65
Pyrolusite	27–47
Stibnite	31–47
Bravoite	31–52
Guettardite	35–42
Tyrrellite	35–45
Horobetsuite	35–50
Vulcanite	35–60
Dufrenoysite	36–41
Jamesonite	36–41
Emplectite	36–42
Seligmannite	36–42
Tintinaite	36–43
Chalcostibite	36–44
Heteromorphite	37–41
Boulangerite	37–42
Sorbyite	37–43
Twinnite	37–43
Launayite	37–44
Kobellite	37–45
Breithauptite	37–48
Geocronite	38–42
Veenite	38–43
Zinkenite	38–43
Bismuthinite	38–50
Hessite	39–41
Nuffieldite	39–45
Aikinite	39–46
Chaoite	40
Hauchecornite	40
Owyheeite	40
Paxite	40
Platynite	40
Robinsonite	40
Semseyite	40
Meneghinite	40–46
Teallite	40–47
Hollingworthite	40–53
Benjaminite	41–43
Gudmundite	41–57
Galenobismutite	42
Madocite	42
Pavonite	42
Ustarasite	42
Violarite	42
Berryite	42–43
Laurite	42–43
Herzenbergite	42–44
Koutekite	42–45
Carrollite	43
Copper	43
Cosalite	43
Parkerite	43–48

Mineral	R
Tellurium	43–54
Galena	44
Pentlandite	44
"Teremkovite"	44
Schapbachite	44–45
Orcelite	44–48
Chalcopyrite	44–49
Argentopyrite	45
Argyropyrite	45
Arsenic	45
Arsenolamprite	45
Betekhtinite	45
Bursaite	45
Csiklovaite	45
Frohbergite	45
Lillianite	45
Paraguanajuatite	45
Siegenite	45
Sternbergite	45
Ullmannite	45
Vysotskite	45
Ferroselite	45–50
Glaucodot	45–50
Linnaeite	45–50
Bonchevite	46
Polydymite	46
Langisite	46–47
Domeykite	47–49
Marcasite	47–53
Gersdorffite	47–54
Irarsite	48
Maucherite	48
Hedleyite	48–51
Oregonite	49–50
Joseite-A	49–53
Niccolite	49–53
Clausthalite	49.5
Reflectance between 50 and 60%	
Niggliite	25–65
Bravoite	31–52
Vulcanite	35–60
Hollingworthite	40–53
Gudmundite	41–57
Tellurium	43–54
Marcasite	47–53
Gersdorffite	47–54
Hedleyite	48–51
Joseite-A	49–53
Niccolite	49–53
Allopalladium	50
Froodite	50
Laitakarite	50
Palladium bismuthide	50
Tungstenite	50
Tetradymite	50–52
Sylvanite	50–58

TABLE II (continued)

Mineral	R	Mineral	R
Maldonite	50-60	Osmium	61-63
Arsenopyrite	51-52	Dyscrasite	61-65
Millerite	51-56	Algodonite	61.5
Joseite-B	51-58	Nickel	63
"Imgreite"	52	Whitneyite	63
Heazlewoodite	52.5	Montbrayite	63-64
Skutterudite	53	Zvyagintsevite	63-64
Moncheite	53-59	Merenskyite	63-65
Cobaltpentlandite	54	Tellurobismuthite	64
Pyrite	54	Calaverite	64-65
Volynskite	54	Geversite	65
Wairauite	54	Iron	65
Stibiopalladinite	54-55	Gold	65-72
Guanajuatite	55	Iridosmium	65-72
Loellingite	55		
Wehrlite	55	Reflectance between 70 and 80%	
Sperrylite	55-56	Gold	65-72
Kostovite	55-60	Iridosmium	65-72
Safflorite	55-60	Palladium	70
Cobaltite	56	Platinum	70
Michenerite	56	Altaite	70-71
Kitkaite	56-58	Allargentum	70-75
Pararammelsbergite	56-58	Antimony	70-75
Melonite	57-61	Krennerite	72
Kotulskite	59-65	Electrum	72-89
Reflectance between 60 and 70%		Reflectance between 80 and 90%	
Niggliite	25-65	Electrum	72-89
Melonite	57-61	Osmiridium	80
Kotulskite	59-65	Iridium	85
Potarite	60	Antimonial silver	85-90
Rammelsbergite	60		
Awaruite	60-61	Reflectance between 90 and 100%	
Bismuth	60-65	Indium	90-95
Lead	60-65	Silver	90-95
Aurostibite	61		

Ore minerals for which no reflectance data are available

Achavalite	Doloresite	Mcconnellite	Sanmartinite
Akaganeite	Feitknechtite	Melanostibite	Sederholmite
Anilite	Freboldite	Metastibnite	Schreibersite
Arsenargentite	Frieseite	Modderite	Selentellurium
Arsenopalladinite	Greigite	Mohsite	Senaite
Arsensulvanite	Grimaldiite	Monteponite	Shandite
Auricuprid	Guadalcazarite	Montesite	Smithite
Aurosmiridium	Guyanaite	Montroseite	"Stannite (?) III"
Avicennite	Häggite	Neyite	"Stannite (?) IV"
Billingsleyite	Hastite	Nolanite	Stannopalladinite
Bornhardtite	Hatchite	Novakite	Stibarsenic
Bracewellite	Hawleyite	Onofrite	Stibiocolumbite
Cafetite	Horsfordite	Parajamesonite	Tin
Cannizzarite	Jaipurite	Paramontroseite	Trechmannite
Cohenite	Kennedyite	Paratenorite	Trogtalite
Crichtonite	Kullerudite	Partridgeite	Trüstedtite
Cuprobismutite	Landauite	Patronite	Ulvöspinel
Derbylite	Magnesiocolumbite	Priderite	Wilkmanite
Dienerite	Mäkinenite	Pyrophanite	Wuestite
Dimorphite	Manjiroite	Redledgeite	Zirconolite

TABLE III
ORE MINERALS LISTED IN ORDER OF INCREASING POLISHING HARDNESS (H)

In this table the ore minerals are listed in order of increasing polishing hardness (H), which can be defined as follows: when increasing the distance between specimen and objective, a bright line of light is seen to move from the harder towards the softer mineral; when decreasing the distance the reverse takes place. Slight deviations from the sequence may occur in view of the large number of new minerals on which there is insufficient information. In some places the reader is referred to special groups of minerals, which are listed separately for convenience in the following tables: selenides (Table IV), Ag-sulphosalts and Ag-Fe-sulphides (Table VI), Pb–Sb-sulphosalts (Table VII), Bi-sulphosalts (Table VIII), Pb–As-sulphosalts (Table IX), and Sn-sulphosalts and Sn-sulphides (Table X).

For some other groups of related minerals the polishing hardness differs so widely that no reference to these minerals is made in Table III. These minerals are also listed separately: tellurides (Table V), platinoid minerals (Table XI), and oxidic manganese minerals (Table XII). The paragenesis of the material studied will usually provide sufficient indication as to the necessity of consulting Table V, XI or XII. The reflectance values given in this table are rounded-off (cf. Table II). More detailed information is given in the descriptions of Part II.

The sequence of the minerals in Table III is the same as that of the mineral descriptions in Part II (p.34–213).

Mineral	Page	Composition	Anisotropy	Reflectance%
Acanthite–argentite	34	Ag_2S	D–I	30–31
Tin	34	Sn	W	?
Bismuth	36	Bi	D–S	60–65
Lead	36	Pb	I	60–65
Jalpaite	38	Ag_3CuS_2	D	30
Patronite	38	VS_4	S	?
SELENIDES	215			
Sulphur	40	S	D	10–15
Realgar	40	AsS	S	20–21
Orpiment	40	As_2S_3	S	25
Getchellite	42	$AsSbS_3$	W	26–27
Kermesite	42	Sb_2S_2O	S	25
Stibnite	42	Sb_2S_3	S	31–47
Metastibnite	44	Sb_2S_3	D	?
Mckinstryite	44	$Cu_{0.8}Ag_{1.2}S$	S	30–35
Imhofite	44	Tl–Cu–As–S	S	28–31
Lorandite	44	$TlAsS_2$	S	31–33
Raguinite	46	$TlFeS_2$	S	25–32
Chalcothallite	46	Cu_3TlS_2	D	30.5
Vrbaite	46	$Tl_4Hg_3Sb_2As_8S_{20}$	D	30–33
Berthierite	48	$FeSb_2S_4$	S	30–40
Jamesonite	48	$4PbS \cdot FeS \cdot 3Sb_2S_3$	S	36–41
Meneghinite	50	$26PbS \cdot Cu_2S \cdot 7Sb_2S_3$	S	40–46
Minium	50	Pb_3O_4	?	20
Massicotite	50	β-PbO	?	20
Lithargite	50	α-PbO	?	20

TABLE III (continued)

Mineral	Page	Composition	Anisotropy	Reflectance %
Ag-SULPHOSALTS	255			
Horobetsuite	52	$(Bi,Sb)_2S_3$	S	35–50
Stromeyerite	52	$CuAgS$	S	26–30
Pb–Sb-SULPHOSALTS	271			
Covellite	54	CuS	S	7–24
Blaubleibender covellite	54	$Cu_{1+x}S$	S	14–25
Idaite	56	Cu_3FeS_4	S	21–31
Chalcocite	58	Cu_2S	W–D	30
Djurleite	60	$Cu_{1.96}S$	W–D	30
Digenite	60	$Cu_{1.8}S$	I	21–24
Parkerite	62	$Ni_3(Bi,Pb)_2S_2$	S	43–48
Shandite	62	$Ni_3Pb_2S_2$	S	?
GALENA	64	PbS	I	44
Bi-SULPHOSALTS	281			
Bismuthinite	66	Bi_2S_3	S	38–50
Sn-SULPHOSALTS (partly)	307			
Bournonite	68	$2PbS.Cu_2S.Sb_2S_3$	W–D	35–39
Lautite	68	$CuAsS$	D	32
Sinnerite	68	$Cu_{1.4}As_{0.9}S_{2.1}$	D–S	29–32
Pb-As-SULPHOSALTS	297			
Gold	70	Au	I	65–72
Electrum	72	(Au,Ag)	I	72–89
Aurostibite	72	$AuSb_2$	I	61
Maldonite	72	Au_2Bi	I	50–60
Silver	74	Ag	I	90–95
Zinc	74	Zn	?	?
Antimony	76	Sb	D	70–75
Onofrite	76	$Hg(S,Se)$	I–W	?
Metacinnabarite	78	HgS	I–W	25
Guadalcazarite	78	$(Hg,Zn)(S,Se)$	I	?
Cinnabar	78	HgS	D	25–29
Livingstonite	80	$HgSb_4S_8$	S	35–40
Antimonial silver	80	(Ag,Sb)	I–W	85–90
Allargentum	82	Ag_6Sb	W–D	70–75
Dyscrasite	84	Ag_3Sb	W–D	61–65
Stibarsenic	84	$AsSb$	D	?
Arsenic	84	As	D	45
Arsenolamprite	86	As	W–S	45
Chalcostibite	86	$Cu_2S.Sb_2S_3$	D	36–44
"Dzhezkazganite"	86	$CuReS_4$ (?)	I	15–30
Bornite	88	Cu_5FeS_4	I	18.5
Mawsonite	90	$Cu_{2+x}Sn_{1-x}FeS_4$	S	24–25
Renierite	90	$Cu_{2.5+x}Ge_{0.5-x}FeS_4$	D–S	23–26
Briartite	90	$Cu_2(Fe,Zn)GeS_4$	W–S	28
CHALCOPYRITE	92	$CuFeS_2$	W	44–49
Talnakhite	94	$CuFeS_2$	I	39
Betekhtinite	94	$Pb_2(Cu,Fe)_{21}S_{15}$	S	45
Cubanite	96	$CuFe_2S_3$	S	39–40
Cubic cubanite	96	$CuFe_2S_3$	I	35

TABLE III (continued)

Mineral	Page	Composition	Anisotropy	Reflectance %
Valleriite	98	$[CuFeS_2] \cdot [(Mg,Al,Fe)(OH)_2]$	S	14–20
Roquesite	98	$CuInS_2$	W	22–23
Sulvanite	100	Cu_3VS_4	I	29–31
Arsensulvanite	100	Cu_3AsS_4	I	?
Doloresite	100	$V_3O_4(OH)_4$	S	?
Montroseite	100	$(V,Fe)OOH$	S	?
Tungstenite	102	WS_2	S	50
Molybdenite	102	MoS_2	S	22–42
"Castaingite"	102	$CuMo_2S_{5-x}$	S	36–39
Graphite	104	C	S	5–20
Chaoite	104	C	?	40
Indium	104	In	W	90–95
Germanite	106	Cu_6FeGeS_8	I	20–33
Goldfieldite	106	$Cu_3(Te,Sb)S_4$	I	32
Freibergite	106	Ag-tetrahedrite	I	30
Schwazite	106	Hg-tetrahedrite	I	30
Tetrahedrite	108	$Cu_3SbS_{3.25}$	I	31–33
Tennantite	108	$Cu_3AsS_{3.25}$	I	27–31
Colusite	108	$Cu_3(As,Sn,V,Fe,Sb)S_4$	I	28–30
Luzonite	110	Cu_3AsS_4	S	25–28
Stibioluzonite	110	Cu_3SbS_4	S	24–26
Enargite	110	Cu_3AsS_4	S	25–29
Gallite	112	$CuGaS_2$	D	22
Novakite	112	$(Cu,Ag)_4As_3$	D	?
Paxite	112	Cu_2As_3	S	40
Koutekite	112	Cu_5As_2	S	42–45
Domeykite	114	Cu_3As	I-D	47–49
Algodonite	114	$Cu_{6-7}As$	S	61.5
Whitneyite	114	(Cu,As)	I	63
Copper	116	Cu	I	43
Delafossite	116	$CuFeO_2$	D-S	20–25
Tenorite	118	CuO	S	20–25
Paratenorite	118	CuO	S	?
Cuprite	118	Cu_2O	S-I	25–30
Sn-SULPHOSALTS (stannite-group)	307			
Nickel	122	Ni	I	63
Heazlewoodite	122	Ni_3S_2	S	52.5
Millerite	122	NiS	S	51–56
Hauchecornite	122	$(Ni,Co)_9(Bi,Sb)_2S_8$	D-S	40
Alabandite	124	MnS	I	25
Greenockite	124	CdS	I	20
Sphalerite	126	$(Zn,Fe)S$	I	17–20
Wurtzite	126	ZnS	I	17–20
Iron	128	Fe	I	65
Wairauite	128	CoFe	I	54
Awaruite	128	(Ni,Fe)	I	60–61
Gudmundite	130	FeSbS	S	41–57
Pentlandite	130	$(Fe,Ni)_9S_8$	I	44
Violarite	132	$(Ni,Fe)_3S_4$	I	42
Bravoite	132	$(Fe,Ni,Co)S_2$	I	31–52
Cattierite	134	CoS_2	I	33
Vaesite	134	NiS_2	I	31
Indite	136	$FeIn_2S_4$	I	29
Dzhalindite	136	$In(OH)_3$	I	8

TABLE III (continued)

Mineral	Page	Composition	Anisotropy	Reflectance %
Cobaltpentlandite	136	$(Co,Fe,Ni)_9S_8$	I	54
Pyrrhotite	138	$Fe_{1-x}S$	S	34–39
Greigite	140	Fe_3S_4	I	?
Smythite	140	Fe_3S_4	S	34.5
Mackinawite	140	FeS	S	22–44
Oregonite	142	Ni_2FeAs_2	W	49–50
Cohenite	142	Fe_3C	W	?
Schreibersite	142	$(Fe,Ni,Co)_3P$	W-D	?
Wulfenite	144	$PbMoO_4$	?	16–18
Descloizite	144	$Pb(Zn,Cu)[OH/VO_4]$	S	15–18
Villamaninite	144	$(Cu,Ni,Co,Fe)S_2$	I	25–30
Linnaeite	146	Co_3S_4	I	45–50
Siegenite	146	$(Ni,Co)_3S_4$	I	45
Polydymite	146	Ni_3S_4	I	46
Carrollite	146	Co_2CuS_4	I	43
Orcelite	148	$Ni_{<5}As_2$	D-S	44–48
Breithauptite	148	NiSb	S	37–48
Langisite	148	$(Co,Ni)As$	D	46–47
Maucherite	148	$Ni_{<3}As_2$	W-D	48
Niccolite	150	NiAs	S	49–53
Skutterudite-series	152	$(Co,Ni,Fe)As_{3-x}$	I	53
Rammelsbergite	154	$(Ni,Co,Fe)As_2$	S	60
Pararammelsbergite	154	$NiAs_2$	S	56–58
Safflorite	156	$(Co,Fe,Ni)As_2$	S	55–60
Loellingite	156	$FeAs_2$	S	55
Gersdorffite	158	$(Ni,Co,Fe)AsS$	I	47–54
Ullmannite	158	NiSbS	I	45
Lepidocrocite	160	γ-FeOOH	S	15–25
Goethite	160	α-FeOOH	D	15–20
Heterogenite	162	CoOOH	S	10–25
Zincite	162	ZnO	?	10
Hauerite	162	MnS_2	I	25
Plattnerite	162	PbO_2	D	17–19
Ludwigite	164	$(Mg,Fe)_2(Fe,Al)BO_5$	S	8–10
Vonsenite	164	$(Fe,Mg)_2(Fe,Al)BO_5$	S	10–15
Avicennite	166	Tl_2O_3	I	?
Wuestite	166	FeO	I	?
Murdochite	166	Cu_6PbO_8	I	17
Magnetite	168	$FeFe_2O_4$	I	21
Titanomagnetite	168		I	17
Maghemite	170	Fe_2O_3	I	26
Ulvöspinel	170	Fe_2TiO_4	I	?
Magnesioferrite	170	$MgFe_2O_4$	I	17.5
Trevorite	172	$NiFe_2O_4$	I	21
Coulsonite	172	FeV_2O_4	I	25
Donathite	172	$(Fe,Mg,Zn)(Cr,Fe)_2O_4$	D	15
Chromite	174	$(Fe,Mg)(Cr,Fe,Al)_2O_4$	I	11–13
Davidite	174	$(Fe,U,Ce,La)_2(Ti,Fe,Cr,V)_5O_{12}$	I-W	15–20
Ilmenite	176	$FeTiO_3$	S	16–20
Geikielite	176	$MgTiO_3$	S	12–15
Anatase	178	TiO_2	?	20
Pseudobrookite	178	Fe_2TiO_5	D	15
Rutile	180	TiO_2	S	20–24
Nb-rutile	180		S	19–20
Ta-rutile	180		S	18–20
Freudenbergite	182	$Na_2Fe_2Ti_6O_{16}$	W	15–20

TABLE III (continued)

Mineral	Page	Composition	Anisotropy	Reflectance %
Perovskite-group	182		I	15
Högbomite	182	$(Fe,Mg)_6(Al,Fe)_{16}TiO_{32}$	D	10
Plumboferrite	184	$PbO.2Fe_2O_3$ (?)	W	15–20
Magnetoplumbite	184	$Pb(Fe,Mn,Al,Ti)_{12}O_{19}$	D	22–24
Scheelite	184	$CaWO_4$	D	10
Wolframite-series	186	$(Fe,Mn)WO_4$	W-S	15–20
Pyrochlore-series	188		I	12–16
Calzirtite	188	$Ca(Ca,Zr)_2Zr_4(Ti,Fe)_2O_{16}$	D	15
Glaucodot	190	$(Co,Fe)AsS$	D	45–50
Arsenopyrite	190	$FeAsS$	S	51–52
Cobaltite	192	$(Co,Fe)AsS$	W-D	56
Nolanite	192	$(U,Fe,Al,Ti)_9O_{18}$	S	?
Pitchblende	194	UO_{2-3}	I	10–15
Uraninite	194	UO_2	I	17
Thorianite	196	$(Th,U,Ce)O_2$	I	15
Brannerite	196	$(U,Th,Ca)[(Ti,Fe)_2O_6]$	I	17
Coffinite	196	$U(SiO_4)_{1-x}(OH)_x$	I-W	7–10
Ilvaite	198	$CaFe_2Fe[OH/O/Si_2O_7]$	S	8–9
Hematite	198	Fe_2O_3	D	26–30
Ixiolite	200	$(Ta,Fe,Sn,Nb,Mn)_4O_8$	I-W	13–14
Wodginite	200	$(Ta,Nb,Fe,Mn,Sn,Zr)_{16}O_{32}$	D	14–15
Tin-tantalite	200	$Mn(Ta,Nb,Sn)_2O_6$	S	15–16
Thoreaulite	200	$Sn(Ta,Nb)_2O_7$	S	16–18
Stibiotantalite	200	$Sb(Ta,Nb)O_4$	W-D	20
Bismutotantalite	200	$Bi(Ta,Nb)O_4$	W-D	20
Columbite-tantalite	202	$(Fe,Mn)(Nb,Ta)_2O_6$	D	15–18
Tapiolite	202	$(Fe,Mn)(Ta,Nb)_2O_6$	S	15–18
Fergusonite	202	$YNbO_4$	I-W	11–14
Formanite	202	$YTaO_4$	I-W	15
Marcasite	204	FeS_2	S	47–53
PYRITE	206	FeS_2	I-W	54
Cassiterite	208	SnO_2	D	11–13
Staringite	208	$(Fe,Mn)_{0.5}Sn_{4.5}(Ta,Nb)_{1.0}O_{12}$	D	12–14
Sukulaite	208	$(Ta,Nb)_2Sn_2O_7$	I	15
Spinel-series	210		I	7–8
Nigerite	210	$(Sn,Mg,Zn,Fe)(Al,Fe)_4(O,OH)_8$	S	7.5
Karelianite	212	V_2O_3	S	18
Eskolaite	212	Cr_2O_3	S	19–21

TABLE IV
SELENIDES

Because of insufficient quantitative information the selenides are here tentatively listed in three different ways: (1) on the basis of anisotropy two groups are distinguished, each group being listed in order of increasing mean reflectance (in green light, in air); (2) in order of increasing minimum (VHN); (3) in order of increasing polishing hardness (H), with reference to some common sulphides.

In Part II (p.215–231) the selenides are described in the same sequence as given in Table IV-1.

TABLE IV-1.

Mineral	Page	Composition	Anisotropy	Reflectance %
Isotropic or weakly anisotropic				
Achavalite	216	FeSe	I (?)	–
Cadmoselite	216	β-CdSe	W	20
Berzelianite	216	$Cu_{2-x}Se$	I	28.5
Stilleite	216	ZnSe	I	30
Tiemannite	216	HgSe	I	30
Aguilarite	218	Ag_4SeS	W	35
Crookesite	218	$(Cu,Tl,Ag)_2Se$	W	35
Penroseite	218	$(Ni,Cu,Co)Se_2$	I	35
Trogtalite	218	$CoSe_2$	I	–
Tyrrellite	220	$(Co,Ni,Cu)_3Se_4$	I	35–45
Trüstedtite	220	Ni_3Se_4	I	–
Bornhardtite	220	Co_3Se_4	I	–
Clausthalite	220	PbSe	I	49.5
Distinctly to strongly anisotropic				
Umangite	222	Cu_3Se_2	S	12.0–16.0
Klockmannite	222	CuSe	S	11.9–35.6
Eskebornite	222	$CuFeSe_2$	S	24.4–34.6
Eucairite	224	α-$Cu_2Se.Ag_2Se$	S	30
Wilkmanite	224	Ni_3Se_4	S	–
Selenium	224	Se	S	25–35
Naumannite	224	Ag_2Se	D	35
Mäkinenite	226	γ-NiSe	S	–
Sederholmite	226	β-NiSe	D-S	–
Kullerudite	226	$NiSe_2$	S	–
Hastite	226	$CoSe_2$	S	–
Freboldite	226	CoSe	S	–
Platynite	228	$Pb_4Bi_7Se_7S_4$	S	40
Paraguanajuatite	"	$Bi_2(Se,S)_3$	D	45
Ferroselite	228	$FeSe_2$	D-S	47–50
Laitakarite	228	Bi_4Se_2S	M	50
Ikunolite	230	$Bi_4(Se,S)_3$	S	50
Guanajuatite	230	$Bi_2(Se,S)_3$	S	55
Kitkaite	230	NiTeSe	D	55.8–58.3

TABLE IV-2

SELENIDES, in order of increasing minimum VHN

Mineral	Page	VHN	Mineral	Page	VHN*
Berzelianite	216	22–99	Achavalite	216	N.d.
Eucairite	224	23–94	Bornhardtite	220	N.d.
Aguilarite	218	25–35	Freboldite	226	N.d.
Tiemannite	216	26–39	Hastite	226	N.d.
Naumannite	224	27–56	Ikunolite	230	N.d.
Paraguanajuatite	228	30–160	Kullerudite	226	N.d.
Laitakarite	228	36–50	Mäkinenite	226	N.d.
Guanajuatite	230	42–210	Platynite	228	N.d.
Clausthalite	220	43–74	Sederholmite	226	N.d.
Klockmannite	222	57–86	Selenium	224	N.d.
Umangite	222	77–112	Stilleite	216	N.d.
Crookesite	218	101–141	Trogtalite	218	N.d.
Kitkaite	230	110–119	Trüstedtite	220	N.d.
Eskebornite	222	141–202	Wilkmanite	224	N.d.
Cadmoselite	216	203–222			
Tyrrellite	220	336–469			
Penroseite	218	407–550			
Ferroselite	228	700–933			

*N.d. = No data

TABLE IV-3

SELENIDES, in order of increasing polishing hardness

Mineral	Page	Mineral	Page	Mineral	Page
Aguilarite	218	Crookesite	218	Stilleite	216
Naumannite	224	Laitakarite	228	Sederholmite	226
Tiemannite	216	Ikunolite	230	Tyrrellite	220
Selenium	224	Paraguanajuatite	228	Trüstedtite	220
Clausthalite	220	Guanajuatite	230	Wilkmanite	224
Mäkinenite	226	Platynite	228	Penroseite	218
Kullerudite	226	(Galena)		Ferroselite	228
Eskebornite	222	Achavalite	216		
Klockmannite	222	Kitkaite	230	Bornhardtite*	220
Eucairite	224	Cadmoselite	216	Freboldite*	226
Berzelianite	216	(Chalcopyrite)		Hastite*	226
Umangite	222	(Tetrahedrite)		Trogtalite*	218

*No H-data available

TABLE V
TELLURIDES

Like the selenides the tellurides are listed tentatively in three different ways: (*1*) on the basis of anisotropy two groups are distinguished each group being listed in order of increasing mean reflectance (in green light, in air); (*2*) in order of increasing minimum VHN; (*3*) in order of increasing polishing hardness (H), with references to some common sulphides.

In Part II (p.233–253) the tellurides are described in the same sequence as given in Table V-1.

TABLE V-1

Mineral	Page	Composition	Anisotropy	Reflectance %
Isotropic or weakly anisotropic				
Weissite	234	$Cu_{2-x}Te$	W-M	30
Coloradoite	234	$HgTe$	I	33.9
Petzite	234	Ag_3AuTe_2	I	37.1
Nagyagite	236	$Au(Pb,Sb,Fe)_8(S,Te)_{11}$ (?)	W-D	38.65
Joseite-A	236	$Bi_{4+x}Te_{1-x}S_2$	W-D	48.8–53.1
Volynskite	238	$AgBi_{1.6}Te_2$	W	54.3
Joseite-B	236	$Bi_{4+x}Te_{2-x}S$	W-D	51.5–58.1
Wehrlite	238	$BiTe$	W-D	55
Montbrayite	238	Au_2Te_3	W-M	63.5
Calaverite	240	$AuTe_2$	W-D	64.4
Altaite	240	$PbTe$	I	70.5
Distinctly to strongly anisotropic				
Rickardite	242	Cu_7Te_5	S	25
Stuetzite	242	$Ag_{5-x}Te_3$	M	37.2–38.9
Hessite	242	Ag_2Te	D-S	38.7–40.85
Empressite	244	$AgTe$	S	34.1–49.9
Csiklovaite	244	$2Bi_2S_3 \cdot Bi_2Te_3$	D	45
Frohbergite	244	$FeTe_2$	S	45.3
Vulcanite	244	$CuTe$	S	35 –60
Tellurium	246	Te	S	43.4–53.5
Hedleyite	246	$Bi_{14}Te_6$	D	48.0–51.2
Tetradymite	248	Bi_2Te_2S	D	49.8–51.7
Sylvanite	248	$AuAgTe_4$	S	49.7–58.2
Kitkaite	248	$NiTeSe$	D	55.8–58.3
Kostovite	250	$AuCuTe_4$	S	54.9–60.1
Melonite	252	$NiTe_2$	D	57.0–60.6
Tellurobismuthite	250	Bi_2Te_3	D	63.6
Krennerite	252	$AuAgTe_4$	S	71.9

TABLE V-2

TELLURIDES, in order of increasing minimum VHN

Mineral	Page	VHN	Mineral	Page	VHN*
Coloradoite	234	23–35	Sylvanite	248	60–250
Hessite	242	24–44	Melonite	252	63–156
Tetradymite	248	25–75	Stuetzite	242	75–90
Tellurium	246	25–87	Wehrlite	238	81
Joseite-B	236	29–67	Empressite	244	108–133
Hedleyite	246	30–89	Kitkaite	248	110–119
Tellurobismuthite	250	32–93	Rickardite	242	133–167
Altaite	240	34–60	Montbrayite	238	198–228
Kostovite	250	35–43	Calaverite	240	198–237
Petzite	234	35–74	Frohbergite	244	250–297
Krennerite	252	36–130	Csiklovaite	244	N.d.
Nagyagite	236	39–129	Vulcanite	244	N.d.
Joseite-A	236	40–87	Weissite	234	N.d.
Volynskite	238	55–103			

*N.d. = No data

TABLE V-3

TELLURIDES, in order of increasing polishing hardness

Mineral	Page	Mineral	Page	Mineral	Page
Hessite	242	Volynskite	238	Krennerite	252
Petzite	234	Altaite	240	Montbrayite	238
Coloradoite	234	Weissite	234	Stuetzite	242
Nagyagite	236	Rickardite	242	Empressite	244
Hedleyite	246	Vulcanite	244	(Galena)	
Wehrlite	238	Kostovite	250	Kitkaite	248
Joseite	236	Sylvanite	248	(Chalcopyrite)	
Csiklovaite	244	Tellurium	246	Melonite	252
Tetradymite	248	(Stibnite)		Frohbergite	244
Tellurobismuthite	250	Calaverite	240	(Tetrahedrite)	

TABLE VI
Ag-SULPHOSALTS AND Ag-Fe-SULPHIDES

The minerals listed below are tentatively placed in order of increasing reflectance in air. The reflectance values in this table must be considered as approximate values since reliable reflectance values, measured with modern equipment, are almost completely lacking for this group of minerals.

A grouping of these minerals on the basis of anisotropy is not very useful, as this feature is often masked by internal reflections.

In Part II (p.255–269) the minerals of this group are described in the same sequence as given in this table.

Mineral	Page	Composition	Anisotropy	VHN	Reflectance %
Argyrodite–					
Canfieldite	256	$4Ag_2S.GeS_2-4Ag_2S.SnS_2$	I–W	133–172	22
Stephanite	256	$5Ag_2S.Sb_2S_3$	S	26–124	25–30
Xanthoconite–					
Pyrostilpnite	258	$3Ag_2S.As_2S_3-3Ag_2S.Sb_2S_3$	S	–	25–30
Proustite–					
Pyrargyrite	258	$3Ag_2S.As_2S_3-3Ag_2S.Sb_2S_3$	S	50–156	25–31
Samsonite	260	$2Ag_2S.MnS.Sb_2S_3$	W	–	28
Andorite	260	$Pb(Ag,Cu)Sb_3S_6$	D	140–206	30
Billingsleyite	260	$Ag_7(As,Sb)S_6$	?	–	30 (?)
Pearceite–					
Polybasite	262	$(Ag,Cu)_{16}(As,Sb)_2S_{11}$	D–S	108–164	30–35
Miargyrite	262	$Ag_2S.Sb_2S_3$	S	88–133	30–35
Aramayoite	264	$Ag_2S.(Sb,Bi)_2S_3$	S	–	30–35
Freieslebenite	264	$4PbS.2Ag_2S.2Sb_2S_3$	D	85–140	35
Diaphorite	264	$4PbS.3Ag_2S.2Sb_2S_3$	W–M	197–242	40
Smithite	266	$AgAsS_2$	M	–	40 (?)
Trechmannite	266	$AgAsS_2$	M	–	40 (?)
Owyheeite	266	$5PbS.Ag_2S.3Sb_2S_3$	S	29–214	40
Argentopyrite	268	$AgFe_2S_3$	S	250–252	45
Sternbergite	268	$AgFe_2S_3$	S	5–74	45
Argyropyrite	268	$Ag_3Fe_7S_{11}$	S	–	45

TABLE VII
Pb-Sb-SULPHOSALTS

The optical and other physical properties of the minerals of this group show too small differences to permit a reliable grouping of these minerals on the basis of one of these properties. All minerals show a white colour with different shades in green or yellow, a distinct to strong bireflectance and a distinct to strong anisotropy: even (VHN) and reflectance values do not offer the possibility of identification. In consequence, these minerals are listed here alphabetically. Identification is usually only possible by X-ray diffraction or electronprobe micro-analysis.

In Part II (p.271–279) the Pb-Sb-sulphosalts are described in the same sequence as given in this table.

Mineral	Page	Composition	Anisotropy	VHN	Reflectance %
Boulangerite	272	$5PbS . 2Sb_2S_3$	D-S	90-183	37.5-41.5
Dadsonite	272	$11PbS . 6Sb_2S_3$	D-S	–	34.9-40.0
Fülöppite	272	$3PbS . 4Sb_2S_3$	M	–	30 -35
Geocronite	274	$27PbS . 7Sb_2S_3$	D	95-206	38.0-41.6
Guettardite	274	$9PbS . 8Sb_2S_3$	S	180-197	34.8-42.0
Heteromorphite	274	$7PbS . 4Sb_2S_3$	S	137-187	37.0-41.0
Launayite	274	$22PbS . 13Sb_2S_3$	S	171-197	36.9-43.8
Madocite	274	$17PbS . 8Sb_2S_3$	S	141-171	max. 42.3
Plagionite	276	$5PbS . 4Sb_2S_3$	D	120-165	34.5-38.0
Playfairite	276	$16PbS . 9Sb_2S_3$	S	150-171	36.4-40.3
Robinsonite	276	$3PbS . 2Sb_2S_3$	S	118-123	40
Semseyite	276	$9PbS . 4Sb_2S_3$	S	109-173	40
Sorbyite	276	$17PbS . 11Sb_2S_3$	S	172-186	37 -43
Sterryite	278	$12PbS . 5Sb_2S_3$	S	–	36.0-38.7
Twinnite	278	$PbS . Sb_2S_3$	S	131-152	36.9-43.0
Veenite	278	$2PbS . Sb_2S_3$	M	156-172	37.5-43.2
Zinkenite	278	$6PbS . 7Sb_2S_3$	M	123-207	37.7-42.7

TABLE VIII
Bi-SULPHOSALTS

The bismuth-sulphosalts show a great similarity in their optical properties which usually prevents identification under the microscope. (VHN)-measurements will only exceptionally be of some help. Besides, some of these minerals are not sharply defined being members of solid solution series. Therefore, these minerals are listed here on the basis of their chemical composition which facilitates comparison. X-ray diffraction and electronprobe micro-analysis will often prove to be the only reliable methods for identification.

In Part II (p.281 - 295) the Bi-sulphosalts are described in the same sequence as given in this table.

Mineral	Page	Composition	Anisotropy	VHN	Reflectance %
Schapbachite	282	$AgBiS_2$	D–S	59–91	43.8–44.9
Pavonite	282	$Ag_2S.3Bi_2S_3$	S	–	42
Schirmerite	282	$2Ag_2S.PbS.2Bi_2S_3$	W	–	35
Benjaminite	284	$Pb_2(Ag,Cu)_2Bi_4S_9$	S	161–194	41 –43
Berryite	284	$Pb_2(Cu,Ag)_3Bi_5S_{11}$	D–S	131–171	41.8–43.0
Neyite	284	$Pb_7(Cu,Ag)_2Bi_6S_{17}$	M	–	–
Galenobismutite	286	$PbS.Bi_2S_3$	S	88–150	42
Cosalite	286	$2PbS.Bi_2S_3$	W	74–161	43
Lillianite	286	$3PbS.Bi_2S_3$	D	120–195	45
Bonchevite	288	$PbS.2Bi_2S_3$	D–S	129–205	46.6
Bursaite	288	$5PbS.2Bi_2S_3$	D	–	45
Cannizzarite	288	$Pb_3Bi_5S_{11}$	?	–	–
Kobellite-Tintinaite	290	$5PbS.4(Bi,Sb)_2S_3$	D	69–173	36.3–45.4
Ustarasite	290	$PbS.3(Bi,Sb)_2S_3$	S	–	42
Giessenite	290	$Pb_9CuBi_6Sb_{1.5}S_{30}$	W–D	65	35
Aikinite	292	$2PbS.Cu_2S.Bi_2S_3$	D–S	165–246	39.2–45.7
Nuffieldite	292	$10PbS.2Cu_2S.5Bi_2S_3$	W	149–178	39.0–44.9
Wittichenite	292	$3Cu_2S.Bi_2S_3$	W–D	161–216	37
Cuprobismutite	294	$CuBiS_2$	?	–	–
Emplectite	294	$CuBiS_2$	S	158–249	36.2–42.2
"Klaprothite"	294	$CuBiS_2$	S	–	32.7–40.2

TABLE IX
Pb-As-SULPHOSALTS

Just like the minerals grouped in Tables VII and VIII, the minerals listed here show very much the same optical and other physical properties which, as a rule, will prevent microscopic identification. X-ray diffraction and electron probe microanalysis will, therefore, also here be the only methods in most cases. An advantage for this group of minerals is, however, that they all were studied in detail by the same person and with the same equipment (Graeser, 1965). This enables us to list these minerals in order of increasing mean reflectance in air. Some of them may contain additional Ag, Cu or Tl.

In Part II (p.297–305) these minerals are described in the same sequence as given in this table.

Mineral	Page	Composition	Anisotropy	VHN	Reflectance % (530 nm)
Hutchinsonite	298	$(Tl,Pb)_2As_5S_9$	D–S	170–171	30.0–31.0
Marrite	298	$PbAgAsS_3$	S	161–171	31.5–34.0
Gratonite	298	$9PbS.2As_2S_3$	D	123–156	33.4–34.4
Liveingite	300	$Pb_{19}As_{13}S_{58}$	D	173–183	34.0–36.0
Lengenbachite	300	$Pb_{37}Ag_7Cu_6As_{23}S_{78}$	W–D	29–40	34.0–37.0
Rathite-I	300	$(Pb,Tl)_3As_4(As,Ag)S_{10}$	S	159–163	34.0–38.5
Baumhauerite	300	acentric: $Pb_{12}As_{16}S_{36}$ centric: $Pb_{11}As_{17}S_{36}$	D–S	128–182	34.0–39.0
Sartorite	302	$PbS.As_2S_3$	D	194–197	35.0–39.0
Dufrenoysite	302	$2PbS.As_2S_3$	S	145–156	36.5–40.5
Jordanite	302	$27PbS.7As_2S_3$	S	149–204	38.0–39.5
Seligmannite	304	$2PbS.Cu_2S.As_2S_3$	S	149–167	36.0–42.0

TABLE X
Sn-SULPHOSALTS AND Sn-SULPHIDES

These minerals are listed in order of increasing mean reflectance in air, as far as possible.

In Part II (p.307–319) these minerals are described in the same sequence as given in this table.

Mineral	Page	Composition	Anisotropy	VHN	Reflectance %
Sakuraiite	308	$Cu_2(In,Sn)ZnS_4$	W	–	20
Hocartite	308	Ag_2SnFeS_4	D–S	–	22.6–24.3
Berndtite	308	SnS_2	?	–	25
Stannite jaune	310	$Cu_{2+x}Sn_{1-x}FeS_4$	S	–	22.7–27.5
Stannoidite	310	$Cu_5Sn(Fe,Zn)_2S_8$	S	232–271	24.4–26.4
Kësterite	310	Cu_2SnZnS_4	W	320–322	24.8–26.2
Stannite	312	Cu_2SnFeS_4	S	140–326	27.4–29.3
Rhodostannite	314	$Cu_2Sn_3FeS_8$	D	243–266	27.8
Cylindrite	316	$6PbS.6SnS_2.Sb_2S_3$ (?)	W–D	31–131	28.2–30.9
Ottemannite	316	Sn_2S_3	S	–	30
Franckeite	316	$5PbS.3SnS_2.Sb_2S_3$	D	13–108	26.6–34.3
Herzenbergite	318	SnS	S	48–114	42.1–44.3
Teallite	318	$PbS.SnS$	D	31–125	40.2–46.6

TABLE XI
PLATINOID MINERALS

To facilitate identification of the platinoid minerals they are divided up into two groups: (*1*) isotropic or weakly anisotropic minerals; and (*2*) distinctly to strongly anisotropic minerals. In each group the minerals are tentatively listed in order of increasing mean reflectance, in air. For unnamed platinoid compounds, see Chudoba (1967).

In Part II (p.321–335) the platinoid minerals are described in the same sequence as given in this table.

Mineral	Page	Composition	Anisotropy	VHN	Reflectance %
Isotropic or weakly anisotropic					
Cooperite	322	PtS	W–D	505–588	39.0
Laurite	322	RuS_2	I	1393–2167	41.8–42.5
Hollingworthite	322	$(Rh,Pt,Pd,Ir,Ru)AsS$	I	657–848	40 –52
Irarsite	322	$(Ir,Ru,Rh,Pt)AsS$	I	976	47.8
Allopalladium	322	Pd	W–D	–	50
Stibiopalladinite	324	Pd_3Sb	W–D	–	54.6
Sperrylite	324	$PtAs_2$	I	960–1277	55.5
Michenerite	324	$(Pd,Pt)BiTe$	I	–	56
Potarite	324	$PdHg$	I	–	60
Zvyagintsevite	326	$(Pd,Pt)_3(Pb,Sn)$	I	241–318	63.6
Geversite	326	$PtSb_2$	I	–	65
Palladium	326	Pd	I	–	70
Platinum	326	Pt	I	329–397	70
Osmiridium	328	(Ir,Os)	I	297–645	80.0
Iridium	328	Ir	I	–	82.1
Distinctly to strongly anisotropic					
Braggite	330	$(Pt,Pd,Ni)S$	S	742–1030	34.5 –35.5
Niggliite	330	$PtSn$	S	306–537	25 –65
Vysotskite	330	$(Pd,Ni,Pt)S$	M	–	45
Froodite	330	$PdBi_2$	S	–	50
Palladium bismuthide	"	$PdBi_3$	S	105–125	50.1
Moncheite	332	$(Pt,Pd)(Te,Bi)_2$	D–S	–	53.2 –58.8
Kotulskite	332	$Pd(Te,Bi)_{1-2}$	S	236	58.7 –64.4
Osmium	334	Os	S	–	60.85–62.8
Merenskyite	332	$(Pd,Pt)(Te,Bi)_2$	D–S	–	63.2 –65.2
Iridosmium	334	(Os,Ir)	W–S	–	65 –72

TABLE XII
OXIDIC MANGANESE MINERALS

Most of these minerals show a wide range in micro-indentation hardness as well as in polishing hardness, depending on grain size or variations in chemical composition. Therefore, they are listed here tentatively in order of increasing mean reflectance although for some minerals, like ramsdellite, this property also varies within wide limits. The range in reflectance values may be due to bireflectance, to variations in chemical composition or to differences in grain size. The anisotropy is given for coarse-grained material. Fine-grained material may appear isotropic.

Microscopic identification is usually only possible on coarse-grained material. X-ray diffraction and electronprobe micro-analysis may in several cases be the only reliable methods for identification.

In Part II (p.337–359) these minerals are described in the same sequence as given in this table.

Mineral	Page	Composition	Anisotropy	VHN	Reflectance %
Gaudefroyite	338	$Ca_4Mn_{3-x}[(BO_3)_3/(CO_3)/(O,OH)_3]$	S	840	10.2–13.1
Rancieite	338	$(Fe,Mg)O.4MnO_2.4H_2O$	S	–	12.5–15.0
Manganosite	338	MnO	I	314–325	14.4
Hetaerolite	338	$ZnMn_2O_4$	S	585–813	13.4–17.5
Groutite	340	$\alpha\text{-}MnOOH$	S	613–813	12.4–20.0
Pyrochroite	340	$Mn(OH)_2$	W–D	224–245	15.2–17.9
Hydrohetaerolite	"	$HZnMn_{2-x}O_4$	D–S	–	15 –20
Marokite	340	$CaMn_2O_4$	S	800	16.3–19.4
Manganite	342	$\gamma\text{-}MnOOH$	S	195–803	14.8–21.4
Franklinite	342	$(Zn,Fe,Mn)(Fe,Mn)_2O_4$	I	667–847	18.4
Lithiophorite	344	$(Al,Li)MnO_2(OH)_2$	S	60–100	10 –25
Chalcophanite	344	$(Zn,Mn,Fe)Mn_3O_7.3H_2O$	S	71–246	9.6–27.3
Aurorite	344	$(Ag,Ba,Ca,Mn,...)Mn_3O_7.3H_2O$	S	–	10 –25
Hausmannite	346	$MnMn_2O_4$	S	466–724	17.2–20.5
α-Vredenburgite	"	$(Mn,Fe)_3O_4$	D	–	18 –20
Quenselite	348	$PbO.MnOOH$	W–D	153–186	17.8–20.8
Jacobsite	348	$(Mn,Fe,Mg)(Fe,Mn)_2O_4$	I	575–875	19.6
Braunite	350	$MnMn_6[O_8/SiO_4]$	W–D	280–1187	20.4–22.4
Todorokite	350	$(H_2O,...)_{\leq 2}(Mn,...)_{\leq 8}(O,OH)_{16}$	S	–	20 –23
Bixbyite	352	$(Mn,Fe)_2O_3$	I	882–1168	22.7
Birnessite	352	$(Ca,Mg,Na,K) \ll_1 (Mn^{4+},Mn^{2+})(O,OH)_2$	W–D	–	25
Ramsdellite	352	$\gamma\text{-}MnO_2$	S	93–1200	10 –40
Woodruffite	352	$(Zn,H_2O)_{\leq 2}(Mn,Zn)_{\leq 8}(O,OH)_{16}$	D	744	26
Cryptomelane	354	$(K,...)_{\leq 2}(Mn,...)_8O_{15}$	S	525–1048	26.7
Psilomelane	354	$(Ba,Mn,...)_3(O,OH)_6Mn_8O_{16}$	S	203–813	15 –20
Cesarolite	354	$PbO.3MnO_2.H_2O$	I	–	28
Hollandite	356	$Ba_{\leq 2}Mn_8O_{16}$	S	272–1048	25.6–32.3
Crednerite	356	$CuMnO_2$	S	200–357	23.6–35.0
Coronadite	356	$Pb_{\leq 2}Mn_8O_{16}$	S	359–840	26.7–32.3
Pyrolusite	358	$\beta\text{-}MnO_2$	S	76–1500	27.2–40.8
Nsutite	358	$Mn_{1-x}Mn_x(OH)_{2x}$	S	350–1288	30 –40

PART II. MINERAL DESCRIPTIONS

Description of minerals from Table III

Name Formula Crystal System	1: Colour 2: Bireflectance 3: Anisotropy 4: Internal reflections	Reflectance in % (Wavelength in nanometres)
ACANTHITE-ARGENTITE* Ag_2S $\quad\quad 177°C$ Mono. $\rightleftharpoons$ Cubic	1. Grey with often a distinct green tint. $\quad\longrightarrow$ galena, much darker and greenish grey. $\quad\longrightarrow$ certain sections of polybasite, no great difference. $\quad\longrightarrow$ silver, distinct green. 2. Very weak, even in oil; only visible on grain boundaries and on twin lamellae. 3. For acanthite: distinct; some spec. may appear isotropic. Sections tarnished in air may appear somewhat more strongly anisotropic. For argentite: isotropic. 4. Not present.	470 nm 33.1-34.3 546 30.4-31.2 589 29.4-30.1 650 28.4-28.9 [194].
TIN β-Sn Tetr.	1. Creamy white. 2. Air: not present. Oil: creamy white to brownish grey (on grain boundaries). 3. Weak, yellowish or brownish grey to bluish grey. 4. Not present.	High.

* The cubic morphology of the high-temperature phase is often preserved in ores, but all room-temperature Ag_2S X-ray diffraction patterns reveal the monoclinic symmetry of acanthite (Ramsdell, 1943). Therefore, usage of argentite as a mineral name refers to a pseudomorph of acanthite after argentite (Taylor, 1969).

Hardness H: polishing hardness VHN: micro-indentation hardness in kg/mm^2	Miscellaneous
H: one of the lowest known: $\sim$ jalpaite. VHN: 20-61 [159, 1113, 1715]. VHN$_{15}$: 22-26 [194].	Occurs as idiomorphic crystals or as aggregates of polygonal grains. Also as cryptocrystalline masses. Twinning lamellae //(001) not uncommon. The presence of twinning cannot be used as a minimum-temperature indicator. Differentially orientated lamellae may be observed under crossed nicols. Occurs as exsolution bodies in galena and as inclusions in pyrite, sphalerite, galena, tetrahedrite. May show intergrowths with covellite, bismuthinite, galena, chalcopyrite, and with many Ag-Sb- or Ag-As-minerals such as pyrargyrite, proustite, freibergite, polybasite. Replaces sulphides such as pyrite, chalcopyrite, covellite, sphalerite, galena, many Ag-minerals, uraninite; these minerals may be found as inclusions in argentite. May be replaced or veined by chalcopyrite, blaubleibender covellite, digenite, electrum, gold, covellite. Physically it is impossible that argentite or acanthite and stromeyerite are formed together. Other ass.: silver, chalcocite, Co-Ni-arsenides. Ref.: 85, 272, 339, 455, 672, 762, 796, 824, 997, 1050, 1153, 1237, 1253, 1343, 1358, 1361, 1392, 1405, 1441, 1513, 1603, 1629, 1679, 1705.
H: $\sim$ argentite. VHN$_{15-25}$: 10-15 [1560].	Occurs as xenomorphic grains. Feathery tongues from the main masses penetrate the calcite host. Twinning //(111) common. Parag.: pyrite, chalcopyrite, bornite, chalcocite, covellite, sphalerite, galena, pitchblende, cassiterite, lead, copper. Ref.: 25, 793, 935, 1395.

Description of minerals from Table III

Name Formula Crystal System	1: Colour 2: Bireflectance 3: Anisotropy 4: Internal reflections	Reflectance in % (Wavelengths in nanometres)
BISMUTH Bi (may contain some Te or As) Hex.	1. White to creamy white; quickly tarnishing to pinkish cream, then turning slowly brownish. ⟶ silver, slightly creamier. ⟶ arsenic, pinkish creamy. ⟶ antimony, brighter. ⟶ niccolite, whiter; niccolite is more pinkish. In contact with bismuth, chalcopyrite is greyish olive-green, pyrrhotite greyish creamy brown, skutterudite, safflorite and rammelsbergite white, bismuthinite greyish white. 2. Weak but distinct, esp. in oil; creamy white to creamy white with a grey tint. More distinct in fresh-polished surfaces: creamy white to pinkish white. 3. Distinct to strong. 4. Not present.	470 nm 55.1-59.8 546 59.9-64.9 589 62.5-68.5 650 62.8-69.3 [558a].
LEAD Pb (may contain some Ag or Sb) Cubic	1. Grey-white, very rapidly tarnishing to grey. ⟶ silver, greyish white, when fresh surface. 2. Not present. 3. Isotropic, sometimes anomalously anisotropic, due to distortion by grinding. 4. Not present.	About 60-65 %.

Hardness H: polishing hardness VHN: micro-indentation hardness in kg/mm^2	Miscellaneous
H< bismuthinite and all other minerals which may accompany bismuth or could be mistaken for bismuth. VHN: 10-19 [159, 1113, 1715] VHN$_{15}$: 9-13 [194] VHN$_{5-10}$: 16-26 [810]	A basal cleavage may be visible. Commonly polysynth. twinned often with parquet-like or feather-like appearance. The twinning is not due to inversion at 75°C. Stresses created within the bismuth on crystallization are apparently sufficient to produce twinning. Intense twinning may also be induced by grinding. Bismuth may be associated with pyrrhotite, pyrite, sphalerite, chalcopyrite, cubanite, bismuthinite, Cu-Bi-sulphides, cosalite, Bi-Te-minerals, gold, Sb-minerals, scheelite, molybdenite, wolframite, stannite, herzenbergite, cassiterite, parkerite, bornite, pentlandite, schapbachite, kobellite, and arsenopyrite, or with: Co-Ni-As-minerals (preferably Co-As-minerals), silver, arsenic, bismuthinite, uraninite, pitchblende. Bismuth may show graphic or myrmekitic intergrowths with galena, bismuthinite, chalcopyrite, and pyrrhotite. Forms a myrmekitic intergrowth with gold, as a product of decomposition of maldonite. May replace galena, tetrahedrite, skutterudite, cobaltite, glaucodot, pitchblende, other sulphides and Bi-tellurides. May be replaced by bismuthinite, schirmerite, silver, niccolite, rammelsbergite, skutterudite, less commonly by arsenic and sulphides. Forms rims around parkerite in pyrrhotite and bornite. Occurs as inclusions in galena, bismuthinite, molybdenite, arsenopyrite, chalcopyrite, schapbachite, Sb-minerals and along cleavages of wolframite. Contains inclusions of joseite, cosalite, bismuthinite. Once observed as idiomorphic cube-like inclusions in chalcopyrite. As skeleton-shaped crystals, dendrites or small cores enclosed in skutterudite, safflorite, rammelsbergite or pitchblende. Ref.: 5, 6, 7, 131, 286, 339, 390, 458, 474, 541, 588, 707, 708, 739, 807, 842, 891, 960, 1045, 1060, 1207, 1237, 1320, 1334, 1356, 1392, 1603, 1608.
H: Very low VHN: 4-6 [1113, 1715] VHN$_{15}$: 5-6 [194]	Usually occurs as granular aggregates and as dendritic, skeletal crystals. Twinning may occur. Polishes without scratches only on a selvyt cloth soaked with a suspension of magnesion oxide in ammonia. Ass.: Mn-oxides, galena, minium, massicotite. Ref.: 79, 794, 811, 1392, 1603.

Description of minerals from Table III

Name / Formula / Crystal System	1: Colour / 2: Bireflectance / 3: Anisotropy / 4: Internal reflections	Reflectance in % (Wavelengths in nanometres)
JALPAITE Ag_3CuS_2 Tetr.	1. Light greyish white (in air). → acanthite, about same shade of grey, may appear as a shade of green. → chalcocite, light yellow-brown. → galena, distinctly darker; jalpaite may show a slight greenish tint. → stromeyerite, purplish. mckinstryite shows a distinct shade of blue against jalpaite. polybasite-pearceite shows a light shade of pink against jalpaite. 2. Only visible in oil, brownish grey to grey. 3. Distinct, stronger than for acanthite, less pronounced than for stromeyerite and mckinstryite. No blue-violet tints as for stromeyerite, colours are blue-green and light green. 4. Not present.	About 30 %.
PATRONITE Varying; two groups corresponding with two groups of synthetic products: $VS_{3.95}$ to $VS_{3.79}$ and $VS_{3.68}$ to $VS_{3.60}$ (Baumann, 1964). Mono.	1. + 2. Varying with chemical composition and orientation. For longitudinal sections: very strong bireflection (similar to that of molybdenite, graphite or valleriite). $VS_{3.68}$ to $VS_{3.60}$: air: ∥ elongation: light grey. ⊥ elongation: brown. oil: ∥ elongation: light grey with yellowish tint. ⊥ elongation: dark brown. $VS_{3.95}$ to $VS_{3.79}$: even stronger: instead of "light grey", read "white". Sections ⊥ elongation have a much weaker bireflection. 3. Very strong. Straight extinction. $VS_{3.68}$ to $VS_{3.60}$, in air: no colour effects; in oil: light grey to blue-green and brown. $VS_{3.95}$ to $VS_{3.79}$: stronger; in oil: white to blue-green and brown. 4. Observed.	No data, but about as for sphalerite.
SELENIDES		

Hardness H: polishing hardness VHN: micro-indentation hardness in kg/mm^2	Miscellaneous
H: ~argentite, <galena. VHN: 23–30 [551].	Forms lamellated and granular aggregates. Good prismatic cleavage. Occurs intergrown with acanthite or pearceite and as inclusions in galena and sphalerite. May contain inclusions of mckinstryite. Veins or replaces pitchblende, chalcopyrite, sphalerite and galena. May be replaced by argentite and silver. Other ass.: stromeyerite, polybasite, tetrahedrite, pyrite, covellite. Difficult to distinguish jalpaite from stromeyerite and mckinstryite under the microscope. Much material described as stromeyerite is in fact jalpaite or mckinstryite, e.g. all stromeyerite in association with argentite or acanthite (Skinner, 1966). Very difficult to distinguish between acanthite and jalpaite. X-ray diffraction should be used for identification. Ref.: 380a, 551, 678, 718, 997, 1237, 1405.
H: very low. VHN: published observations very doubtful.	Forms aggregates of very fine short-prismatic needle-like crystals. Grain size very small (2-10 micron). A basal cleavage is present. Twinning not observed. Parag.: bravoite, pyrite, molybdenite, chalcopyrite and marcasite. Ref.: 27, 89, 1237.
H: generally very low.	For the description, see p. 215–231.

Description of minerals from Table III

Name Formula Crystal System	1: Colour 2: Bireflectance 3: Anisotropy 4: Internal reflections	Reflectance in % (Wavelengths in in nanometres)

SULPHUR
α-S Ortho.
γ-S Mono.
	1. Air: greyish.
	   Oil: much darker, dull grey.
	   ⟶ sphalerite, much darker, especially in oil.
	2. Distinct in air; in oil masked by the intense int. refl.
	3. Distinct, only observable in air, with nicols not completely crossed; in oil masked by the int. refl.
	4. Abundant and intense: white to light yellowish.
		About 10-15 %.

REALGAR
AsS
Mono.
	1. Dull grey, in air much lighter.
	   ⟶ orpiment, slightly darker.
	   ⟶ sphalerite, almost the same tint, in air lighter.
	   ⟶ cinnabar, much darker.
	2. Weak, but distinct: grey with reddish to grey with bluish tint.
	3. Strong, but not perceptible in oil because of strong int. refl.
	4. Abundant and intense, yellowish red.
		530 nm 20.0-21.1 [523].

ORPIMENT
As_2S_3
Mono.
	1. Grey, in air much lighter.
	   ⟶ realgar, slightly lighter.
	   ⟶ sphalerite, in air lighter; in oil about the same, some sections darker.
	2. Strong: in air about as molybdenite.
	   in air: in oil:
	   ∥ a white greyish white
	   ∥ b dull grey with darkest, velvety
	      reddish tint
	   ∥ c dull grey-white dull grey-white
	                          with reddish tint.
	3. Strong, in oil masked by int. refl.
	4. Abundant and intense, white or light yellow.
		About 25 %.

Hardness H: polishing hardness VHN: micro-indentation hardness in kg/mm^2	Miscellaneous
H: very low. VHN: 30-66 [1715]. VHN$_{5-10}$: 24-35 [810].	When occurring as an alteration product of sulphides, usually fine-granular. Easily recognized. Parag.: occasionally with sulphides in zones of cementation. Ref.: 1209, 1237.
H: <orpiment, ≪stibnite. VHN: 47-60 [159, 1715]. VHN$_{15}$: 51-57 [194]. VHN$_{10}$: 50-57 [810].	Usually occurs as cave-fillings between other ore minerals. As inclusions in orpiment. Forms orientated intergrowths with orpiment. Parag.: nearly always orpiment, commonly also stibnite, arsenopyrite, pyrite, sphalerite, arsenic, arsenical sulphosalts, lorandite, tennantite, loellingite, enargite, proustite, Fe-hydroxides. Ref.: 523, 682, 967, 1237, 1389, 1390, 1392, 1603, 1629. Note: Dimorphite, As$_4$S$_3$, Ortho. No optical data. Ref.: 1237, 1475.
H: slightly >realgar. VHN: 22-52 [159, 1715]. VHN$_{15}$: 29-58 [194]. VHN$_{10}$: 26-33 [810].	May occur as needle-formed or tabular crystalline masses, often sheaf-like or radiating, but usually as alteration product around realgar. Forms oriented intergrowths with realgar. Parag.: as realgar, nearly always with realgar. Ref.: 967, 1237, 1392, 1629.

Name Formula Crystal System	1: Colour 2: Bireflectance 3: Anisotropy 4: Internal reflections	Reflectance in % (Wavelengths in nanometres)
GETCHELLITE $AsSbS_3$ Mono.	1. Greyish white with a slight blue tint. → orpiment, realgar, lighter; orpiment is violet-grey-brown. → stibnite, darker; stibnite is yellowish white. 2. Distinct (only in oil). 3. Weak, obscured by the int. refl. 4. Strong, blood-red.	470 nm 28.4-30.7 546 25.9-27.3 589 25.1-26.5 650 24.3-25.6 [194].
KERMESITE Sb_2S_2O Tricl.	1. Grey. 2. Distinct: grey to brownish or greenish grey. 3. Strong: violet to blue-green; in oil masked by intense int. refl. 4. Intense, deep-red with a violet tint.	About 25 %.
STIBNITE Sb_2S_3 Ortho.	1. White to greyish white. → galena, slightly darker with a creamy tint. → bournonite, much lighter. → chalcostibite, very similar. → bismuthinite, darker, less creamy. → antimony, slightly grey. 2. Strong: ∥ a dull grey white; ∥ b brownish grey; ∥ c pure white. 3. Very strong: 45°: blue, grey-white, brown, pinkish brown. Straight extinction. Undulose extinction very common. 4. Not present.	R_p R_m R_g 470 nm 31.5 42.5 50.4 546 31.4 40.9 46.8 589 30.4 39.6 44.3 650 30.2 39.6 41.3 [194].

Hardness H: polishing hardness VHN: micro-indentation hardness in kg/mm^2	Miscellaneous
H: slightly > orpiment. VHN$_{15}$: 30-50 [194].	Perfect cleavage // (001). Occurs as inclusions in orpiment and realgar. May contain inclusions of pyrite and stibnite. Other ass.: cinnabar. Ref.: 74, 1669.
H: no data. VHN$_{25}$: 36-99 [1715].	Occurs as irregular grains or as radiated aggregates. Also as single needle-like crystals. Produced by alteration of stibnite. Diff.: proustite and pyrargyrite show a higher refl. and never a fibrous texture. Parag.: nearly always associated with stibnite and sulphosalts. Ref.: 219, 1237, 1392, 1509, 1732.
H: slightly varying with orientation. > orpiment, realgar, ≦ berthierite, galena, bournonite, ≪ chalcopyrite. VHN: 42-129 [159, 1113, 1715]. VHN$_{50}$: 72-138 [194-1715]. VHN$_{10-20}$: 65-153 [810].	Occurs as irregular granular masses or as radiated aggregates. Pressure-twins, crumpling-lamellae and deformations very common. May show fine growth-zoning (when etched). May be intergrown with berthierite, boulangerite, pyrargyrite, kermesite. Replaces pyrite, arsenopyrite, sphalerite, fahl-ore, pyrrhotite. Replaced by cinnabar. Other ass.: Fe-, Cu-, Zn-, Pb-sulphides, enargite, stibioluzonite, gold, stibarsenic, antimony, realgar, orpiment, sulphur, lorandite, vrbaite, livingstonite, bravoite, vaesite, pitchblende, wolframite, Pb-Sb-sulphosalts, molybdenite, scheelite, berthierite. Diff.: jamesonite shows a less distinct birefl. and anisotr. Bismuthinite has a higher refl., is light yellowish and has a more pronounced cleavage. Bournonite is harder, has a weaker birefl. and anisotr. Ref.: 11, 219, 261, 339, 602, 666, 967, 1237, 1281, 1379, 1392, 1429a, 1603, 1629, 1732.

44 Description of minerals from Table III

Name Formula Crystal System	1: Colour 2: Bireflectance 3: Anisotropy 4: Internal reflections	Reflectance in % (Wavelengths in nanometres)
METASTIBNITE Sb_2S_3 ?	1. White with bluish tint. $\longrightarrow$ stibnite, distinctly darker. 2. Not observable. 3. Distinct (in air); in oil masked by int. refl. 4. Intense, deep-red. Not as abundant as in proustite, pyrargyrite or cinnabar.	Distinctly lower than for stibnite.
MCKINSTRYITE $Cu_{0.8+x}Ag_{1.2-x}S$ with $x \leq 0.02$ Ortho.	1. Light greyish white. $\longrightarrow$ stromeyerite, whiter. $\longrightarrow$ jalpaite, distinctly blue. 2. Distinct. 3. Strong, grey, pale greyish blue and light tan. Less pronounced than for stromeyerite, stronger than for jalpaite. 4. Not present.	About 30-35 %.
IMHOFITE Tl-As-sulphosalt. Mono.	1. Pure white. $\longrightarrow$ galena, cream coloured. 2. Not perceptible in air and oil. 3. Very strong. 4. Abundant, bright red.	530 nm 28.0-31.0 [198].
LORANDITE $TlAsS_2$ Mono.	1. Grey-white with a bluish tint. $\longrightarrow$ galena, distinct bluish tint. $\longrightarrow$ realgar, much lighter. $\longrightarrow$ pyrargyrite, in air very similar. 2. Weak, but noticeable. 3. Strong, but obscured by the abundant int. refl. 4. Abundant and strong, dark-red; realgar has more abundant int. refl. (orange-yellow).	530 nm 31.4-32.6 [523].

Hardness H: polishing hardness VHN: micro-indentation hardness in kg/mm^2	Miscellaneous
H: no data. VHN: no data.	Commonly with colloform texture. Under crossed nicols the sphaerolitic or cauliflower-like texture becomes more distinct. Parag.: stibnite, kermesite. Ref.: 7, 17, 841, 1209.
H: no data. VHN: 60 [551].	Occurs as coarse-grained aggregates of intergrown crystals. Forms intergrowths with jalpaite. Parag.: silver, arsenopyrite, chalcopyrite, stromeyerite. Difficult to distinguish from stromeyerite under the microscope. Many spec. that have been described as "stromeyerite", e.g. all stromeyerite in ass. with argentite or acanthite, is in fact mckinstryite or jalpaite. Ref.: 551, 678, 1405, 1407.
H: ≪ hutchinsonite. VHN$_3$: 38 [198].	Optical properties much like hutchinsonite, but much lower hardness. Parag.: realgar. Ref.: 198, 523, 1018.
H: distinctly >realgar, metacinnabarite,< cinnabar. VHN$_{30}$: 39 [523]. VHN$_{10}$: 40-57 [810].	Cleavage visible only in larger crystals. Twinning not observed. Replaces raguinite. Parag.: cinnabar, realgar, orpiment, marcasite, melnicovitepyrite, pyrite, imhofite, As-rich sulphosalts. Ref.: 73, 75, 523, 790, 967, 1237, 1343, 1392, 1615, 1730.

Description of minerals from Table III

Name / Formula / Crystal System	1: Colour / 2: Bireflectance / 3: Anisotropy / 4: Internal reflections	Reflectance in % (Wavelength in nanometres)
RAGUINITE $TlFeS_2$ Ortho.	1. See at birefl. 2. Strong: R_g ($\parallel$ elongation of fibres): distinct pink; R_p' ($\perp$ elongation of fibres): greyish white with creamy tint. 3. Strong, with vivid interference colours, orange dominating. Straight extinction. 4. Not observed.	480 nm 24.9-31.9 540 25.4-31.9 580 25.7-35.7 640 26.0-40.6 Dispersion curves: [790].
CHALCOTHALLITE Cu_3TlS_2 Not indicated; for X-ray powder patterns, see [1380]	1. Light grey. $\longrightarrow$ chalcocite, weak pinkish-lavender tint. 2. Air: noticeable: R_g = colourless or weakly bluish; R_p = pinkish lavender. 3. Distinct, orange-brown tints. 4. Not present.	480 nm 32.7 540 30.5 580 29.5 640 29.1 Dispersion curve: [1380].
VRBAITE $Tl_4Hg_3Sb_2As_8S_{20}$ Ortho.	1. Greyish white with a bluish tint. $\longrightarrow$ lorandite, somewhat more bluish. $\longrightarrow$ cuprite, very similar. 2. No data. 3. Distinct, not as strong as for lorandite: blue or bluish green to reddish yellow. 4. Abundant, esp. in oil: red.	480 nm 32.8-36.4 540 30.0-33.4 580 27.6-30.7 640 26.2-28.5 Dispersion curves: [232].

Hardness H: polishing hardness VHN: micro-indentation hardness in kg/mm^2	Miscellaneous
H: no data. VHN: no data.	Occurs as fibres intimately intergrown with pyrite, and forming pseudomorphs of pseudo-hexagonal outlines after an unidentified mineral. Replaced by realgar and lorandite. Other ass.: orpiment, vrbaite. Ref.: 681, 790.
H: < vrbaite. VHN$_?$: 61-90 [1380].	Occurs as lamellar aggregates in veins of ussingite intersecting syenite. Shows a lamellated structure. Pronounced longitudinal cleavage. A transverse cleavage is distinct. The lamellae are deformed by bending. Replaced by chalcocite. Contains inclusions of silver, chalcocite, vrbaite, avicennite(?). Ref.: 1380.
H: > chalcothallite, ≪ cuprite VHN: no data.	Forms no intergrowths with other minerals. Occurs as single grains, and as inclusions in chalcothallite. Ref.: 232, 442, 967, 1019, 1380, 1392.

Name Formula Crystal System	1: Colour 2: Bireflectance 3: Anisotropy 4: Internal reflections	Reflectance in % (Wavelength in nanometres)

BERTHIERITE
$FeSb_2S_4$
Ortho.
	1. White-grey with a pinkish or brownish tint.
	2. Strong and characteristic:
	    ∥ a brownish pink (about same as for pyrrhotite).
	    ∥ b grey-white.
	    ∥ c white.
	3. Very strong, about same as for stibnite, but colours still more vivid. Sharp extinction.
	4. Not present.
		About 30-40 %.

JAMESONITE
$4PbS \cdot FeS \cdot 3Sb_2S_3$; may contain Bi up to Bi:Sb = 1.07:1.
Mono.
	1. White.
	  → galena, according to orientation: about the same or slightly greenish.
	  → stibnite, distinctly lighter, some sections more greenish.
	2. Air and oil: strong.
	  Oil: ∥ c bright white with slight yellow-green tint.
	    ∥ b greyish yellow-green.
	    ∥ a dark yellow-green or olive.
	3. Strong: grey, tan, brown, light blue, dark blue. Basal sections are nearly isotropic.
	4. Bi-jamesonite has reddish int. refl.
		$R_g \parallel c$ $R_p \perp c$
		472 nm 39.7 36.3
		550 40.6 36.5
		579 38.7 34.7
		640 36.3 32.5
		Dispersion curves: [959].

Hardness H: polishing hardness VHN: micro-indentation hardness in kg/mm^2	Miscellaneous
H: $\geq$ stibnite, pyrargyrite, $\ll$ sphalerite. VHN: 102-213 [159, 194, 1715]. VHN$_{50}$: 134-206 [1086, 1715].	Commonly radially developed or with fibrous or spathic texture. Needle-like crystals, sometimes in aggregates. Forms orientated intergrowths with stibnite and myrmekitic ones with chalcopyrite. Replaces arsenopyrite, pyrite, pyrrhotite, gudmundite, sphalerite. May be replaced by stibnite. Inclusions of pyrite, arsenopyrite, marcasite or pyrrhotite not uncommon. Residual patches of berthierite may be included in stibnite. Other ass.: sphalerite, galena, breithauptite, niccolite, Ni-Co-Fe-arsenides, silver, pyrargyrite. Ref.: 182, 184, 219, 275, 348, 477, 1086, 1164, 1237, 1281, 1379, 1392, 1531, 1540, 1546, 1629, 1732, 1734.
H: $<$ galena. VHN: 67-126 [159, 194, 1113, 1715]. VHN$_{20}$: 104-123 [959].	Occurs as needle-shaped crystals and in aggregates of these. Cleavage $\perp$ elongation usually well visible as distinct from boulangerite; absence of transverse cleavage, however, does not prove boulangerite. Twinning lamellae very common and characteristic, unlike those of stibnite; developed $/\!/$ elongation, (100); very thin crystals may be untwinned. Zonal texture not observed. May be intimately intergrown with galena, pyrargyrite and boulangerite. Zonal replacement of pyrite by jamesonite occurs. Replaces sphalerite, pyrite, freibergite, arsenopyrite, semseyite. Replaced by chalcopyrite and galena. Occurs as inclusions in sphalerite, arsenopyrite, chalcopyrite, tetrahedrite. May contain inclusions of antimony. Occurs as decomposition product of geocronite. Other ass.: pyrrhotite, marcasite, gel-pyrite, cassiterite, stannite, franckeite, stibnite, silver, gold, andorite, proustite, polybasite, realgar, sternbergite, Pb-Sb-minerals. Diff.: stibnite shows a stronger birefl. and anistr. Boulangerite shows another birefl.; no basal cleavage, no twins. Ref.: 110, 224, 227, 239, 275, 339, 628, 682, 741, 776a, 891, 967, 1056, 1237, 1281, 1316, 1379, 1392, 1490, 1529, 1647, 1653, 1666, 1732. Note: Parajamesonite: has the same chem. comp., but another cryst. Forms columnar crystals. Ass.: sphalerite, galena, pyrrhotite, chalcopyrite, tetrahedrite. Ref.: 1733.

Description of minerals from Table III

Name / Formula / Crystal System	1: Colour / 2: Bireflectance / 3: Anisotropy / 4: Internal reflections	Reflectance in % (Wavelength in nanometres)
MENEGHINITE $Cu_2S \cdot 26PbS \cdot 7Sb_2S_3$ Ortho.	1. White. ⟶ galena, slightly bluish; galena shows a pinkish tint. 2. Weak, much weaker than for jamesonite. White with a brownish yellow tint to greyish white with a reddish or faint greenish tint. Pale $\parallel$ c, darker $\perp$ c. 3. $\parallel$ elongation: maximum, strong. $\perp$ elongation: minimum, moderately strong. Light tan, brown, blue-grey; in oil stronger than jamesonite and boulangerite. Straight extinction. 4. Rare, red.	480 nm 40.7-46.7 540 40.0-45.9 580 39.6-45.6 640 38.6-43.0 Dispersion curves: [197], see also: [958, 959].
MINIUM Pb_3O_4 Tetr.	1. Influenced by strong int. refl., it has a pinkish yellow-grey colour. The normal colour is bluish. 2. Weak, yellowish grey or orange-grey to grey-bluish. 3. Obscured by int. refl. 4. Abundant and strong, see at colour.	About 20 %.
MASSICOTITE β-PbO Ortho.	1. Grey-white. 2. Not discernible. 3. Obscured by int. refl. 4. Abundant, white.	About 20 %.
LITHARGITE α-PbO Tetr.	1. Grey-white with numerous int. refl. 2. Not discernible. 3. Strongly obscured by int. refl. 4. Abundant, purpur-red.	About 20 %.
Ag-SULPHOSALTS		Between 20 and 40 %.

Hardness H: polishing hardness VHN: micro-indentation hardness in kg/mm^2	Miscellaneous
H: slightly < galena*, distinctly < chalcopyrite. VHN: 113-155 [1715]. VHN$_{25}$: 135-183 [1715]. VHN$_{20}$: 119-170 [958, 959]. VHN$_{15}$: 137-149 [194].	Occurs as acicular or needle-like idiomorphic crystals and as granular aggregates. One perfect cleavage ∥ (010). Warren (1947) reports a well-marked twinning, apparently ⊥ cleavage. Ramdohr (1950 c) and Williams (1960 b) report a parquet-like twinning due to deformation resembling that of jamesonite. Zonal texture not observed. Found as irregular inclusions in galena. May contain exsolution lamellae of geocronite. Forms intergrowths with galena, bournonite, chalcopyrite, sphalerite, boulangerite. Replaces pyrite, sphalerite, galena, arsenopyrite; may be replaced by galena, chalcopyrite, boulangerite. Very difficult to distinguish from boulangerite, zinkenite, jamesonite. Parag.: pyrite, chalcopyrite, sphalerite, galena, jamesonite, boulangerite, stannite, arsenopyrite, gold, tetrahedrite, geocronite, antimony, stibnite, bournonite, jordanite, wurtzite, covellite, freibergite, pyrargyrite, pyrrhotite, marcasite. Sobotka, 1961, gives a complete list of occurrences. Ref.: 18, 121, 197, 339, 368, 418a, 720, 958, 1104, 1227, 1237, 1382, 1392, 1416, 1531, 1647, 1695.
H: < galena, lithargite, massicotite. VHN: no data.	Generally extremely fine-grained. Replaces and forms pseudomorphs after galena and cerussite. Parag.: massicotite, lithargite, galena, cerussite, wulfenite, secondary Zn-minerals. Ref., 524, 751, 1408.
H: > minium. VHN: no data.	Often shows alteration to lithargite. Parag.: minium, lithargite, galena, cerussite, wulfenite, secondary Zn-minérals. Ref.: 524.
H: no data. VHN: no data.	Parag.: minium, massicotite, galena, cerussite, wulfenite, secondary Zn-minerals. Ref.: 524.
H: usually distinctly < galena.	For the description, see p.255–269.

* Burnol et al. (1965) and Aicard et al. (1968) report H > galena.

Description of minerals from Table III

Name / Formula / Crystal System	1: Colour / 2: Bireflectance / 3: Anisotropy / 4: Internal reflections	Reflectance in % (Wavelength in nanometres)
HOROBETSUITE $(Bi_{0.45}Sb_{0.55})_2S_3$ Ortho.	1. See at birefl. 2. Not distinct: very pale brown to pale blue. 3. In contrast to bismuthinite, yellow to dark brown; straight extinction. 4. Not present.	Intermediate between bismuthinite and stibnite.
STROMEYERITE $Cu_{1+x}Ag_{1-x}S$ Ortho. $\stackrel{94°}{\rightleftharpoons}$ cubic.	1. Grey with a violet-pinkish tint. $\longrightarrow$ chalcocite, lavender-grey. $\longrightarrow$ jalpaite and mckinstryite, very similar. 2. Weak but distinct (in oil): grey-brown to light grey with a slight bluish green or pinkish tint. 3. Strong: light violet, purple, brown, orange and yellow. 4. Not present.	470 nm 27.8-35.4 546 25.8-29.6 589 25.2-28.4 650 25.0-27.6 [194].
Pb-Sb-SULPHOSALTS	1. White with greenish or yellowish tints. 2. Distinct to strong. 3. Distinct to strong.	Between 35 and 45 %.

Hardness H: polishing hardness VHN: micro-indentation hardness in kg/mm^2	Miscellaneous
H: no data. VHN: no data.	Sb-rich end member in the naturally occurring solid solution series between Bi_2S_3 and Sb_2S_3. Occurs as prismatic crystals ($/\!\!/$ c). Ratio Bi_2S_3:Sb_2S_3 varies from 9:11 to 13:17. Parag.: sulphur, Fe-sulphides. Ref.: 595, 1429a.
H: <galena, chalcocite. VHN: 38-44 [159]. VHN_{15}: 27-62 [194].	Needle-like idiomorphic crystals are rare. In these crystals the anisotropy may be less distinct than in aggregates or intergrowths of stromeyerite with other minerals. An oleander-leaf texture is not uncommon, due to change of texture on heating. Occurs as exsolution intergrowths with chalcocite; may be intergrown with silver, freibergite, tennantite, tetrahedrite, bornite, chalcopyrite, galena. Replaces bornite, pyrite, tennantite, tetrahedrite, chalcopyrite, chalcocite, galena, pitchblende; may be replaced by covellite under supergene alteration. Occurs as reaction rims between silver and chalcocite. Observed as inclusions in chalcocite, galena. Other ass.: sphalerite, stibioluzonite, enargite, sternbergite, seligmannite, polybasite, copper. Physically it is impossible that argentite (or acanthite) and stromeyerite are formed together. Stromeyerite described in association with argentite (or acanthite) is either jalpaite or mckinstryite. Difficult to distinguish under the microscope stromeyerite from jalpaite or mckinstryite. Ref.: 6, 55, 339, 342, 458, 553, 718, 840, 918, 967, 1054, 1237, 1260, 1310, 1343, 1358, 1361, 1365, 1367, 1392, 1405, 1441, 1603.
H: slightly <galena. VHN: between 100 and 200.	For the description, see p. 271–279.

Name Formula Crystal System	1: Colour 2: Bireflectance 3: Anisotropy 4: Internal reflections	Reflectance in % (Wavelength in nanometres)
COVELLITE CuS Hex.	1. Basal sections indigo-blue; cross sections show extremely high birefl., both in air and in oil. 2. Extraordinary high; most characteristic feature. In air: O deep blue with violet tint. E bluish white; ⟶ chalcocite, slightly lighter and pinkish. In oil: O Purple to violet-red; scratches may appear as orange-red light lines. E blue-grey, slightly pinkish. Orientation: E∥c, O⊥c. 3. Extremely high; 45°: fiery orange, reddish brown. 4. Not present.	R_o R_e 486 nm 11.3 27.5 546 7.2 24.3 589 4.5 21.7 656 5.9 21.8 Dispersion curves: [485].
BLAUBLEIBENDER COVELLITE $Cu_{1+x}S$ (in comparison with normal covellite a slight Cu-excess of 1.5-2 %) Hex.	1. The name blaubleibender covellite refers to a Cu-sulphide without the characteristic purplish or violet-red colours of covellite for O in oil, but instead showing the same colour as in air. 2. Extremely high: air: O : blue E : bluish white oil: O : deep blue, often with a slight violet tint. E : same as in air, only darker. 3. Strong, but the reddish interference colours are not as vivid as for normal covellite. 4. Not present.	R_o 486 nm 17.5 548 14.0 591 10.8 657 8.0 Dispersion curve: [485]

Hardness H: polishing hardness VHN: micro-indentation hardness in kg/mm^2	Miscellaneous
H:$\gg$ argentite, slightly <galena,<chalcopyrite; basal sections slightly <chalcocite, cross sections slightly>chalcocite. VHN: 59-129 [159, 1113, 1715]. VHN$_{50}$: 92-110 [194].	Usually occurs as idiomorphic tabular crystals. Perfect basal cleavage. No twinning or zoning observed. Occurs as an alteration product of many Cu-minerals. Replaces chalcopyrite, stromeyerite, enargite, tennantite, tetrahedrite, bornite, chalcocite, emplectite, stannite, sphalerite, galena, pyrite, and most of the associated minerals. May be replaced by digenite, chalcocite, bornite, chalcopyrite. Forms intergrowths, often orientated, with chalcocite, digenite, bornite, sphalerite, chalcopyrite, galena, pyrite, marcasite, aikinite, cuprite. Other ass.: stibioluzonite, delafossite, tenorite, silver. Ref.: 87, 117, 187, 227, 339, 891, 967, 1237, 1343, 1358, 1392, 1603, 1629.
H:$\sim$ covellite. VHN: no data.	Blaubleibender covellite is an independant mineral. There is no solid solution with stoechiometric CuS (normal covellite). Occurs less than normal covellite. Blaubleibender covellite is an intermediate alteration product. Originates from orthorhombic chalcocite or digenite by the action of acid water. When replacing chalcocite, nearly always intergrown ∥ (001) of chalcocite. Replaces digenite, bornite, chalcopyrite. Ass.: normal covellite, other Cu-minerals, sphalerite, galena, acanthite, scheelite, pyrite. Ref.: 424, 824, 933, 1237, 1616.

Description of minerals from Table III

Name / Formula / Crystal System	1: Colour / 2: Bireflectance / 3: Anisotropy / 4: Internal reflections	Reflectance in % (Wavelength in nanometres)

IDAITE
Cu_3FeS_4*
Probably tetr.*
	1. See at birefl.
	2. Very strong: air: O: reddish orange to red-brown.
	    E: bright yellowish grey.
	    oil: same, but more vivid.
	3. Extremely strong:
	  - as decomposition product in bornite: in 45° position, vivid green.
	  - as crystals of hypogene origin: in 45° position: greyish green.
	4. Not present.
		R_o
		480 nm 17.9–22.9
		540 23.6–31.2
		580 27.5–33.4
		640 29.6–35.1
		
		R'_e
		480 nm 14.8–22.6
		540 21.0–29.2
		580 27.0–33.6
		640 29.6–35.9
		
		The dispersion curves [832] cross at about 580 nm.

* Frenzel (1959 a and b) originally gave Cu_5FeS_6 as formula because the mineral was very similar to the hexagonal synthetic compound Cu_5FeS_6. Lévy (1967) gave the new formula based on microprobe-analyses. The crystal structure may be the same as mawsonite, which is considered as an intermediate member between stannite and idaite.

Hardness H: polishing hardness VHN: micro-indentation hardness in kg/mm^2	Miscellaneous
H: somewhat > covellite. VHN: 176-202 [1715]. VHN$_{20-50}$: 216-260 [810].	Two modes of occurrence: (a) supergene, as alteration product in bornite: lattices, lamellae and veinlets $/\!/$ (100) and (111) of bornite, often together with chalcopyrite in the same form. The formation of idaite may be preceeded by spindles of somewhat higher refl. and more yellow-orange colour than bornite. In other cases bornite shows a "fracture-disease" previous to idaite forming. As a supergene mineral it may also form films between chalcopyrite and chalcocite, or replace pyrite in company of chalcocite, digenite and covellite. (b) hypogene: large hexagonal tabular crystals. Parallel intergrowths with covellite may occur. Forms inclusions in pyrite, wurtzite and bornite. Alters to covellite and pyrite. Ref.: 424, 425, 429, 431, 483, 760, 832, 1503, 1569, 1720.

Name Formula Crystal System	1: Colour 2: Bireflectance 3: Anisotropy 4: Internal reflections	Reflectance in % (Wavelength in nanometres)

CHALCOCITE
Cu_2S
 	
Ortho. $\xrightleftharpoons{103°}$ hex.
	1. Bluish white. Bluish streaks may be due to polishing.
	   ⟶ galena, distinctly bluish.
	   ⟶ pyrite, bluish grey.
	   ⟶ covellite, white, no pinkish tint.
	   ⟶ tetrahedrite, distinctly bluish.
	   ⟶ bornite, bluish white.
	   ⟶ copper, bluish grey.
	2. Very weak.
	3. Weak to distinct; emerald-green to light pinkish (strong illumination); straight extinction. Apparant isotropism may be a result of too great pressure while polishing.
	4. Not present.
		About 30 %.

Hardness H: polishing hardness VHN: micro-indentation hardness in kg/mm^2	Miscellaneous
H: $\gg$argentite,~digenite, galena,$>$basal sections of covellite,$<$prismatic sections of covellite, $<$bornite, tetrahedrite. VHN: 58-98 [159, 194, 1113, 1715]. VHN$_{15-25}$: 62-99 [194, 810, 1715].	Chalcocite formed above 103°C is generally coarse-grained and shows intergrowths with digenite. Lamellar twinning very common: partly broad, lancet-shaped (proving formation above 103°C), partly fine-lamellar (then most probably formed above 103°C); absence of lancet-like twins does not prove a formation below 103°C. Coarse-grained pseudomorphs after bornite, digenite or pyrite very common. Quickly crystallized, this type may be finer grained, resembling supergene chalcocite. Exsolutions are numerous, especially from Cu_2S- or Cu_5FeS_4-rich digenites, e.g. "lamellar chalcocite", a regular intergrowth of chalcocite, digenite, covellite and bornite. Lamellar texture, however, may also be the result of replacement of bornite // (111). Supergene chalcocite is generally fine-grained. May be replaced by digenite. "Steely chalcocite" principally consists of this type. "Sooty chalcocite" commonly consists of a mixture of very fine-grained supergene chalcocite, supergene digenite and covellite. Etching brings out a granular or irregular pattern. Extremely fine-grained aggregates are apparently isotropic; etching leaves an "inactive" pattern. May occur together with chalcopyrite formed by decomposition of bornite. Replaces most of the associated minerals, such as: pyrite (then numerous minute remnants may be enclosed), chalcopyrite (bornite usually forming reaction rims), sphalerite, galena, enargite, tetrahedrite, stannite, bornite, digenite, algodonite, domeykite, niccolite. High temperature chalcocite may be replaced by chalcopyrite, bornite, digenite, covellite, argentite, niccolite, silver, copper. Chalcocite may show eutectic or pseudo-eutectic intergrowths with digenite, covellite, bornite, chalcopyrite, stromeyerite, tetrahedrite, galena. Other ass.: occasionally iron oxides, tellurides, selenides, wittichenite. Ref.: 83, 87, 187, 278, 297, 333, 334, 339, 361, 536, 658, 1044, 1214, 1219, 1237, 1280, 1333, 1356, 1358, 1361, 1364, 1365, 1367, 1392, 1442, 1603, 1632. Note: So called "pinkish grey" chalcocite is greyish white with pinkish tint to pale pinkish grey (in oil bluish violet-grey). Birefl. very weak. Anisotropy is stronger than for orthorhombic chalcocite: pinkish and greenish blue. Occurs in digenite and in "lamellar chalcocite". Ref.: 1237.

Name Formula Crystal System	1: Colour 2: Bireflectance 3: Anisotropy 4: Internal reflections	Reflectance in %
DJURLEITE $Cu_{1.96}S$ Ortho.	1. Very similar to chalcocite. 2. Very similar to chalcocite. 3. Very similar to chalcocite. 4. Not present.	Similar to chalcocite.
DIGENITE $Cu_{1.765}S$ to $Cu_{1.79}S$*, may contain some Ag. Cubic.	1. Greyish blue, varying with quantity Cu_2S or CuS dissolved. → galena, clear blue. → bornite, clear blue. → chalcocite, distinct blue, darker. Some spec. have lost their blue colour more or less completely due to admixtures, but remain isotropic; the colour, then, has turned greyish white, distinctly darker than that of chalcocite. 2. Not present. 3. Isotropic, remains isotropic at normal temperatures; may sometimes show a weak anomalous anisotr. 4. Not present.	480 nm 25.2 540 21.3 580 20.0 640 19.3 Dispersion curve: [1066].

* Above 83°C digenite may contain increasing amounts of Cu_2S or CuS in solid solution. Oosterbosch et al., 1964 described digenite from Katanga containing up to 18% Se. It shows no bluish colour, but more the colour of chalcocite. VHN_{15}: 30-38.

Hardness H: polishing hardness VHN: micro-indentation hardness in kg/mm^2	Miscellaneous
H: = chalcocite. VHN_{50}: 74–83 [194].	Cannot exist above 93°C: breaks down to hexagonal chalcocite and high digenite. Because this process is reversible, the presence of djurleite provides no evidence of deposition below 93°C, unless it occurs as crystals or as large pure masses. Parag.: digenite, bornite, pyrite, chalcocite, chalcopyrite, covellite, cuprite, tenorite, goethite, delafossite. Diag.: very similar to chalcocite in all aspects. Diff.: djurleite behaves differently from chalcocite during polishing: individual grains of djurleite show many different shades of blue and grey. However, continued polishing ultimately results in a uniform colour. Ref.: 946, 1287, 1289, 1307, 1394a.
H: ~ chalcocite, galena. VHN: 56–83 [159, 194, 1715]. VHN_{15-25}: 30-74 [194, 1066, 1715].	Octahedral cleavage often visible. As exsolution products chalcocite or covellite lamellae, sometimes also chalcopyrite blebs or spindles, may occur, proving a formation above 83°C; absence of such lamellae, however, does not prove a formation at lower temperatures. The exsolution of digenite may sometimes result in so called "lamellar chalcocite", a regular intergrowth of chalcocite with digenite and covellite or bornite or both. Irregular or granular intergrowths of these minerals also occur. Extremely fine lamellar intergrowths with chalcocite of covellite (due to exsolution) may suggest an "anisotropic digenite". Formed below 83°C. digenite often encloses bornite. Sometimes it occurs reticulated in bornite, chalcopyrite, pyrite, tetrahedrite, enargite, etc. or as fine lamellae in chalcocite. For other replacements, see chalcocite. Ref.: 187, 278, 310, 950, 1066, 1219, 1237, 1289, 1392, 1500, 1616. Note: Anilite, Cu_7S_4, is orthorhombic. Occurs as prismatic or platy crystals intimately intergrown or associated with djurleite. No optical data can be given because mixtures of anilite and djurleite transform to a homogeneous digenite-like phase at the surface in the process of polishing. Ref.: 948a.

Description of minerals from Table III

Name Formula Crystal System	1: Colour 2: Bireflectance 3: Anisotropy 4: Internal reflections	Reflectance in % (Wavelength in nanometres)
PARKERITE α-Ni$_3$(Bi,Pb)$_2$S$_2$ Ortho.	1. Creamy white with a faint mauvish tint. $\longrightarrow$ galena, sharply contrasting. 2. Distinct, creamy white to greyish creamy white. 3. Strong, greenish grey to yellowish brown or slate-blue to salmon-pink, depending on adjustment of the nicols. 4. Not present.	470 nm 42.0-44.7 to 44.4-47.7 546 43.5-46.1 to 46.2-48.0 589 44.0-46.5 to 47.6-48.6 650 45.2-47.6 to 48.0-50.1 [1155]

SHANDITE
β-Ni$_3$Pb$_2$S$_2$
Hex.
	1. White with a slight creamy tint.
	2. Strong, E: similar to heazlewoodite.
	    O: much darker and bluish grey.
	    $\longrightarrow$ heazlewoodite, bluish green.
	3. Very strong, grey blue to yellow-brown.
	4. Not present.
		No data.

Hardness H: polishing hardness VHN: micro-indentation hardness in kg/mm^2	Miscellaneous
H: slightly $<$ or $\sim$ galena. VHN$_{15}$: 111-142 [1155].	Usually occurs as xenomorphic grains; also as irregular, rounded and subhedral particles; as veinlets and stringers. Usually polysynthetically twinned, especially well visible in oil; may be transversed by less pronounced lamellae. Notched cleavage traces. As a result of the twinning a continuous cleavage trace suffers displacement at the junction of alternate lamellae. This zig-zag trace may be sufficiently characteristic to permit of recognition of parkerite. Occurs as inclusions in bismuth, galena, niccolite, siegenite, cobalt-pentlandite. Rimmed by bismuth when occurring as inclusions in pyrrhotite and bornite. Surrounds bismuth and forms intergrowths with bismuthinite. Other ass.: chalcopyrite, sphalerite, cubanite, sperrylite, arsenides, tellurides, gold, schapbachite, rammelsbergite, bravoite, maucherite, safflorite, langisite, pyrite, marcasite. Ref.: 588, 908, 1137, 1155, 1342, 1531.
H: no data. VHN: no data.	Practically always occurs in orientated intergrowths with heazlewoodite: lamellae of shandite $/\!/$ heazlewoodite (0001). Replaced by heazlewoodite. May contain lamellae of sphalerite$/\!/$ (0001). Parag.: sphalerite, chromite, magnetite, heazlewoodite. Ref.: 1137, 1225, 1237, 1629, 1692.

Name Formula Crystal System	1: Colour 2: Bireflectance 3: Anisotropy 4: Internal reflections	Reflectance in % (Wavelength in nanometres)
GALENA PbS (may contain some Te or Se, and Ag in solid solution).* Cubic	1. Bright white, sometimes with a pinkish tint. Traces of tellurium change the colour from white to purple. Extremely fine-grained galena is much darker than coarse-grained galena: tan to grey. ⟶ sphalerite, white. ⟶ boulangerite, pinkish. ⟶ bismuthinite, pinkish. ⟶ tellurium, slightly greyer. ⟶ arsenic, slightly grey. ⟶ altaite and clausthalite, slightly darker and pinkish. ⟶ stibnite, distinctly lighter. ⟶ tennantite, pinkish. 2. Not present. 3. Isotropic; a weak anomalous anisotr. is not uncommon: medium grey to grey-black. 4. Not present.	470 nm 47.9 546 43.8 589 43.1 650 43.7 Dispersion curve: [300]; see also [43].
Bi-SULPHOSALTS	1. Greyish white. 2. Weak. 3. Distinct to strong.	About 40-45 %.

* For minor elements in galena, see Fleischer (1955).

Hardness H: polishing hardness VHN: micro-indentation hardness in kg mm^2	Miscellaneous
H: $\gg$argentite, $>$proustite, stephanite, covellite, slightly$>$boulangerite, jamesonite, $\sim$chalcocite, bismuthinite, slightly $<$bournonite, $<$bornite, chalcopyrite*, tetrahedrite VHN: 56-116 [159, 194, 1113, 1715]. VHN$_{10-25}$: 64-110 [810, 1715].	Occurs as granular aggregates and as skeletal crystals. Very often developed with crystallographic boundaries. A perfect cubic cleavage is nearly always visible. Triangular pits along the cleavage lines are very typical. Twinning caused by mechanical deformation or pressure may occur. Zonal texture not uncommon; may be brought out by etching. Replacements are numerous, partly due to its latent plasticity. Replaces arsenopyrite, pyrite, marcasite, pyrrhotite, sphalerite, chalcopyrite, tetrahedrite, bornite, enargite. May be replaced by covellite, chalcocite, silver, electrum, Ag-Au-tellurides, argentite, arsenopyrite, stannite, magnetite, sphalerite, pyrite, bismuth, uraninite, stephanite. Inclusions of other minerals very common, e.g. sphalerite, tetrahedrite, freibergite, boulangerite, stannite, electrum, silver, argentite, dyscrasite, pyrargyrite, polybasite, and other Ag-minerals, bismuthinite, tetradymite, bournonite, antimony, galenobismutite. The silver content of many galena spec. may partly be due to solid solution, partly to minute inclusions of Ag-minerals, such as argentite, freibergite, pyrargyrite, polybasite, silver, dyscrasite, stromeyerite, pearceite, stephanite, schapbachite. May occur as inclusions in sphalerite, chalcopyrite, uraninite, geocronite, galenobismutite. Forms pseudomorphs after cylindrite, franckeite and teallite in graphic-like intergrowths with cassiterite and boulangerite. Forms coarse- or fine-grained intergrowths with pyrite, sphalerite, chalcopyrite, bournonite, boulangerite, geocronite. Forms myrmekitic (pseudo-eutectic) intergrowths with tetrahedrite, chalcopyrite, freibergite, covellite, chalcocite, stromeyerite, stephanite, polybasite, pyrargyrite, pearceite, proustite, bismuth. Forms orientated intergrowths with chalcopyrite, tetrahedrite, pyrrhotite, millerite, antimony, bismuth, tetradymite, chalcocite, argentite, altaite, schapbachite, covellite, polybasite, bournonite, pyrargyrite, jamesonite, geocronite, galenobismutite, pyrite, ullmannite, arsenopyrite. Other ass.: Co-Ni-arsenides, tellurides, Bi-minerals, pentlandite, sperrylite, cubanite. Ref.: 223, 227, 332, 333, 339, 588, 599, 726, 805, 806, 891, 1058, 1192, 1237, 1328, 1343, 1346, 1347, 1358, 1379, 1392, 1393, 1479, 1603, 1608, 1629, 1662, 1679, 1705, 1732.
	For the description, see p.281-295.

---

* Dunn (1937b) observed a galena spec. with unusually great hardness: slightly$<$chalcopyrite.

Description of minerals from Table III

Name Formula Crystal System	1: Colour 2: Bireflectance 3: Anisotropy 4: Internal reflections	Reflectance in % (Wavelength in nanometres)
BISMUTHINITE Bi_2S_3. Forms a solid solution series with stibnite (see horobetsuite) and with aikinite (see aikinite). Ortho.	1. White (in air); in oil with a bluish grey tint. ⟶ bismuth, darker, bluish grey. ⟶ chalcopyrite, bluish grey. ⟶ galena, lighter, creamy white. ⟶ stibnite, much lighter. ⟶ emplectite, lighter and whiter. Emplectite, wittichenite, and annivite are more or less greyish brown compared with bismuthinite. 2. Weak to distinct; much weaker than for stibnite. c = creamy white. a = bluish grey-white. b = grey-white. 3. Very strong, esp. in oil, but weaker than for stibnite. Slate-grey to yellowish brown or grey-violet. Straight extinction. Large crystals commonly show undulatory extinction. 4. Not present.	470 nm 39.5-50.6 546 37.8-50.2 589 37.4-49.2 650 36.8-47.2 [194].
Sn-SULPHOSALTS (partly)		

Hardness H: polishing hardness VHN: micro-indentation hardness in kg/mm^2	Miscellaneous
H; $\geq$ galena; slightly $>$ emplectite; $<$ chalcopyrite. VHN: 67-216 [159, 1113, 1715]. VHN$_{50}$: 68-190 [194, 810, 1715].	Usually radial-fibrous developed, seldom granular. Idiomorphic acicular and needle-like crystals are rare. Cleavage $/\!/$ (010) not uncommon. Lamellar twinning or a "spindle-like" texture, due to stress, is reported. Replaces bismuth, emplectite, stibioluzonite, arsenopyrite, pyrite, pyrrhotite, sphalerite, chalcopyrite, magnetite, uraninite, parkerite, May be replaced by Bi-tellurides, Au-Ag-tellurides, stannite, cassiterite (rare), tetrahedrite, sulphosalts, dignite. May show inclusions of bismuth, gold, chalcopyrite, cosalite. May occur as inclusions in stannite, chalcopyrite, bismuth, ikunolite, tetrahedrite, molybdenite, cassiterite, pyrrhotite. Shows intergrowths with emplectite, chalcopyrite, galena, argentite, stannite, chalcocite, gold, tetradymite, tetrahedrite, parkerite. Other ass.: herzenbergite, franckeite, cylindrite, teallite, wolframite, huebnerite, scheelite, ferberite, jamesonite, stibnite, cubanite, cosalite, arsenic, wittichenite, annivite, Ni-Co-arsenides, marcasite, joseite, niccolite, langisite, siegenite. Diff.: stibnite shows a stronger birefl. Ref.: 5, 6, 7, 11, 12, 84, 227, 339, 344, 458, 588, 632, 739, 803, 806, 844, 899, 1122, 1155, 1204, 1207, 1237, 1320, 1322, 1334, 1356, 1358, 1392, 1429a, 1451, 1558, 1603, 1629.
	For the description, see p.307-319

Name Formula Crystal System	1: Colour 2: Bireflectance 3: Anisotropy 4: Internal reflections	Reflectance in % (Wavelength in nanometres)
BOURNONITE $2PbS \cdot Cu_2S \cdot Sb_2S_3$ Ortho.	1. Greyish white with a distinct blue or bluish green tint. ⟶ boulangerite, darker. ⟶ tetrahedrite, lighter, bluish. ⟶ galena, darker, bluish green. 2. Air: weak; only visible on grain boundaries and on twin lamellae. Oil: distinct: white with a bluish green tint to white with an olive-brown tint. 3. Air: weak. Oil: distinct (esp. when twinned): pale blue, greenish grey, brownish yellow, dark brown, purplish. 4. Not present.	470 nm 36.4-37.3 to 37.5-40.0 546 34.6-35.2 to 35.9-39.2 589 34.3-34.7 to 35.4-39.4 650 33.3 to 34.6-39.9 [558a].
LAUTITE CuAsS (some Ag may be present) Ortho.	1. Greyish white with a brownish pink tint (in oil). ⟶ enargite, very similar. ⟶ galena, brownish. ⟶ pyrrhotite, less brown. 2. Weak, only visible in oil on grain boundaries: brownish pink to somewhat more bluish. 3. Distinct, much weaker than for enargite: bluish green or bluish violet to violet-brown. 4. Very weak, sometimes visible.	515 nm 31.8-32.5 [644].
SINNERITE $Cu_{1.4}As_{0.9}S_{2.1}$ Tricl.	1. ⟶ galena, distinct yellow-brown. 2. Not observable. 3. Air: distinct. Oil: strong, grey-brown to grey-blue. 4. Not present.	530 nm 29.5-31.5 [522].
Pb-As-SULPHOSALTS	1. White with different tints. 2. Weak to distinct. 3. Distinct to strong.	Between 30 and 40 %.

Hardness H: polishing hardness VHN: micro-indentation hardness in kg/mm^2	Miscellaneous
H:$>$boulangerite, jamesonite, stibnite, slightly $>$galena, $<$chalcopyrite, sphalerite, tetrahedrite. VHN: 132-213 [159, 1113, 1715]. VHN$_{20-50}$: 166-212 [194, 810, 1715].	Occurs usually as aggregates of polygonal grains. No apparent cleavage. May be transversed by minute cracks. Twinning after (110) very common and characteristic, often parquet-like developed. Occurs as inclusions in galena, ullmannite, bornite. May contain inclusions of boulangerite, chalcopyrite, galena. Veins tetrahedrite, pitchblende, pyrite. Replaces galena, pyrargyrite, arsenopyrite, pyrite, meneghinite, chalcopyrite. May be replaced by galena, tetrahedrite, covellite, stibioluzonite. Forms myrmekitic intergrowths with galena. Forms reaction rims between tetrahedrite, chalcopyrite, geocronite and galena, between chalcopyrite and boulangerite, and between chalcopyrite and chalcostibite. Other ass.: semseyite, stibnite, jordanite, chalcocite, gold, silver, bismuth, sphalerite. Diff.: Twinning, colour and anisotropy may allow identification from similar minerals. Ref.: 18, 31, 33, 239, 322, 339, 346, 556, 598, 738, 797, 891, 1174, 1232, 1237, 1270, 1281, 1317, 1379, 1416, 1452, 1454, 1456, 1545, 1646, 1687.
H: $\ll$enargite, $<$arsenic. VHN$_{50}$: 142-147 [810].	Occurs as bundle-like aggregates of idiomorphic crystals. Cleavage not observed (as distinct from enargite). Twinning may occur. Occurs as inclusions in digenite. Diff.: enargite shows a much stronger anisotr. and often a distinct cleavage; enargite is harder than arsenic, tennantite, chalcopyrite. Parag.: arsenic, enargite, stibioluzonite, tennantite, covellite, pyrite, chalcopyrite, galena, loellingite, proustite, bismuth, bornite, digenite, chalcocite. Ref. 285, 635, 644, 873, 896, 1237, 1494.
H: no data. VHN: 357-390 [522].	Under crossed nicols abundant twinning is visible which shows a remote similarity to the twinning of arsenopyrite. Parag.: tennantite, sphalerite. Ref.: 522, 873, 1018, 1028. Note: Nowackiite, $Cu_6Zn_3As_4S_{12-13}$, hexagonal. Forms twins $\parallel$ (0001). Ass.: sphalerite. VHN$_{30}$: 480-500; Refl. about 30 %. Ref.: 872, 876, 1018, 1020.
H:$\sim$galena. VHN: usually around 150.	For the description, see p.297-305.

Description of minerals from Table III

Name Formula Crystal System	1: Colour 2: Bireflectance 3: Anisotropy 4: Internal reflections	Reflectance in % (Wavelength in nanometres)

GOLD
Au (may contain Ag, Pd, Cu, Bi, Pt, Hg, Rh).
Cubic
	1. Bright or "golden" yellow, varying with the content of admixed metals.
	⟶ electrum, yellower.
	⟶ chalcopyrite, much paler and brighter yellow.
	⟶ silver, yellow.
	⟶ bismuth, more yellow.
	⟶ platinum, much more yellow.
	⟶ all sulphides, much lighter.
	Pd-rich gold is nearly creamy white, Ag-rich gold is pale yellow, and Cu-rich gold looks pink to reddish.
	2. Not present.
	3. Isotropic; no complete extinction under crossed nicols, but typical greenish tints.
	4. Not present.
		Depending on method and quality of polishing, and on Ag-content. For pure gold, values estimated from the dispersion curve [230]:
		
		470 nm 35
		550 66
		590 71
		650 82
		
		Reflectance with varying Ag-content:
		
		              470 nm 541 nm
		1000 fine 36.4 71.6
		900 fine 43.5 77.9
		800 fine 56.0 83.1
		700 fine 66.8 86.2
		600 fine 75.1 88.0
		500 fine 81.5 89.4
		                               [327a].

Hardness H: polishing hardness VHN: micro-indentation hardness in kg/mm^2	Miscellaneous
H: >galena, <tetrahedrite, sphalerite, ~chalcopyrite. VHN: 41-94 [159, 1113, 1715]. VHN$_{10-25}$: 42-88 [194, 810, 1715].	Related minerals: Electrum: (Au, Ag) with Ag-content higher than 25%. Rhodite: (Au, Rh). Porpezite: (Au, Pd). Auricuprid: AuCu$_3$: occurs in exsolution intergrowths as lamellae // (100) of gold, and as grains coated by gold. H>gold. Colour: violet-pink. Ref.: 1240. As mentioned the colour changes with varying metal-content. The different phases do not only occur as separate grains, but are frequently seen to be intergrown with each other in intricate patterns. Pure gold can be rimmed by more pinkish looking Cu-rich gold. Growth zoning can be observed by the different shades of colour. Rarely occurs as euhedral pentagonal or dodecahedral crystals but normally as isolated grains of different size; as fine veinlets; skeleton-shaped; clustered; sheaf-like; as pseudo-hexagonal crystals; fine acicular; as crystal aggregates; in colloidal form or as solid solution in pyrite, arsenopyrite and sphalerite; forms rims on pyrite, chalcopyrite and galena. As globular specks or tiny blebs in many ore minerals, such as: arsenopyrite, pyrite, marcasite, galena, sphalerite, chalcopyrite, bismuth, bismuthinite, wittichenite, pyrrhotite, pentlandite, tetrahedrite, freibergite, argentite, many tellurides, polybasite, stephanite, stibnite, copper, skutterudite, safflorite, bournonite, magnetite. Secondary gold occurs as fillings in covellite, chalcocite, digenite, "limonite", hematite. Replaces pyrrhotite, pyrite, arsenopyrite, galena and other sulphides, tellurides of Ag, Au and Hg, niccolite, maucherite, gersdorffite. Replaces cubanite, millerite. May be intergrown with tetradymite, clausthalite, altaite and other tellurides. May form a myrmekitic intergrowth with bismuth, due to the decomposition of maldonite. Some grains may show coatings consisting of very fine-grained sulphides, bismuth or other material. Other ass.: electrum, bornite, enargite, stibioluzonite, Se-minerals, goethite, dyscrasite, galenobismutite, cosalite, aikinite, stibnite, jamesonite, orpiment, realgar, cinnabar, valleriite, parkerite, magnetite, ilmenite, stannite. Ref.: 44, 145, 161, 274, 275, 280, 333, 337, 339, 343, 396, 438, 464, 466, 557, 588, 602, 604, 684, 712, 739, 765, 850, 884, 899, 906, 916, 953, 978, 1046, 1047, 1070, 1076, 1205, 1229, 1237, 1343, 1366, 1379, 1392, 1452, 1453, 1454, 1456, 1459, 1461, 1463, 1465, 1475, 1485, 1494, 1498, 1545, 1603, 1644, 1652, 1679.

Name Formula Crystal System	1: Colour 2: Bireflectance 3: Anisotropy 4: Internal reflections	Reflectance in % (Wavelength in nanometres)
ELECTRUM (Au, Ag) 25 % and more Ag. Cubic	1. Creamy white or faintly yellow. ⟶ chalcopyrite, paler yellow. ⟶ gold, less yellow. 2. Not present. 3. Isotropic. No complete extinction under crossed nicols; greenish brown tints may appear. 4. Not present.	See at gold.
AUROSTIBITE $AuSb_2$ Cubic	1. White, tarnishes very quickly to a bornite-pink under certain conditions. ⟶ galena, pinkish tint is present. In oil, the decrease in refl. is less than for galena. 2 + 3. Isotropic. 4. Not present.	546 nm 61.0 [194].
MALDONITE Au_2Bi Cubic	1. ⟶ bismuth, air: darker, greenish grey. oil: bluish grey. 2 + 3. Isotropic. 4. Not present.	About 50-60 %.

Hardness H: polishing hardness VHN: micro-indentation hardness in kg/mm^2	Miscellaneous
H: see gold. VHN: 34-44 [159]. VHN$_{10-20}$: 68-82 [810].	Occurrence similar to that of gold; commonly as irregular grains of different size and as veinlets; enclosed in sulphides, tellurides, etc. Replaces pyrite, sphalerite, chalcopyrite, galena, Ag-Sb-minerals, argentite, "limonite", enargite. Parag.: see gold. Ref.: see gold, more in particular: 333, 557, 588, 1342a, 1343, 1485, 1679.
H: slightly $>$ gold. VHN$_{15}$: 248-262 [194].	Occurs as angular, xenomorphic grains. No apparent cleavage. Practically always associated with gold. Often as reaction product of gold and stibnite. Decomposes to gold and stibnite in a myrmekitic intergrowth of these minerals. Replaces pyrrhotite, arsenopyrite, gold. Forms coatings on gold particles. May contain some tetrahedrite. As inclusions in stibnite, Parag.: gold, freibergite, stibnite, jamesonite, chalcostibite, arsenopyrite, pyrite, chalcopyrite, sphalerite, galena, tennantite, tetrahedrite, gersdorffite, antimony, silver, copper. Diagn.: when occurring without gold and stibnite, difficult to distinguish from galena. Ref.: 274, 275, 326, 527, 1414.
H: $\gg$ bismuth, somewhat $>$ galena and gold. VHN: no data.	Supply of bismuth causes desintegration to a myrmekitic intergrowth of bismuth and gold. The associated bismuth has no twin lamellae. May be found unaltered as inclusions in arsenopyrite. Parag.: bismuth, bismuthinite, gold, galena, arsenopyrite. Ref.: 160, 1229, 1237, 1581.

74 Description of minerals from Table III

Name Formula Crystal System	1: Colour 2: Bireflectance 3: Anisotropy 4: Internal reflections	Reflectance in % (Wavelength in nanometres)
SILVER Ag (may contain minor amounts of Au, Hg, As, Sb, Pt, Ni, Pb, and Fe). Cubic	1. Bright white with a creamy tint; usually tarnishes quickly in air to a more creamy or pinkish colour which turns pinkish to brown, sometimes irid. ⟶ antimony, brighter and creamy white. ⟶ arsenic, much brighter and creamy. ⟶ dyscrasite, yellowish. ⟶ tellurium, creamy white, higher refl. ⟶ copper, white, lighter. ⟶ iron, creamy white. ⟶ platinum, creamy white, lighter. ⟶ bismuth, brighter white. 2. Not present. 3. Isotropic; scratches and pits give false light effects. 4. Not present.	About 90-95 %.
ZINC Zn Hex.	No data.	No data.

Hardness H: polishing hardness VHN: micro-indentation hardness in kg/mm^2	Miscellaneous
H: $\gg$proustite, $>$galena, slightly $<$dyscrasite; $<$arsenic, tetrahedrite, $\ll$sphalerite. VHN: 46-118 [159, 824, 1113, 1715]. VHN$_{10-20}$: 40-57 [194, 810].	Related minerals: Huntilite: silver containing some As. Chilenite: silver containing up to 10% Bi. Twinning and zoning not uncommon. Cleavage not observed. Occurs as dendrites; as skeleton- or cross-shaped crystals surrounded by niccolite, rammelsbergite or Ni-skutterudite. As irregular masses; as desseminated grains or leaves; as tubercle-like grains; as inclusions in argentite, pyrargyrite, rammelsbergite, bornite, chalcopyrite, pyrrhotite, arsenic, dyscrasite. Forms exsolution-intergrowths with allargentum; in this case the silver usually contains some Sb. May be formed by decomposition of argentite, freibergite or Ag-galena. Replaces argentite, chalcopyrite, galena, bornite, chalcocite, allargentum, bismuth, stephanite, tennantite, arsenopyrite, pitchblende, niccolite, maucherite, stromeyerite, pearceite. Forms pseudomorphs after safflorite, rammelsbergite, pitchblende-dendrites, argentite, proustite, pyrargyrite, stephanite. Forms intergrowths with galena, bornite, chalcocite. Enclosed by chalcocite, but mostly separated from it by stromeyerite. Replaced by argentite, galena, rammelsbergite and younger sulphides and arsenides. Other ass.: bismuthinite, graphite, pyrite, tetrahedrite, polybasite, electrum, cobaltite, domeykite, algodonite; of the Co-Ni-Fe-arsenides, silver prefers association with the Ni-richer members. Ref.: 6, 84, 162, 339, 342, 458, 475, 588, 632, 659, 707, 718, 762, 763, 837, 938, 960, 997, 1001, 1033, 1050, 1124, 1164, 1237, 1248, 1343, 1358, 1392, 1475, 1537, 1538, 1603, 1655, 1732. Note: Kongsbergite, α-(Ag,Hg), and moschellandsbergite, γ-(Ag,Hg) are cubic and form cubic crystals, grains and aggregates. VHN = 48-81 [1715]. Ref.: 102, 1001, 1248.
H: no data. VHN: 130 [1560].	Occurs in the oxidized parts of Pb-Zn-Ag-deposits, in basalts, in gravel deposits. Ass.: silver, Fe-hydroxides, Mn-oxides, sulphur, sphalerite, freibergite, galena, gold, platinum. Ref.: 79, 163.

Name Formula Crystal System	1: Colour 2: Bireflectance 3: Anisotropy 4: Internal reflections	Reflectance in % (Wavelength in nanometres)
ANTIMONY Sb (usually some As is present)* Hex.	1. White. ⟶ arsenic, slightly brighter white. ⟶ galena, brighter white. ⟶ platinum, slightly greyer. ⟶ silver, not so bright; silver is creamy white. ⟶ dyscrasite, no great difference. ⟶ bismuth, not so bright. 2. Weak, weaker than for bismuth and arsenic; about as for dyscrasite. 3. Distinct: yellowish grey, brownish, bluish grey; sometimes only slight variations. 4. Not present.	About 70-75 %.
ONOFRITE Hg(S,Se) Cubic	1. Similar to metacinnabarite. 2. Not present. 3. Isotropic, sometimes slightly anisotropic in spots. 4. Not present.	Lower than for cinnabar.

*Volborth (1960) reported antimony from Eräjärvi, Finland, which contained 15% Bi in solid solution, here referred to as bismuthian antimony.

Hardness H: polishing hardness VHN: micro-indentation hardness in kg/mm^2	Miscellaneous
H: > stibnite, < dyscrasite, stibarsenic, arsenic. VHN: 45-101 [159, 194, 1113]. VHN$_{20-50}$: 51-135 [194, 810, 1086, 1715].	Commonly occurs as fine-granular aggregates. Idiomorphic cube-like crystals rarely occur enclosed in galena. Cleavage commonly visible. Twinning lamellae, often polysynthetic, usually occur. Bismuthian antimony is not twinned. Replaces stibnite. May be bordered by stibnite. Forms crusts around dyscrasite. May occur as minute grains in arsenic, stibnite and geocronite. Observed as a decomposition product in stibiotantalite. May contain minute inclusions of stibnite and kermesite. Forms graphic intergrowths with stibarsenic (so called "allemontite I"); the components can be identified by etching and by hardness. Diff.: stibnite is strongly anisotr.; skutterudite and safflorite are much harder, the former being isotropic; dyscrasite and allargentum show a less pronounced cleavage and are harder. Parag.: arsenic, stibarsenic, pyrite, arsenopyrite, Co-Ni-arsenides, stibnite, kermesite, galena, pyrargyrite, Cu- and Ag-tellurides or -selenides, copper, realgar, gold, cinnabar, breithauptite, niccolite, gudmundite, fahl-ore, scheelite, bournonite, boulangerite. Ref.: 219, 339, 891, 997, 1048, 1086, 1174, 1237, 1603, 1622, 1629, 1647, 1704.
H: < cinnabar. VHN: no data.	Occurs as equigranular aggregates. No cleavage observed. Twinning is present in spots. Contains relics and lamellae of cinnabar / (111). Parag.: cinnabar, pyrite, chalcopyrite, sphalerite. Ref.: 1237, 1596a, 1629.

Description of minerals from Table III

Name Formula Crystal System	1: Colour 2: Bireflectance 3: Anisotropy 4: Internal reflections	Reflectance in % (Wavelength in nanometres)
METACINNABARITE HgS (may contain some Zn, Cd and/or Se). Cubic	1. In air: greyish white. ⟶ cinnabar's O, much darker. ⟶ cinnabar's E, very similar, slightly darker. In oil: much darker with a brownish grey tint. ⟶ cinnabar's O, darker and brownish pink. ⟶ cinnabar's E, very similar; cinnabar is more bluish. 2. Usually not present. 3. Isotropic; often shows anomalous anisotr.; then a very weak birefl. may be visible at grain boundaries. 4. Not present.	For Cd-metacinnabarite: 522 nm 25.0 594 25.6 658 26.1 [1595].
GUADALCAZARITE (Hg,Zn)(S,Se) Cubic	1. Almost white in air: distinct brownish grey tint in oil. 2. Not present. 3. Isotropic. 4. Not present.	No data.
CINNABAR HgS Hex.	1. In air: white with a bluish grey tint. In oil: bluish grey. ⟶ galena, darker, bluish. 2. Distinct in oil. 3. Distinct, commonly masked by the int. refl. 4. Intense and abundant, bright red.	460 nm 27.4-29.9 540 25.0-29.3 580 24.5-28.2 640 23.9-26.6 [1382a].

Hardness H: polishing hardness VHN: micro-indentation hardness in kg/mm^2	Miscellaneous
H: <cinnabar; Cd-metacinnabarite is harder than metacinnabarite, but still<cinnabar. VHN_{10-20}: 73-86 [194, 810]. VHN_{20}: 149-161 [1595, for Cd-metacinnabarite].	Occurs as single crystals and as granular aggregates. Cleavage not distinct. Twin lamellae $\parallel$ (111) and $\parallel$ (211) are nearly always present. Very often intergrown with, or partly altered to cinnabar. Cinnabar may take over the twinned texture of metacinnabarite. Parag.: cinnabar, stibnite, pyrite, marcasite, hematite, chalcopyrite. Diff.: cinnabar shows intensive red int. refl. and no twins; schwazite is much harder and not twinned. Ref.: 80, 307, 604, 1166, 1211, 1237, 1392, 1397, 1595, 1629. Note: The name "saukovite", given by Vasilyev (1966) to Cd-metacinnabarite, was rejected before publication by the IMA-CNMMN.
H: <cinnabar. $VHN_?$: 66-89 [1593].	Occurs as veinlets in cinnabar. Ass.: pyrite. Ref.: 1397, 1593.
H: slightly > antimony, metacinnabarite, ≪ schwazite, cuprite. VHN_{10-25}: 51-98 [194, 810, 1715].	Occurs as idiomorphic crystals. Commonly as granular aggregates, sometimes intercrystallized. Replaces stibnite, pyrite, chalcopyrite, tetrahedrite, metacinnabarite. Forms pseudomorphs after, and may contain inclusions of stibnite. May be replaced by marcasite, chalcopyrite. Alters to metacinnabarite and livingstonite. Occurs as lamellae along (111) of onofrite. Other ass.: schwazite, bornite, chalcocite, covellite, silver, gold, realgar, orpiment, galena, cassiterite, arsenopyrite, Fe-hydroxides, enargite, stibioluzonite, sulphides of Zn and Cu. Ref.: 10, 105, 307, 317, 465, 604, 1237, 1355, 1374, 1392, 1438, 1603, 1629.

Description of minerals from Table III

Name Formula Crystal System	1: Colour 2: Bireflectance 3: Anisotropy 4: Internal reflections	Reflectance in % (Wavelength in nanometres)
LIVINGSTONITE $HgSb_4S_8$* Mono	1. Light grey with a creamy tint. 2. Distinct: creamy grey to creamy brown with a violet tint. 3. Strong: dark violet to light green-grey (nicols not completely crossed). 4. Occurs in places, blood-red; not as abundant as in proustite or cinnabar.	About 35–40 %.
ANTIMONIAL SILVER (Ag,Sb) α-phase in the system Ag-Sb** Cubic; hex. at higher Sb-contents.	1. Pure white. ⟶ silver, slightly darker. ⟶ allargentum, slightly whiter. 2. Not present. 3. Anisotropic at higher Sb-contents. 4. Not present.	Somewhat lower than for silver.

* According to Gorman (1951), livingstonite from Guerrero is triclinic and has the composition $HgSb_4S_7$.

** The phases in the Ag-Sb system are indicated as α, ζ, and ϵ, respectively, in accordance with the nomenclature of Hansen and Anderko (1958).

Hardness H: polishing hardness VHN: micro-indentation hardness in kg/mm^2	Miscellaneous
H: >galena, <chalco-pyrite. VHN$_{15-25}$: 74-131 [194, 1715].	Occurs as prismatic crystals. Cleavage usually visible. A parallel displacement often occurs causing a wedge-shaped or digitate texture with differently orientated lamellae. Alters to cinnabar and Sb-oxides; pseudomorphosed by cinnabar. Diff.: cinnabar is more bluish, shows a much weaker anisotr. and more abundant int. refl. Miargyrite looks very similar, but usually shows more scratches, is slightly more bluish, and has another paragenesis. Parag.: stibnite, cinnabar, sulphur, metacinnabarite, onofrite, guadalcazarite, tiemannite. Ref.: 380, 516, 1237, 1264, 1392.
H: slightly > silver, slightly < allargentum. VHN: no data.	Twinning may be present. Occurs as groundmass for the oleander-leaf-shaped exsolution lamellae of allargentum at Cobalt, Ont. See also at allargentum. Ref.: 567, 1124, 1419.

Description of minerals from Table III

Name Formula Crystal System	1: Colour 2: Bireflectance 3: Anisotropy 4: Internal reflections	Reflectance in % (Wavelength in nanometres)
ALLARGENTUM Ag_6Sb (reported analyses range from 11.18 to 16.17 wt % Sb). ζ-phase in the Ag-Sb system. (see footnote ** on p. 80).	1. White with a slight grey tint. $\longrightarrow$ silver, slightly greyer. $\longrightarrow$ antimonial silver, very similar. 2. Not present. 3. Weak to distinct. 4. Not present.	Somewhat higher than for dyscrasite.

Hardness H: polishing hardness VHN: micro-indentation hardness in kg/mm^2	Miscellaneous
H: somewhat > antimonial silver. VHN: 143-157 [194].	Occurs as orientated lancet- or oleander-leaf-shaped exsolution lamellae in a groundmass of antimonial silver. Also in equigranular mosaics with antimonial silver. The "dyscrasite" reported by many authors from Cobalt, Ont. is in fact Ag_6Sb. Some of the dyscrasite from Andreasberg, Harz, is also Ag_6Sb. The ζ-phase is only stable with antimonial silver or with dyscrasite. Intergrowths of three Ag-Sb phases (antimonial silver, Ag_6Sb, and dyscrasite) were reported by Ramdohr (1960 from Cobalt, Ont., and by Markham and Lawrence (1962) from Broken Hill, Australia. Ramdohr described exsolution lamellae of dyscrasite in a groundmass of a mineral supposed to be Ag_6Sb and which he called "allargentum". Lamellae of antimonial silver cut dyscrasite and "allargentum". Halls et al. (1967) described and analysed the same type of intergrowth form Cobalt, Ont. The "allargentum" of Ramdohr consisted of a Ag-phase with 4.9-8.7 wt. % Sb and 0.7-4.3 wt. % Hg. This phase has no synthetic equivalent in the Ag-Sb system, but a stable phase of this composition might be possible in a ternary Ag-Sb-Hg system. The oleander-leaf-shaped lamellae called dyscrasite by Ramdohr were proved to be the ζ-phase with 12.2-13.8 wt. % Sb and some Hg. The lamellae of "antimonial silver" contained 2.1-4.1 wt. % Sb and 0.8-8.1 wt. % Hg. Petruk et al. (1969) obtained the same results as Halls et al. (1967). The groundmass described by Ramdohr as "allargentum" is slightly anisotropic and yields a cubic powder diagram. Consequently it can be considered as antimonial silver. The information summarized above shows that the identification of "allargentum" by Ramdohr is incorrect, and it was suggested that the term "allargentum" should be discarded (Halls et al., 1967). Nevertheless Petruk et al. (1969) proposed to give the name allargentum to the lamellae of Ag_6Sb. This regrettable proposal, which only increases the confusion around the name "allargentum", has been approved by the IMA-CNMMN. Markham and Lawrence (1962) observed dyscrasite, "allargentum" and antimonial silver in all possible combinations. If their identification is correct it may be concluded that (1) some are disequilibrium assemblages, and (2) the ζ-phase is possibly unstable at 25°C. (Somanchi and Clark, 1966).

Name Formula Crystal System	1: Colour 2: Bireflectance 3: Anisotropy 4: Internal reflections	Reflectance in % (Wavelength in nanometres)
DYSCRASITE Ag_3Sb (reported analyses range from 22.00 to 27.88 wt. % Sb). Ortho. (ϵ-phase in the Ag-Sb system; see also footnote ** on p.80).	1. Pure white (in air); in oil slightly greyer. ⟶ galena, creamy white. ⟶ silver, slightly greyer. ⟶ antimony, slightly creamy. 2. Weak, white to creamy white. 3. Weak to distinct. 4. Not present.	470 nm 59.9-61.5 to 62.6-63.6 546 61.5-63.3 to 64.0-65.2 589 62.2-64.0 to 65.1-66.2 650 62.1-63.9 to 68.5-69.5 Dispersion curves: [43].
STIBARSENIC AsSb Hex.	1. White; tarnishes in air less quickly than arsenic and more quickly than antimony. 2 + 3. Between antimony and arsenic. 4. Not present.	Intermediate between antimony and arsenic.
ARSENIC As (usually some Sb is present). Hex.	1. White; tarnishes in air in one day, more quickly than stibarsenic. ⟶ antimony, light grey, darker. ⟶ skutterudite and safflorite, slightly greyer and darker. ⟶ galena, white with a creamy tint (in air); in oil about the same. Silver is much brighter and creamy against arsenic. Bismuth is pinkish cream against arsenic and proustite bluish grey. 2. In oil distinct: greyish white with a yellow tint (O), or light bluish grey (E). 3. Very distinct, stronger than for antimony or stibarsenic. Steel-grey, yellow-grey or dark grey tints. 4. Not present.	Values for R_o estimated from the dispersion curve: [230]. 470 nm 47 550 45 590 45 650 44

Hardness H: polishing hardness VHN: micro-indentation hardness in kg/mm^2	Miscellaneous
H: $\gg$ galena, slightly $>$ silver and antimony. slightly $<$ arsenic and chalcopyrite. VHN: 152-178 [159, 1715].	May occur idiomorphically (with prismatic, square or rhombic sections) or allotriomorphically. Cleavage rarely visible. Usually untwinned; irregular jig-saw twinning occurs. Occurs as inclusions in arsenic, galena, cobaltite, pyrite. The beautiful exsolution lamellae of "dyscrasite" reported from Cobalt, Ont., have been shown to consist of allargentum: Ag_6Sb. Ref. (among which several apply to allargentum): 14, 339, 342, 475, 567, 997, 1124, 1310, 1358, 1419, 1450, 1456, 1464, 1603, 1675, 1732. Note: Arsenargentite, Ag_3As, forms needles in a groundmass of arsenic. Ref.: 568, 1475.
H: $>$ antimony, $<$ arsenic. VHN: 170-202 [1715].	Intergrowths with graphic or concentric texture of stibarsenic with antimony are called "allemontite I"; those of stibarsenic with arsenic are called "allemontite III", the last type occurring more often than the first one. The different intergrowths may be identified by hardness and by the etch tests as described by Wretblad (1941). Occasionally contains small crystals of loellingite. Ref.: 637, 866, 1183, 1603, 1704.
H: $\gg$ bismuth, $>$ silver, antimony, slightly $>$ dyscrasite, stibarsenic. VHN: 57-167 [159, 1715]. VHN_{20-50}: 69-137 [194, 810, 1715].	May show a fine to coarse crystalline texture; smaller grains tending to be equigranular, coarser crystals showing a plume- or sheaf-like appearance. Usually, however, arsenic shows a colloform texture with concentric layers or spheroids with radiated texture. Idiomorphic crystals may occur as an alteration product of geocronite. A basal cleavage is often visible. Twinning lamellae very common. Zonal colloform textures are developed by exposure to air for some days. May show inclusions of silver, antimony or dyscrasite. Forms graphic intergrowths or concentric aggregates with stibarsenic (so called "allemontite III"). May contain small spherical inclusions of rammelsbergite. Replaces skutterudite, rammelsbergite, bismuth. Replaced by proustite; surrounded by loellingite. Other ass.: arsenopyrite, pyrrhotite, marcasite, pyrite, sphalerite, galena, chalcopyrite, tetrahedrite, tennantite, lautite, niccolite, Ag-minerals, cinnabar, stibnite. Ref.: 53, 250, 339, 475, 632, 835, 997, 1109, 1110, 1237, 1392, 1445, 1603, 1704.

Description of minerals from Table III

Name Formula Crystal System	1: Colour 2: Bireflectance 3: Anisotropy 4: Internal reflections	Reflectance in % (Wavelength in nanometres)

ARSENOLAMPRITE
As
Ortho.
	1. Very similar to arsenic.
	2. Distinct in air, stronger in oil:
	   ∥ (001) similar to arsenic,
	   ⊥ (001) darker.
	3. Weak in air, strong in oil.
	4. Not present.
		Close to that of arsenic.

CHALCOSTIBITE
$Cu_2S \cdot Sb_2S_3$
Ortho.
	1. White with a pinkish grey tint.
	  ⟶ silver, greyish.
	  ⟶ sphalerite, pinkish.
	  ⟶ bournonite, lighter, less pinkish.
	  ⟶ galena, darker, yellowish grey.
	  ⟶ tetrahedrite, much lighter, faintly blue.
	2. Distinct in oil, esp. in granular aggregates: creamy white to light brown, darker ⊥ elongation.
	3. Distinct, more vivid than for bournonite: pinkish, greenish grey, bluish grey. Extinction ∥ elongation.
	4. Pale red, rare.
		$R_\alpha \parallel c$
		472 nm 37.5-37.7
		550 36.0-37.5
		579 35.0-36.5
		640 33.0-34.0
		
		$R_\beta \parallel a$
		472 nm 38.5-38.8
		550 39.0-39.2
		579 38.0-38.4
		640 35.5-36.0
		
		$R_\gamma \parallel b$
		472 nm 51.0-51.2
		550 43.5-43.8
		579 42.0-42.3
		640 39.0-39.1
		
		Dispersion curves: [150].

"DZHEZKAZGANITE" Tentative formula: $CuReS_4$ X-ray amorphous.	1. Brownish grey; in oil the brown tint is more distinct. 2. Not present. 3. Isotropic. 4. Not present.	About 15-30 %.

Hardness H: polishing hardness VHN: micro-indentation hardness in kg/mm^2	Miscellaneous
H:∼arsenic. VHN: no data.	Forms radiating aggregates of thick tabular crystals with perfect cleavage. The crystals are more or less curved. May contain lamellae of arsenic. Replaced by arsenic. Ass.: silver, loellingite, safflorite, pyrite. Diagn.: in polished section, arsenic is etched much quicker in air than arsenolamprite. Ref.: 674, 1237.
H: distinctly > silver, slightly < chalcopyrite and sphalerite. VHN: 193-285 [159, 1715]. VHN_{50}: 212-249 [194]. VHN_5: 183-287 [150].	Occurs as allotriomorphic grains, rarely as euhedral tabular prismatic crystals. Cleavage ∥ (001) commonly visible; other cleavage systems are rarely visible; triangular pits may be shown. Forms intergrowths with enargite. Replaces pyrite, sphalerite, chalcopyrite. May be replaced by silver, tenorite, bournonite, chalcocite, covellite, galena, boulangerite, jamesonite, pyrrhotite, the replacing minerals following sometimes the cleavage system and occasionally producing a fine network. Occurs as inclusions in tetrahedrite. Other ass.: arsenopyrite, marcasite, pyrite, wurtzite, gudmundite, tetrahedrite, cinnabar, stannite, argyrodite, canfieldite, andorite, argentite, stibnite, Ag- and Pb-sulphosalts, bismuth, bismuthinite, gold, zinkenite. Ref.: 7, 14, 16, 150, 227, 238, 239, 563, 891, 1070, 1212, 1237, 1392, 1415, 1596, 1653.
H:∼bornite. VHN_{20}: 230 [1172].	Occurs as colloform aggregates and as fine veinlets and minute segregations in bornite and digenite, locally forming networks. May replace bornite almost completely. Parag.: galena, digenite, bornite, covellite, tennantite, arsenopyrite. Ref.: 1172, 1258, 1325. Note: Morris and Short (1966) suggest that the mineral probably might be a Re-Mo-sulphide in stead of a Re-Cu-sulphide. The name "dzhezkazganite" has been disapproved by the IMA-CNMMN.

Name Formula Crystal System	1: Colour 2: Bireflectance 3: Anisotropy 4: Internal reflections	Reflectance in % (Wavelength in nanometres)
BORNITE Cu_5FeS_4 $228°C$ Tetr. $\rightleftharpoons$ Cubic	1. Pinkish brown to orange, soon tarnishing purplish, violet or iridescent. $\longrightarrow$ enargite, much darker, more variegated. $\longrightarrow$ stibioluzonite, more orange and darker. $\longrightarrow$ germanite, much darker and duller. $\longrightarrow$ renierite and mawsonite, less yellow and darker. 2. Often a slight birefl. is perceptible on grain boundaries and on twin lamellae. 3. Usually shows a weak anisotropy. Only very fine-grained aggregates may appear isotropic. 4. Not present.	470 nm 16.6 546 18.5 589 21.5 650 25.8 [484].

Hardness H: polishing hardness VHN: micro-indentation hardness in kg/mm^2	Miscellaneous
H: >galena, chalcocite, slightly<chalcopyrite. VHN: 68-105 [159, 194, 1113, 1715]. VHN$_{20-25}$: 97-124 [810, 1715].	Commonly occurs as aggregates of rounded grains. Idiomorphic crystals are very rare. Cleavage often visible; two directions, $\parallel$ (100) and $\parallel$ (111), may occur. Twinning often present. Fine-grained aggregates usually do not show twins. Zonal texture not observed. May be replaced by chalcocite, covellite, chalcopyrite, digenite, argentite. Replaces chalcopyrite, pyrite, tetrahedrite, galena, parkerite, rammelsbergite, pitchblende, hematite. Forms exsolution intergrowths with chalcopyrite, chalcocite and tetrahedrite; if cooled rapidly, the subordinate mineral is orientated along the crystallographic planes of the host mineral (e.g. chalcopyrite spindles or blades form a triangular pattern $\parallel$ cleavage lines of bornite); on slow cooling a rapid solid diffusion may enable exsolution bodies to segregate, producing a granular aggregate. Exsolution and replacement textures may closely resemble each other. Bornite often shows inclusions of the replaced minerals: enargite, cubanite, linnaeite, cobaltite, tetradymite, magnetite, sphalerite, melonite, hessite, sylvanite, coloradoite, altaite, colusite, germanite, renierite, mawsonite, and skeleton-shaped stannite jaune may also be enclosed. Occurs intimately intergrown with wittichenite, linnaeite, chalcocite, digenite, chalcopyrite, covellite, hematite, magnetite, stromeyerite, pearceite, freibergite, polybasite. Bornite may alter to a lamellar intergrowth of chalcopyrite and idaite. Previous to the formation of idaite bornite may show a characteristic "fracture-disease". The so called "orange bornites" have been shown to consist of renierite, mawsonite or luzonite. Other ass.: arsenopyrite, galena, pyrrhotite, gold, silver, bismuth, many sulphides, tellurides and selenides. Ref.: 86, 156, 166, 297, 333, 338, 339, 424, 425, 588, 658, 718, 764, 773, 867, 949, 1214, 1219, 1237, 1319, 1333, 1342, 1345, 1357, 1358, 1364, 1365, 1371, 1392, 1603, 1607, 1616, 1629.

Notes:
1. Some bornites contain patches and layers of a slightly lighter-coloured bornite. These inclusions consist of bornite with some sulphur (0.4 wt. %) in excess. The X-ray powder patterns of normal bornite and the lighter-coloured "X-bornite" are similar, but diagnostic reflections for each phase are present.
   Ref.: 166, 948, 1722.
2. "Galenobornite" (Satpayeva et al., 1964) is a plumbian variety of bornite, with brownish grey to grey colour and a weak anisotropy. Occurs as elongated crystals similar in form to betekhtinite. Desintegrates to bornite, galena and chalcopyrite.
   Ass.: bornite, chalcopyrite, chalcocite, galena, betekhtinite.
   The name has been desapproved by the IMA-CNMMN.

Description of minerals from Table III

Name Formula Crystal System	1: Colour 2: Bireflectance 3: Anisotropy 4: Internal reflections	Reflectance in % (Wavelength in nanometres)
MAWSONITE $Cu_{2+x}Sn_{1-x}FeS_4$; $0.5 < x < 1$ Intermediate member in the stannite-idaite series. Tetr.	1. Brownish orange. 2. Strong: orange to brown with a slight orange tint. In the latter position it is almost identical to hexastannite and stannite jaune. 3. Very strong: bright straw-yellow to bright royal blue and to dark grey-blue near extinction. 4. Not present.	R_o 480 nm 20.6–20.9 540 24.3–24.8 580 26.7–27.2 640 30.2–30.9 R'_e 480 nm 17.8 540 23.7–24.5 580 28.4–29.6 640 34.2–36.1 The dispersion curves [237, 832] cross at about 550 nm.
RENIERITE $Cu_{3-x}Ge_xFeS_4$; $x < 0.5$. Tetr.	1. Orange-brown, similar to bornite, but does not tarnish. 2. Distinct: orange to bronze with a violet tint. 3. Distinct to strong: yellow-brown to dark brown or greyish yellow to bluish grey. 4. Not present.	R_o 480 nm 18.3–19.4 540 23.4–25.2 580 27.0–29.4 640 29.0–31.4 R'_e 480 nm 18.5–19.7 540 25.2–26.4 580 29.4–30.2 640 31.6–32.5 The dispersion curves [832] cross at about 460 nm.
BRIARTITE $Cu_2(Fe,Zn)GeS_4$ Tetr.	1. Blue-grey. 2. Not observable. 3. Air: weak; oil: distinct to strong: grey-green and reddish tints. 4. Not present.	480 nm 26.2 540 27.6 580 27.4 640 26.0 Dispersion curve: [414, 832].

Hardness H: polishing hardness VHN: micro-indentation hardness in kg/mm^2	Miscellaneous
H: slightly > bornite. VHN$_?$: 166-210 [867].	No cleavage or twinning observed. Occurs in bornite-rich ore as rounded to irregular inclusions in bornite. Together with hexastannite or stannite jaune as reaction rims between cassiterite and bornite. Most of the "orange bornite" described in the literature is mawsonite. According to their chemical composition, "stannite jaune-orange", "stannite brun-jaune" and "stannite orangée" described by Picot, Troly and Vincienne (1963) also belong to the mawsonite series. Parag.: chalcocite, chalcopyrite, tetrahedrite, pyrite, galena, tennantite, enargite, cassiterite, hexastannite or stannite jaune, colusite, arsenopyrite. Ref. 237, 832, 867, 1427.
H: >bornite, ≧ chalcopyrite, <tennantite, sphalerite. VHN: 295-425 [194, 1398, 1715].	Occurs as granular aggregates and as single irregular grains. Also as idiomorphic grains in chalcopyrite, galena, sphalerite, tennantite, bornite, chalcocite. Polysynthetic and parquet-like twinning have been observed. As disseminated grains in bornite and germanite. Forms exsolution lamellae and bodies in chalcopyrite and sphalerite. Forms intergranular networks in aggregates of stannite and germanite. Occurs as lamellae along the cubic and octahedral planes of germanite, forming a regular intergrowth with this mineral. Contains inclusions of tennantite and briartite. Rarely observed as inclusions in galena and pyrite. Forms shells of variable thickness around germanite ovoids in bornite, and around gallite inclusions in germanite. Replaces germanite, from incipient to complete. Part of "orange bornite" described in the literature is renierite. Other ass.: enargite, digenite. Ref.: 785, 792, 832, 834, 1160, 1371, 1429b, 1470, 1588, 1607.
H: = chalcopyrite. VHN: no data.	May show lighter- and darker-coloured zones. The darker zones have a higher Zn-content. Under crossed nicols, in oil, shows a polysynthetic twinning, in parallel laths, in two directions, perpendicular to each other. Occurs as inclusions in chalcopyrite, tennantite, renierite, sphalerite, and forms a reticular texture with chalcopyrite. Also as rounded grains in sphalerite, sometimes together with gallite. May contain inclusions of renierite and chalcopyrite. Other ass.: pyrite, arsenopyrite, galena, chalcocite, bornite, germanite. Diff.: can only be mistaken for gallite, which has a slightly lower refl. Ref.: 414, 832, 1083, 1429b, 1607.

Name Formula Crystal System	1: Colour 2: Bireflectance 3: Anisotropy 4: Internal reflections	Reflectance in % (Wavelength in nanometres)
CHALCOPYRITE* $CuFeS_2$ Tetr.	1. Brassy yellow; tarnishes after some time. Small Se-contents turn colour to brown. ⟶ cubanite, hard bright yellow. ⟶ bismuth, greyish olive-green. ⟶ gold, olive-green when the chalcopyrite blebs are smaller than the gold flakes; brass-yellow when small gold spots occur in large areas of chalcopyrite. ⟶ galena, darker, yellow. ⟶ pyrrhotite, slightly darker, no creamy or brownish tints. ⟶ pyrite, darker more yellow. ⟶ sphalerite, magnetite, bright yellow-white. ⟶ tetrahedrite, stannite, bright yellow. ⟶ digenite, bright yellowish white, if the blebs of chalcopyrite are small. 2. Weak, occasionally observable. 3. Usually weak, but distinct: grey-blue and greenish yellow; if more Fe than usual is present, the anisotr. may be stronger. 4. Not present.	470 nm 34.0-37.6 to 35.6-38.6 546 43.8-48.0 to 46.4-48.8 589 45.7-48.9 to 48.0-50.5 650 46.7-50.1 to 49.4-51.4 Dispersion curves: [43].

* For minor elements in chalcopyrite, see Fleischer (1955).

Hardness H: polishing hardness VHN: micro-indentation hardness in kg/mm^2	Miscellaneous
H:>galena,<sphalerite, pentlandite, pyrrhotite. VHN: 174-219 [159, 194, 1113, 1715].	Twinning very common; lamellae may be differentially developed: very fine lamellar or broad, "irregularly" distributed or showing different systems. As a result of exsolution, chalcopyrite may show inclusions of cubanite, valleriite or mackinawite (orientated lath-shaped or worm-like bodies), pyrrhotite (hair-like wisps), stannite, sphalerite and tetrahedrite (curving lamellae or star-shaped bodies). Also contains inclusions of stephanite and ullmannite. Pseudo-eutectic intergrowths with these minerals and with pyrargyrite, chalcocite, bornite, sphalerite, magnetite not uncommon. Forms orientated intergrowths with digenite, bornite, pentlandite, sphalerite, tetrahedrite, cubanite, covellite, valleriite, mackinawite, bournonite, linnaeite, pyrite, marcasite, magnetite, stannite, pyrrhotite, polybasite, cobaltite. Replaces pyrite, pyrrhotite, sphalerite, magnetite, stannite, tetrahedrite, galena, polybasite, Sb-minerals (then bournonite may form reaction rims), cubanite, niccolite, maucherite, pentlandite, pitchblende. May be replaced by chalcocite, digenite, covellite, Ag-minerals, electrum, hematite, magnetite, goethite. As an exsolution product chalcopyrite may occur as fine inclusions in sphalerite (often orientated), tetrahedrite, bornite (as blades ⫽111), digenite, chalcocite (spindle-like), enargite and as lamellae in cubanite and in talnakhite (see at these minerals). May be present in several generations. May be formed together with chalcocite or idaite by desintegration of bornite. Other ass.: arsenopyrite, scheelite, ferberite, molybdenite, bismuth, Bi-sulphides, galenobismutite, tetradymite, proustite, miargyrite, diaphorite, stromeyerite, sternbergite, cinnabar, gold, copper, cuprite, tenorite, gersdorffite, sperrylite, parkerite, hessite, heazlewoodite, silver. Ref.: 262, 288, 297, 321, 333, 337, 338, 339, 413, 425, 492, 543, 588, 686, 891, 916, 1000, 1045, 1153, 1204, 1237, 1263, 1270, 1319, 1328, 1342, 1343, 1357, 1358, 1367, 1379, 1385, 1392, 1444, 1479, 1603, 1616, 1629, 1732.

Description of minerals from Table III

Name Formula Crystal System	1: Colour 2: Bireflectance 3: Anisotropy 4: Internal reflections	Reflectance in % (Wavelength in nanometres)
TALNAKHITE $CuFeS_2$ Cubic	1. Dark yellow with a distinct pinkish tint in comparison to chalcopyrite. 2. + 3. Isotropic. 4. Not present.	546 nm 38.6 589 40.2 Dispersion curve: [492].
BETEKHTINITE $Pb_2(Cu,Fe)_{21}S_{15}$ Ortho.	1. ⟶ galena, more yellow, distinct cream. ⟶ chalcocite, chalcocite is more greenish grey. ⟶ bornite, weakly reddish. ⟶ tetrahedrite, similar. 2. Air: distinct: ∥ c light cream, ⊥ c more yellow-cream. Oil: stronger, the colours having a more yellow tinge. 3. Air: strong: ∥ c dark grey-black, ⊥ c dark yellow-brown and dark blue-green. Oil: stronger, more vivid colours. 4. Not present.	Similar, but somewhat higher than galena.

Hardness H: polishing hardness VHN: micro-indentation hardness in kg/mm^2	Miscellaneous
H: no data. VHN: no data.	Originally described as "cubic chalcopyrite" by Budko and Kulagov (1963). Crystallographic and thermal studies by Cabri (1967) showed that the so called cubic chalcopyrite cannot be considered as a cubic modification of chalcopyrite, but rather is a separate mineral species. On suggestion of the IMA-CNMMN, Budko and Kulagov renamed the mineral (Cabri, 1967). Occurs as intergrowths with fine plates and mesh structures of chalcopyrite. Fills the spaces between these plates and mesh structures. The plates of chalcopyrite form mesh structures, the plates wedging out at the intersections. Enlargement of the plates is observed in the peripheral parts of the chalcopyrite grains, esp. at boundaries with cubic cubanite. In places the plates pass into narrow continuous rims, sometimes surrounding narrow rims of cubic cubanite. At the same time, at the boundaries with ortho. cubanite, there are no changes either in amount or size of the plates of chalcopyrite. The grains which consist of the intergrowths of talnakhite and chalcopyrite are twinned in places. The features of these intergrowths indicate their origin: the result of the decomposition of a solid solution. The intergrowths may contain lamellae of pentlandite. Parag.: chalcopyrite, cubic and ortho. cubanite, pentlandite, magnetite, sphalerite, pyrrhotite. Ref.: 176, 212, 492, 803.
H:>bornite. VHN: no data.	Occurs as tabular, long prismatic or needle-like crystals. May form granular aggregates. Alters to bornite and galena. Replaced by chalcopyrite, galena, bornite and silver. Replaces chalcocite. Forms intimate intergrowths with chalcocite and galena, due to replacement. The betekhtinite-relics are practically always rimmed by a narrow seam of bornite or chalcopyrite. May contain exsolution bodies of native silver. Other ass.: sphalerite, digenite, annivite, tennantite, mawsonite, enargite. Diagn.: similar to annivite, but anisotropic; wittichenite is very similar, but less anisotropic. Ref.: 55, 313, 634, 868, 961, 1237, 1326, 1354.

Description of minerals from Table III

Name Formula Crystal System	1: Colour 2: Bireflectance 3: Anisotropy 4: Internal reflections	Reflectance in % (Wavelength in nanometres)

CUBANITE
$CuFe_2S_3$
Ortho.
	1. Creamy grey or pale brown.
	   ⟶ pyrrhotite, very similar, more yellow, less pinkish.
	   ⟶ chalcopyrite, pinkish grey.
	2. Distinct (in oil): creamy grey to brownish grey.
	3. Strong (but not as strong as for pyrrhotite): pinkish brown, greyish blue; basal sections are less anisotr. Sections ⊥ (001) show on close examination very fine lamellae, proving the polysynthetic structure of the cubanite blades.
	4. Not present.
		Only maximum values:
		546 nm 39.2-40.4
		589 40.8-42.2 [492].

CUBIC CUBANITE $CuFe_2S_3$ Cubic	1. Distinct pinkish tint and somewhat darker in contrast to ortho. cubanite which has a yellowish colour with only a slight reddish-pink tint. 2. + 3. Isotropic. 4. Not present.	546 nm 35.0 589 37.1-37.5 [492].

Hardness H: polishing hardness VHN: micro-indentation hardness in kg/mm^2	Miscellaneous
H: slightly >chalcopyrite, <sphalerite, ≪pyrrhotite. VHN: 150-264 [159, 194, 1113, 1715].	Occurs as irregular granular or polygonal aggregates; also lamellar; intimately intergrown with chalcopyrite, bornite, or pyrrhotite, commonly due to exsolution, developed as blade-like or prismatic lamellae. May show inclusions of these minerals and of sphalerite, galena, valleriite, pentlandite. Occurs as inclusions in galena, pyrrhotite and as narrow rims on the contact between chalcopyrite and pyrrhotite. Replaces chalcopyrite and pyrrhotite. May be replaced by cubic cubanite and talnakhite. Desintegrates to intergrowths which may contain the following minerals: chalcopyrite, pyrrhotite, magnetite, chalcocite, pyrite, marcasite, valleriite. Other ass.: arsenopyrite, stannite, cassiterite, wolframite, hematite, molybdenite, tetrahedrite. Diff.: pyrrhotite is much harder; enargite shows a much lower refl., esp. in oil. Ref.: 30, 52, 144, 172, 249, 262, 333, 339, 353, 492, 588, 1000, 1045, 1200, 1202, 1204, 1237, 1254, 1319, 1342, 1358, 1385, 1392, 1450, 1501, 1589, 1603. Note: Rao and Rao (1968) described a type of cubanite which probably may be identical to tetragonal cubanite. Its optical properties are more or less in between these of cubic and orthorhombic cubanite.
H: no data. VHN: no data.	Occurs in close intergrowth with ortho. cubanite. Occurs as fine rims on the periphery of exsolution lamellae of ortho. cubanite in chalcopyrite. Originates from ortho. cubanite by replacement, which starts along the edges of the lamellae and along their fissures. It may end in an almost complete replacement of ortho. cubanite by cubic cubanite. Then the cubic cubanite shows fine parallel fissures, apparently indicating the decrease in volume. The mineral had been described already by Ramdohr (1928) as "cubanite II". Parag.: talnakhite, chalcopyrite, ortho. cubanite, pyrrhotite, pentlandite, galena, sphalerite, magnetite. Ref.: 492, 1203, 1254.

Description of minerals from Table III

Name Formula Crystal System	1: Colour 2: Bireflectance 3: Anisotropy 4: Internal reflections	Reflectance in % (Wavelength in nanometres)

VALLERIITE
$[CuFeS_2] \cdot [(Mg,Al,Fe)(OH)_2]$
Hex.
	1. Varies strongly, even in well-polished sections; usually brown or bronze-coloured (in contrast to mackinawite).
	2. Very strong, not as strong as for graphite or mackinawite.
	   O: light grey-brown or creamy bronze to darker greyish pink-violet.
	   E: bluish grey to dark grey.
	3. Extreme: white to grey-bronze.
	4. Not present.
		450 nm 14.0-14.3
		550 14.2-20.2
		589 14.2-22.0
		650 14.0-23.6
		           [1382a].

ROQUESITE $CuInS_2$ Tetr.	1. Grey with a slight bluish tint. ⟶ sphalerite, somewhat similar. 2. Not indicated. 3. Weak, polysynthetic twinning is visible. 4. Not present.	471 nm 23.5-24.2 554 22.3-23.4 580 21.8-23.0 624 21.4-22.4 Dispersion curves: [1157].

Hardness H: polishing hardness VHN: micro-indentation hardness in kg/mm^2	Miscellaneous
H: >chalcopyrite, ~cubanite, ≪pyrrhotite. VHN$_{50}$: 30 [244].	Appears as inclusions in chalcopyrite, pyrrhotite and pentlandite, often orientated and probably due to exsolution, showing highly varying shapes: needles, threads, worm-like masses, laths, blebs, stars. Also occurs as fine-laminated or pisolitic aggregates and as contacts between chalcopyrite and pyrrhotite. May occur as inclusions in chromite and magnetite. Replaces most of the accompanying minerals. Other ass.: cubanite, mackinawite, graphite, pyrite. The numerous literature references to valleriite are ambiguous and could equally well apply to mackinawite; actually much of the valleriite mentioned in the literature has been proved to be mackinawite. It is practically impossible to distinguish valleriite from mackinawite under the microscope; identification is only possible by X-ray diffraction or electron-microprobe techniques. It appears that mackinawite is a common and widespread mineral, while valleriite is only found in contact-metamorphic environments associated with carbonates and serpentine. Ref.: 41, 140, 144, 244, 370, 372, 411, 749, 1045, 1106, 1243, 1426, 1464. The following references could refer as well to valleriite as to mackinawite: 339, 543, 588, 739, 845, 888, 891, 1044, 1047, 1120, 1202, 1502.
H: >bornite, <tetrahedrite. VHN: 241 [1157].	Occurs as inclusions in bornite and chalcopyrite. Other ass.: wittichenite, chalcocite, covellite, sphalerite, idaite, stromeyerite, silver. Ref.: 703b, 1157, 1489a.

Name Formula Crystal System	1: Colour 2: Bireflectance 3: Anisotropy 4: Internal reflections	Reflectance in % (Wavelength in nanometres)
SULVANITE Cu_3VS_4 (forms a complete solid solution series with arsensulvanite, Cu_3AsS_4). Cubic	1. Air: light yellow or creamy yellow. ⟶ galena, light yellow. ⟶ chalcopyrite, lighter yellow. Oil: much darker yellow or creamy brown. 2. + 3. Isotropic. 4. Not present.	480 nm 25.8-27.9 540 29.2-31.2 580 29.1-30.4 640 33.0-34.8 Dispersion curves: [832]
ARSENSULVANITE Cu_3AsS_4 (V may substitute As). Cubic	1. Distinct yellow-brown tint. In the series sulvanite-arsensulvanite it seems that the colour becomes more brown with increasing As-content. 2. + 3. Isotropic. 4. Not present.	No data.
DOLORESITE $V_3^{4+}O_4(OH)_4$ Mono.	1. Grey, variable along the elongation of the crystals. 2. Present. 3. Very strong. 4. Not indicated.	No data.
MONTROSEITE (V,Fe)OOH Ortho.	1. White to grey. ⟶ karelianite, somewhat darker, bluish grey. 2. Weak. 3. Strong: bright greyish yellow to dark brown. 4. Not indicated.	No data.

Hardness H: polishing hardness VHN: micro-indentation hardness in kg/mm^2	Miscellaneous
H: >chalcocite, covellite, somewhat >bornite. VHN_{15-50}: 152-165 [194, 810].	Forms idiomorphic cube-like crystals. Also occurs as needle-like crystals and as irregular orthogonal grains. Cubic cleavage occasionally visible; triangular pits may occur. May be replaced completely by chalcocite and covellite. Similar to galena, but the yellow tint and the much lower refl. may be decisive. Ref.: 40, 337, 418, 832, 1117, 1237, 1332, 1410, 1567, 1723.
H: no data. VHN: no data.	Parag.: enargite, luzonite, covellite, pyrite. Ref.: 124, 912. Note: Sclar and Drovenik (1960) reported arsensulvanite with only traces of V; they called it "lazarevićite"; this name has been disapproved by the IMA-CNMMN.
H: no data. VHN: no data.	Occurs massive with radiating botryoidal texture. Fibrous cleavage; lamellar twinning $/\!/$ (100) is always present. Intergrown with häggite on a submicroscopic scale in parallel orientation. Replaces paramontroseite. Parag.: other V-oxides, coffinite, uraninite, clausthalite. Ref.: 374, 375, 1443. Note: Häggite, $V^{3+}V^{4+}O_2(OH)_3$, mono. Occurs intergrown with doloresite. Ref.: 374, 1665.
H: no data. VHN_{10}: 266-300 [170].	Occurs as bladed crystals. Replaces karelianite. Alters in air to paramontroseite. Parag.: pitchblende, pyrite, galena, other sulphides, other V-oxides. Ref.: 170, 371, 373, 479, 544, 1443, 1664. Note: Paramontroseite, VO_2, Ortho. Alteration product of montroseite. White colour in polished section. Parag.: V-oxides, coffinite, uraninite. Ref.: 373, 1443, 1665.

Description of minerals from Table III

Name Formula Crystal System	1: Colour 2: Bireflectance 3: Anisotropy 4: Internal reflections	Reflectance in % (Wavelength in nanometres)
TUNGSTENITE WS_2 Hex.	1. Similar to molybdenite; the decrease in refl. in oil is not as strong as for molybdenite. 2. + 3. Similar to molybdenite. 4. Not present.	On basal sections: 589 nm $R_o = 50$ % [521].
MOLYBDENITE MoS_2 (may contain some Re). Molybdenite-2H: hexagonal Molybdenite-3R: rhombohedral	1. See at bireflection. 2. Extremely strong and characteristic. O: white. $\longrightarrow$ galena, very similar. $\longrightarrow$ graphite (O), much lighter, without brownish tint. E: dull grey with dark bluish tint. $\longrightarrow$ graphite (E), much lighter. 3. Very strong; 45°: white with pinkish tint; nicols not completely crossed: dark blue, very characteristic. 4. Not present.	R_o R_e 480 nm 47.0 – 550 41.7 22.0 590 40.0 20.9 650 40.0 20.1 Dispersion curves: [825].
"CASTAINGITE" $CuMo_2S_{5-x}$ (may contain some Fe, Pb, and Bi). Hex.	1. Pale yellowish grey. $\longrightarrow$ molybdenite, more yellowish. 2. Weak (only visible in oil). 3. Strong, but less than molybdenite. 4. Not present.	546 nm 36-39 [271b].

Hardness H: polishing hardness VHN: micro-indentation hardness in kg/mm^2	Miscellaneous
H: >galena. VHN: (on basal sections): 14.6-15.6 [521].	Occurs as fine scaly aggregates. Also as inclusions in safflorite and skutterudite. Forms pseudomorphs after scheelite. Parag.: galena, sphalerite, tetrahedrite, wolframite, scheelite, rutile, bismuth, bismuthinite. Impossible to distinguish tungstenite from molybdenite in polished section. Ref.: 521, 807, 1237, 1269, 1392.
H: in fine-grained aggregates slightly >chalcopyrite, <graphite. VHN$_{15}$: 16-101 [194]. VHN$_{50}$ (on basal planes): molybd.-2H: 24-32 [521] molybd.-3R: 18-23 [521].	Often more or less curved plates or crystals with undulatory extinction. May show colloform texture. Forms rosette-shaped aggregates. Cleavage ∥ (0001) nearly always visible. Parallel displacement very common, often producing a twinning-like texture. Polysynthetic twinning of primary origin is indicated by marked variations in polarization effects in sections ∥ c. Observed as minute inclusions in many sulphides, such as pyrite, chalcopyrite, arsenopyrite, bismuthinite, tetrahedrite; interstitial in cassiterite and wolframite; as tiny spots in scheelite. May contain inclusions of bismuth, bismuthinite, ikunolite, joseite, gold, galena. May be replaced by powellite (which shows abundant int. refl.), arsenopyrite, chalcopyrite. Replaces pyrite, chalcopyrite, arsenopyrite, magnetite. Other ass.: sphalerite, cubanite, pyrrhotite, tennantite, uraninite, giessenite, zinkenite, rutile. Molybdenite-2H cannot be distinguished from molybdenite-3R under the microscope. Molybdenite-3R is rare and may occur associated or intergrown with molybdenite-2H. Ref.: 264, 323, 333, 339, 433, 521, 540, 714, 758, 803, 806, 825, 848, 890, 1045, 1167, 1237, 1310, 1392, 1420, 1447, 1506, 1562, 1608, 1624, 1693, 1695, 1716. Notes: 1. The mineral jordisite, X-ray amorphous MoS$_2$, occurs as spherolites. It shows grey to black colours, similar to those of anthracites and other coaly substances. Usually isotropic, but sometimes anisotropic. VHN$_?$: 178 [309]. May contain blades of molybdenite; recrystallizes to molybdenite. Alters to supergene Mo-minerals, such as ilsemannite and wulfenite. Parag.: molybdenite, sphalerite, galena, hematite, arsenopyrite, pyrite, chalcopyrite. Ref.: 91, 309, 530, 902, 1053, 1436. 2. "Femolite" (Skvortsova et al., 1964) has been proposed as a ferriferous modification of molybdenite with slightly different optical features. The name has been disapproved by the IMA-CNMMN because the material is similar to molybdenite.
H: >molybdenite. VHN: 90-160 [271b].	Occurs in radiating aggregates, like a crystallized gel, the core consisting of castaingite and the outer layers of molybdenite. Also as reaction product between primary molybdenite and supergene djurleite. The name has been disapproved by the IMA-CNMMN because of the lack of data in the first publication. New convincing data were presented by Clark and Sillitoe (1969). Ref.: 271b, 1353.

Name Formula Crystal System	1: Colour 2: Bireflectance 3: Anisotropy 4: Internal reflections	Reflectance in % (Wavelength in nanometres)

GRAPHITE
C
Graphite-2H: hexagonal
Graphite-3R: rhombohedral
	1. See at bireflection.
	2. Extremely strong and characteristic.
	   O: brownish.
	      ⟶ sphalerite, much lighter.
	      ⟶ molybdenite (O), much darker.
	      ⟶ chalcopyrite and pyrrhotite, distinctly darker and brown.
	   E: much darker, almost completely black (in air bluish grey).
	      ⟶ sphalerite, much darker.
	      ⟶ molybdenite (E), much darker.
	3. Very strong: straw-yellow to dark brown or violet-grey. The anisotropy is strongest for yellow light. Basal sections appear isotropic.
	4. Not present.
		About 5 to 20 %.

CHAOITE C Hex.	1. Metallic grey to white. 2. Not present. 3. Not observed (probably on account of the extremely small grain size). 4. Not present.	About 40 %.
INDIUM In Tetr.	1. Pinkish white. 2. Not present. 3. Weak. 4. Not present.	About 90-95 %.

Hardness H: polishing hardness VHN: micro-indentation hardness in kg/mm^2	Miscellaneous
H: slightly >chalcopyrite and molybdenite. VHN_{5-15}: 7-12 [194, 810].	Forms small plates, blades or laths, and sheaf-like aggregates. Rarely radial-fibrous shells. The basal cleavage is usually well visible. Twinning and zonal texture were not observed. Parallel displacements and curved or wave-like texture very common. As inclusions in sphalerite, pyrite. May contain inclusions of silver, pyrite, galena, brannerite. Other ass.: arsenic, realgar, arsenopyrite, magnetite, hematite, pyrolusite, psilomelane, pyrrhotite. Ref.: 128, 138, 227, 543, 769, 837, 1098, 1237, 1310, 1337, 1605.
H: slightly > graphite. VHN: no data.	Occurs as relatively thin lamellae (3 to 15 microns wide) alternating with graphite and perpendicular to (0001) of graphite. Ref.: 354.
H: no data. VHN_{20}: 130-159 [1615].	Occurs in close association with native lead in greisenized and albitized granite. Ass.: silver. Ref.: 1615. Note: According to McLellan (1945) pure indium is the softest metal known. Its scratch- and Brinell-hardness is even lower than for lead. Vlasov (1966), however, gives a VHN for artificial metallic indium of 137-165.

Description of minerals from Table III

Name / Formula / Crystal System	1: Colour / 2: Bireflectance / 3: Anisotropy / 4: Internal reflections	Reflectance in % (Wavelength in nanometres)
GERMANITE Cu_6FeGeS_8 (may contain some Ga, Zn, As, W, Mo, and V). Cubic	1. Pinkish violet-grey. ⟶ enargite (in darkest position), similar. ⟶ bornite, no yellow tint; slightly lighter. The colour turns bluish grey due to W, light brown due to Mo, and yellow due to V. 2. + 3. Isotropic. 4. Not present.	Range due to differences in chem. comp. 480 nm 19.0–24.6 540 19.7–25.5 580 21.4–26.2 640 24.2–26.9 Dispersion curves: [832].
GOLDFIELDITE $Cu_3(Te,Sb)S_4$ Cubic	1. Greyish white with a faint brownish tint (air). ⟶ tetrahedrite, very similar. ⟶ bismuthinite, darker. ⟶ sphalerite, lighter. 2. + 3. Isotropic. 4. Not present.	480 nm 31.8 540 32.0 580 31.9 640 31.5 Dispersion curve: [832].
FREIBERGITE Ag-tetrahedrite, Ag-content may be higher than 20 wt. %. Cubic	1. Grey; in oil with a faint yellowish brown tint. ⟶ proustite, brownish. ⟶ galena, greyish brown tint. ⟶ sphalerite, much lighter. 2. + 3. Isotropic. 4. Brownish red, not uncommon, but scarcely spread.	469 nm 29.9 559 30.0 589 29.4 669 28.0 Dispersion curve: [1551].
SCHWAZITE Hg-tetrahedrite Cubic	1. Grey-white with a yellowish brown tint. ⟶ tetrahedrite, much lighter. 2. + 3. Isotropic. 4. Not present.	About 30 %.

Hardness H: polishing hardness VHN: micro-indentation hardness in kg/mm^2	Miscellaneous
H: >galena, slightly <tetrahedrite, tennantite, galena, <sphalerite. VHN: 372-450 [194, 1113, 1715].	Occurs as fine-granular aggregates. Also as ovoids in a mass of tennantite, galena and enargite. May show a distinct zonation or mottling, due to a paler and greyer colour variant of the same polishing hardness and optical behaviour. The zonation is due to different contents of W, Mo, V. The colour variants may occur as irregular cores or concentric spherical shells within the ovoids. Some germanite grains are wholly composed of the paler colour variants. Greyish zones in germanite may be due to incomplete exsolution of briartite. Moulds and replaces pyrite; also replaces chalcopyrite and bornite. Replaced by tennantite, tetrahedrite, chalcocite, renierite, luzonite. Contains exsolutions of renierite and briartite. Other ass.: galena, sphalerite, gallite. Ref.: 414, 697, 832, 951, 969, 1158, 1160, 1237, 1307, 1371, 1392, 1411, 1429b, 1439, 1476, 1517, 1518, 1607. Note: Mitreyeva et al. (1968) described a V-As-bearing germanite: $Cu_3(As_{0.5}Ge_{0.3}V_{0.2})S_4$. It shows a pinkish colour with a creamy tint. Isotropic. VHN= 508. Refl. is about 27 to 33 %.
H: ~stibioluzonite. VHN: no data.	Zonal texture common. Replaces stibioluzonite, bismuthinite. Replaced by gold. Other ass.: marcasite, sphalerite, Au-Ag-tellurides. Ref.: 832, 1428, 1558.
H: varying with Ag-content, >silver-sulpho-salts, ≪sphalerite; for Ag-rich spec. even <galena. VHN: 272-375 [159, 194, 1715]. VHN_{50}: 252-310 [194, 1551].	Cleavage rarely visible. May be replaced by proustite, argentite, galena, zinkenite. Other ass.: silver, chalcopyrite, gold, Co-Ni-Fe-arsenides, franckeite, stannite, andorite, Ag-Sb-sulphosalts, arsenic, jamesonite, stibnite, realgar. Ref.: 85, 224, 239, 475, 1237, 1392, 1428, 1551, 1646, 1732.
H: >chalcopyrite. VHN: 262-373 [194, 1594].	Replaced by chalcocite, covellite, cinnabar, cuprite, tenorite, copper. Other ass.: pyrite, chalcopyrite, bornite, Ag-minerals. Ref.: 686, 1428, 1594, 1650.

Name Formula Crystal System	1: Colour 2: Bireflectance 3: Anisotropy 4: Internal reflections	Reflectance in % (Wavelength in nanometres)
TETRAHEDRITE $Cu_3SbS_{3.25}$ (may contain some Fe, Zn, Hg, Bi, Te, Pb). Forms a complete solid solution series with tennantite. The pure end-member does not exist in nature. Cubic	1. Grey, varying with admixed elements, commonly with olive or brownish tint. ⟶ galena, brownish grey, sometimes greenish. ⟶ chalcopyrite, bluish grey. ⟶ bournonite, darker and brownish. ⟶ stannite, light grey. ⟶ sphalerite, lighter. 2. + 3. Isotropic. 4. Brownish red, may be visible.	480 nm 31.1-32.2 540 31.4-33.1 580 31.1-33.2 640 29.0-31.4 Dispersion curves: [832].
TENNANTITE $Cu_3AsS_{3.25}$ (may contain some Fe, Zn, Ag). Forms a complete solid solution series with tetrahedrite. Cubic	1. Grey, varying with admixed elements, commonly with bluish green tint. ⟶ galena, distinctly greenish. ⟶ pearceite, very similar. ⟶ chalcocite, greenish. ⟶ chalcopyrite, bluish grey. 2. + 3. Isotropic. 4. Commonly visible, various shades of red.	480 nm 28.4-31.2 540 27.3-31.4 580 26.1-30.8 640 23.2-29.3 Dispersion curves: [832].
COLUSITE $Cu_3(As,Sn,V,Fe,Sb)S_4$* Cubic	1. Coppery cream to creamy tan. ⟶ pyrite, creamy tan. ⟶ tennantite, brass-yellow. ⟶ enargite, cream. ⟶ chalcopyrite, tan. 2. + 3. Isotropic. 4. Not present.	480 nm 25.9-26.5 540 28.2-29.9 580 30.1-31.6 640 30.5-31.4 Dispersion curves: [832].

* The Te-content in colusite reported in the literature is due to another mineral of the tetrahedrite group (goldfieldite) which occurs intergrown with colusite (Lévy, 1967, and Springer, 1969a).

Hardness H: polishing hardness VHN: micro-indentation hardness in kg/mm^2	Miscellaneous
H:$\gg$ galena,$\sim$chalcopyrite,$<$ sphalerite. For the tetrahedrite-tennantite series: VHN: 251-425 [159, 194, 522, 810, 920, 1715].	Cleavage not always distinct. A zonal texture may be developed by etching. May show minute inclusions of sphalerite, chalcopyrite, stannite or pyrrhotite due to exsolution; tiny blebs of arsenopyrite, pyrite, galena, gold, pyrargyrite and of tellurides may also be enclosed. Replaces arsenopyrite, gersdorffite, pyrite (sometimes separated from pyrite by a reaction rim of bornite), marcasite, sphalerite; may be replaced by chalcopyrite, bournonite, galena, Ag-sulphosalts, silver. Rarely shows graphic exsolution intergrowths with galena, chalcopyrite, bornite (arranged as blades $/\!/$ (111) in bornite). Forms more or less rounded grains, curving lamellae or star-shaped bodies enclosed in chalcopyrite, pyrrhotite, stannite, sphalerite, galena, bornite. Other ass.: gudmundite, zinkenite, jamesonite, meneghinite, chalcocite, bismuth, bismuthinite, huebnerite, wolframite, cassiterite, molybdenite. Ref.: 28, 34, 37, 322, 333, 336, 338, 339, 474, 488, 832, 891, 920, 943, 1074, 1204, 1237, 1379, 1392, 1428, 1450, 1451, 1452, 1454, 1603, 1646, 1690, 1732. Note: Annivite is Bi-tetrahedrite. Shows a higher refl. (slightly lower than for chalcopyrite) and a more creamy-white tint than tetrahedrite. The extinction is not always complete. Ref.: 488, 900.
H:$\gg$ galena,$\sim$chalcopyrite,$<$sphalerite. For the tetrahedrite-tennantite series: VHN: 251-425 [159, 194, 522, 810, 920, 1715].	Forms myrmekitic intergrowths with galena, stromeyerite, chalcopyrite, pyrargyrite. Replaces pyrite, enargite. May be replaced by galena (a reaction rim of seligmannite and some arsenopyrite may be present), stibioluzonite, enargite, chalcopyrite, stromeyerite. Cleavage not always discernable. Parag.: about as for tetrahedrite; besides the named minerals: sphalerite, bornite, covellite, chalcocite, lautite, wittichenite, sternbergite, frieseite, argentite, pearceite, silver, gold, sartorite, baumhauerite, realgar, linnaeite. Ref.: 6, 57, 322, 333, 505, 522, 832, 835, 840, 891, 1118, 1153, 1237, 1392, 1428, 1494, 1708.
H:$\gg$ bornite, slightly $>$tennantite,$<$enargite. VHN: 296-376 [194, 1398, 1715].	Occurs as isometric crystals or as aggregates. Cleavage not observed. Zonal texture very common, visible by differences in colour shade and polishing hardness; these growth-zones correspond to varying chemical composition. May be replaced by tetrahedrite, tennantite, enargite, bornite, chalcocite, chalcopyrite. May be surrounded by several orange, brown and yellow varieties of stannite, which probably are members of the stannite jaune - mawsonite series. Other ass.: pyrite, tellurides. Ref.: 101, 571, 786, 832, 937, 969, 993, 1161, 1398, 1517, 1728.

Name / Formula / Crystal System	1: Colour / 2: Bireflectance / 3: Anisotropy / 4: Internal reflections	Reflectance in % (Wavelength in nanometres)
LUZONITE-STIBIOLUZONITE* $Cu_3(As,Sb)S_4$ complete solid solution series. Tetr.	1. Somewhat lighter than enargite. Luzonite is pinkish orange and stibioluzonite is pinkish, without orange tints. 2. Distinct to strong (in oil), the lamellar twinning which is always present causing beautiful colour effects: orange-brown to greyish violet. 3. Very strong: dark brown, greyish green; accentuated by compound twinning. Commonly oblique extinction with regard to the twin lamellae. Stibioluzonite shows more vivid polarization colours than luzonite. 4. Not present.	Luzonite: R_o 480 nm 24.3-24.9 540 24.8-25.6 580 25.9-26.0 640 26.7-27.2 R'_e 480 nm 25.0-25.6 540 26.1-27.9 580 26.8-29.4 640 27.2-30.5 Stibioluzonite: R_o 480 nm 23.7-23.8 540 23.7-23.9 580 24.0-24.2 640 26.0-26.8 R'_e 480 nm 24.2-24.5 540 25.4-26.2 580 26.4-27.4 640 28.1-29.0 Dispersion curves: [832].

ENARGITE-STIBIOENARGITE** $Cu_3(As,Sb)S_4$ Ortho.	1. Pinkish grey or light pinkish brown (in air); in oil much darker, more violet-grey or brownish grey in darkest position.
	 → bornite, pinkish white.
	 → chalcocite, pinkish brown.
	 → galena, greyish brown.
	 → luzonite, less brown, bluish pink.
	 → tennantite, darker pink.
	 → krennerite, similar, but much lower refl.
	 2. Distinct (in oil): // a greyish pink with yellow tint
	 // b pinkish grey
	 // c greyish violet.
	 3. Strong: polarization colours in bluish, greenish, reddish and orange tints.
	 4. A deep-red int. refl. may occur.
		//a //b
		 470 nm 27.42 26.91
		 546 25.15 25.88
		 589 24.41 26.07
		 650 25.67 28.07
		 //c
		 470 nm 29.22
		 546 28.72
		 589 28.66
		 650 28.36
		
		 Dispersion curves: [849].

* The pure end-members and intermediate members have been observed in nature. Gaines (1957) proved that the original material described by Stelzner in 1873 was tetr. Cu_3SbS_4, and thus the name "famatinite" should have priority over stibioluzonite. However, there has been much confusion in the literature about the name "famatinite" because it has been given as well to tetr. as to ortho. Cu_3SbS_4. For these reasons the nomenclature of Strunz (1966) is followed here.

** The maximum Sb-content ever found in enargite was about 6 wt. %. There is no evidence for the existence of natural stibioenargite. The name is used by Strunz (1966) to replace the doubts around the name "famatinite".

Hardness H: polishing hardness VHN: micro-indentation hardness in kg/mm^2	Miscellaneous
H: distinctly>bornite, chalcopyrite, slightly >tetrahedrite, ≦enargite, slightly<sphalerite. VHN: 205-397 [159, 194, 1113, 1715].	Forms irregular, rounded and isometric grains. Cleavage not observed. Twinning (polysynthetic and compound) is always present (as distinct from enargite); star-shaped and pseudo-hexagonal twins also occur. The twin lamellae in luzonite are more regular than in stibioluzonite. Replaces chalcopyrite, tetrahedrite, sphalerite, chalcocite; may be replaced by bismuthinite, enargite, tennantite. Enargite originates from luzonite and vice-versa. Concentric textures with enargite occur. Forms intergrowths with stannite, chalcopyrite, tetrahedrite, enargite. Occurs as exsolution bodies in chalcopyrite. Parag.: nearly always enargite, chalcopyrite; moreover: members of the tetrahedrite group, pyrite, marcasite, sphalerite, galena, seligmannite, covellite, bismuthinite, gold, electrum, silver, argentite, hessite, sylvanite, stephanite, proustite, pyrargyrite, stannite. Ref.: 418, 431, 460, 575, 654, 832, 896, 1057, 1237, 1372, 1392, 1410, 1429, 1494, 1558, 1679. Note: Moh and Ottemann (1962) described several luzonites with a low Sn-content and named them "stannoluzonite". This name has been disapproved by the IMA-CNMMN.
H: >galena, chalcocite, bornite, chalcostibite, chalcopyrite, ≧tennantite, luzonite,<sphalerite. VHN: 133-383 [159, 194, 522, 810, 1113, 1715].	Occurs as prismatic crystals and as allotriomorphic or rounded grains. Cleavage // (110) nearly always visible. Usually untwinned (as distinct from luzonite-stibioluzonite); twinning due to stress may occur; enargite not influenced by stress only rarely shows twins or star-shaped trillings. Often shows zonal texture when etched. Replaces pyrite, chalcopyrite, tennantite, sphalerite, pitchblende; may be replaced by bornite, galena, chalcocite, covellite, pyrite. Gradually alters to tennantite ("grüne Enargit" of Schneiderhöhn, and "mottled enargite" of Graton and Murdoch). May originate from luzonite, and vice-versa. Forms intergrowths with jordanite, klockmannite, wurtzite, sphalerite, luzonite. Commonly associated with luzonite. Other ass.: marcasite, arsenopyrite, seligmannite, chalcostibite, lautite, meneghinite, jamesonite, pyrargyrite, Bi-minerals, selenides, tellurides, gold, silver, electrum, millerite, realgar, stannite, coffinite, digenite. Ref.: 227, 333, 418, 460, 481, 522, 572, 575, 625, 688, 832, 848, 1119, 1122, 1237, 1372, 1392, 1429, 1450, 1452, 1494, 1496, 1568, 1629. Note: Moh and Ottemann (1962) described an enargite from Ohkubo, Hokkaido, Japan, which contained more Sn than Sb + As, and called it "stannoenargite". The vote on this name by the IMA-CNMMN was inconclusive.

Description of minerals from Table III

Name Formula Crystal System	1: Colour 2: Bireflectance 3: Anisotropy 4: Internal reflections	Reflectance in % (Wavelength in nanometres)
GALLITE $CuGaS_2$ Tetr.	1. Grey*. $\longrightarrow$ chalcocite, distinct brownish grey to violet-grey tint. $\longrightarrow$ briartite, distinct greenish tint. $\longrightarrow$ tetrahedrite, very similar. 2. Oil: distinct on grain boundaries and twin lamellae: brownish grey to violet-grey. 3. Distinct: grey-blue tints. 4. Not present.	480 nm 22.8 540 21.9 580 21.4 640 21.2 Dispersion curve: [832].
NOVAKITE $(Cu,Ag)_4As_3$ Tetr.	1. Light cream. $\longrightarrow$ arsenic, slightly lighter. 2. Not observed. 3. Medium, dark-blue-grey and lighter brown-ocher colour effects in marginal positions. 4. Not present.	No data.
PAXITE Cu_2As_3 Ortho.	1. White. $\longrightarrow$ arsenic, greyer. 2. Not present, even in oil. 3. Strong, light green-grey to dark brown with a violet tint. 4. Not present.	Somewhat lower than for arsenic.
KOUTEKITE Cu_5As_2 Hex.	1. Bluish grey; tarnishes grey after some time. $\longrightarrow$ cuprite, very similar, but no int. refl. 2. Weak, lighter and darker blue-grey. 3. Strong, blue to light pinkish brown. 4. Not present.	Values estimated from the dispersion curves: 550 nm 42-45 % [1163].

* Vlasov (1966) described two varieties of gallite: a lighter one, distinctly anisotropic, and a darker one, apparently isotropic.

Hardness H: polishing hardness VHN: micro-indentation hardness in kg/mm^2	Miscellaneous
H: distinctly $>$ galena, renierite, slightly $>$germanite, slightly $<$sphalerite. VHN$_{25}$: 446-471 [194].	Occurs as exsolution bodies in sphalerite and germanite and as rounded grains or crusts in germanite-rich ores. Twinned $/\!/$ (112) and (111), also lamellar. When exsolved in sphalerite (up to 15 %), forms basal lamellae along the (100) and (111) planes of sphalerite. Sphalerite exsolution needles may occur along the (001) planes of gallite. The sphalerite needles contain in turn gallite exsolution bodies. As exsolution grains in germanite and renierite. Replaced by tennantite. Difficult to distinguish from briartite. Parag.: germanite, renierite, briartite, enargite, chalcopyrite, galena, sphalerite, pyrite. Ref.: 414, 1237, 1429b, 1470, 1476, 1517.
H: no data. VHN: 387-398 [194].	Most frequently intergrown with arsenic, wherein it forms metasomatic veinlets. Loellingite occurs in the center of the veinlets. On account of the atmospheric oxidation of the polished section, a reaction rim between novakite and arsenic is visible. Arsenic may be replaced to a high extent by novakite. Occurring without arsenic, novakite forms irregular aggregates surrounded by loellingite. These aggregates show an oleander-leaf texture due to unmixing of the solid solution. Parag.: arsenic, arsenolamprite, koutekite, paxite, silver, loellingite, chalcocite, skutterudite, chalcopyrite, bornite, pitchblende. Ref.: 679, 680.
H: no data. VHN: no data.	Under crossed nicols, displays a distinctly lamellar texture. Intergrown with novakite, koutekite, and especially arsenic. The intergrowth with arsenic is the result of a gel decomposition. Parag.: arsenic, arsenolamprite, silver, loellingite, niccolite, skutterudite, koutekite, novakite, chalcocite, bornite, chalcopyrite, tiemannite, clausthalite, uraninite, hematite. Ref.: 677.
H: no data. VHN$_{15}$: 114-120 [194]. VHN$_?$: 147 [1163].	Occurs as rounded single crystals in paxite. Also as polycrystalline aggregates of very small crystals. May show a distinct lamellated texture under crossed nicols. Domeykite may be penetrated by a network of very fine veinlets of koutekite. These veinlets may join each other, giving the illusion of a single crystal. Replaced by loellingite and paxite. May contain very fine inclusions of arsenic. Occurs as inclusions in paxite and arsenic. Parag.: arsenic, silver, skutterudite, loellingite, chalcocite, novakite, paxite, domeykite, algodonite, copper, hematite. Ref.: 565, 673, 675, 1163.

Description of minerals from Table III

Name Formula Crystal System	1: Colour 2: Bireflectance 3: Anisotropy 4: Internal reflections	Reflectance in % (Wavelength in nanometres)

DOMEYKITE
α-Cu_3As: cubic
β-Cu_3As: hexagonal
	1. α-domeykite: creamy yellow.
	   β-domeykite: somewhat bluish grey in contrast.
	   ⟶ algodonite, darker and greyish yellow.
	   ⟶ whitneyite, darker and yellowish grey.
	2. α-domeykite: isotropic.
	   β-domeykite: distinct.
	3. α-domeykite: isotropic.
	   β-domeykite: very distinct, orange tints.
	4. Not present.
		Values estimated from the dispersion curves:
		550 nm 47.5-48.7 [1163].

ALGODONITE
$Cu_{6-7}As$ (β-algodonite contains somewhat more Cu than α-algodonite).
Hex.
	1. Yellowish to greenish grey (α-algodonite) or pinkish yellow (β-algodonite).
	   ⟶ whitneyite, somewhat whiter and darker.
	   ⟶ domeykite, whiter and lighter.
	2. Distinct.
	3. Both varieties are strongly anisotropic, without colour effects.
	4. Not present.
		Value estimated from the dispersion curve:
		550 nm 61.5 [1163].

WHITNEYITE
(Cu,As) As-content up to 12 %.
Cubic
	1. Creamy white. As-poor whitneyite shows yellow to orange tints. Tarnishes very quickly.
	   ⟶ algodonite, more yellow and somewhat lighter.
	   ⟶ domeykite, more white and lighter.
	2. + 3. Isotropic.
	4. Not present.
		Value estimated from the dispersion curve:
		550 nm 63.0 [1163].

Hardness H: polishing hardness VHN: micro-indentation hardness in kg/mm^2	Miscellaneous
H:>chalcocite,~algodonite,<breithauptite. VHN?: domeykite: 206 [1163]. β-domeykite: 250 [1417].	Never idiomorphically developed. Forms granular aggregates composed of minute, more or less rounded grains. Forms intergrowths with algodonite: algodonite may occur as needles or oleander-leaf textures in domeykite and vice-versa. Replaces algodonite. May be replaced by algodonite, huntilite, chalcocite. Forms crusts on niccolite, chalcocite and rammelsbergite. Lamellae of α-domeykite may be found in a groundmass of algodonite and whitneyite. α-domeykite may contain lamellae of β-domeykite. Other ass.: copper, silver, cuprite, tenorite. Ref.: 203, 339, 896, 1159, 1163, 1237, 1250, 1356, 1392, 1417.
H:~domeykite,>chalcocite,<breithauptite. VHN$_{15}$: 248-255 [194]. VHN?: 217 [1163].	Does not show idiomorphism, only known as granular aggregates. Forms intergrowths with whitneyite and domeykite: algodonite may occur as needles or oleander-leaf textures in whitneyite or domeykite and vice-versa. Chalcocite and covellite form pseudomorphs after algodonite. Forms intimate intergrowths with niccolite. Replaced by domeykite. α-algodonite only occurs as exsolution lamellae in β-algodonite. Other ass.: copper, silver, cuprite, tenorite. Ref.: 148, 203, 339, 860, 1159, 1163, 1237, 1250, 1356, 1392. Note: Horsfordite, Cu_6Sb (?), crystal system not indicated. Shows a creamy white colour in comparison to galena, and is bluish grey-white against silver. Ref.: 967.
H:>chalcocite,<breithauptite. VHN?: 58 [1163].	Whitneyite is a compound with varying composition and colour. Always as xenomorphic grains, forming crust-like or massive aggregates. Occurs on the borders of native copper, the contact being progressive and not distinct. Sometimes on the contacts between copper and domeykite. Often bordered by a rim of algodonite associated with silver. Forms intergrowths with algodonite: algodonite may occur as needles or oleander-leaf textures in whitneyite and vice-versa. Lamellae of domeykite may occur in a groundmass of whitneyite and algodonite. Ref.: 1163, 1237, 1417.

Description of minerals from Table III

Name Formula Crystal System	1: Colour 2: Bireflectance 3: Anisotropy 4: Internal reflections	Reflectance in % (Wavelength in nanometres)
COPPER Cu (may contain some Ag or As). Cubic	1. Pink, soon tarnishing brownish. ⟶ silver, pink. ⟶ chalcocite, much lighter; chalcocite is dull blue-grey. ⟶ cuprite, much lighter; cuprite is dull bluish green-grey. 2. Not present. 3. Isotropic. No complete extinction; scratches may give false light effects. 4. Not present.	Values estimated from the dispersion curve: 470 nm 38 550 43 590 68 650 79 [230].
DELAFOSSITE $CuFeO_2$* Hex.	1. + 2. Bireflectance very distinct. In air: O yellowish rose-brown to creamy brown. E rose-brown. In oil: O dull pinkish grey to yellowish grey. E much darker, light cacao-brown to pinkish brown-grey. ⟶ enargite and tenorite, very similar colours; tenorite shows a more distinct yellow tint. 3. Very distinct to strong; 45°: bluish grey. Straight extinction. Basal sections appear isotropic. 4. Not present.	About 20-25 %.

* Buist et al. (1966) failed in their attempts to synthetize $CuFeO_2$. They reported $3Cu_2O.Fe_3O_4$ as formula for delafossite. According to Wiedersich et al. (1968) the probable cause of the discrepancies may have been the use of platinum crucible liners by Buist et al. Platinum can easily extract copper from Cu-bearing oxides at elevated temperatures.

Hardness H: polishing hardness VHN: micro-indentation hardness in kg/mm^2	Miscellaneous
H: distinctly >chalcocite, chalcopyrite,<cuprite. VHN: 48-143 [159, 194, 1715].	Occurs coarse- and fine-grained (as concretions); with allotriomorphic or panidiomorphic texture; as crystal aggregates. Supergenic copper may show a dendritic or spear-like form. Cleavage not observed. After etching a lamellar twinning is always visible. Zonal texture not uncommon. Occurs as minute inclusions in cuprite, enargite, bornite, chalcocite, pyrrhotite, pentlandite. Forms fringes on the periphery of aggregates of iron and cohenite. Contains orientated intergrowths of cubanite and cuprite. As threads or minute irregular blebs in serpentine surrounded by magnetite. Replaces chalcocite; may be replaced by, or oxidized to cuprite. Other ass.: tenorite, chalcopyrite, covellite, delafossite, "limonite", silver, domeykite. Ref.: 8, 83, 245, 283, 322, 339, 1122, 1237, 1360, 1392, 1437, 1603, 1629.
H: distinctly <cuprite, goethite. VHN: no data.	Forms aggregates of subparallel crystals, also sheaf-like or coarse-laminated aggregates, or very fine-grained inclusions in goethite, etc., with concentric or botryoidal texture and then very similar to tenorite. In this last form rather common; the coarser crystalline modification is rare. Cleavage may be visible. Parag.: goethite, "limonite", cuprite, tenorite, copper, pyrite, marcasite, bornite, chalcocite, djurleite, covellite, galena, tennantite. Diff.: tenorite shows a higher refl. and stronger anisotr. Ref.: 189, 436, 624, 891, 1087, 1210, 1392, 1394a, 1603, 1686.

Name Formula Crystal System	1: Colour 2: Bireflectance 3: Anisotropy 4: Internal reflections	Reflectance in % (Wavelength in nanometres)
TENORITE CuO Mono.	1. In air: grey to greyish white. In oil: much darker and strongly pleochroic: ⟶ creamy grey to dull brownish grey. ⟶ cuprite, brownish, without bluish tint. ⟶ chalcocite, brownish. ⟶ goethite, lighter and yellowish grey. 2. See at colour. 3. Strong: blue to light creamy grey; extinction not complete and oblique. 4. Not present.	About 20-25 %.
PARATENORITE CuO Tetr.	1. White with a faint pinkish brown tint (in air). ⟶ tenorite, some resemblance but much whiter. 2. Weak, white to pinkish brown. 3. Strong. 4. Not present.	No data.
CUPRITE Cu_2O (may contain some V). Cubic	1. In air: light grey with a bluish tint. In oil: much darker and more distinctly bluish. ⟶ copper, much darker greyish blue. ⟶ chalcocite, darker and greenish. ⟶ hematite, darker, greenish blue tint. 2. Visible on close examination. 3. Always shows strong anomalous anisotr.: deep grey-blue to olive-green (a spec. from Burra-Burra, S. Australia, showed beautiful colours from violet to emerald-green). Due to the abundant int. refl. better visible in air than in oil. 4. Deep-red, always visible and highly characteristic.	About 25-30 %.

Hardness H: polishing hardness VHN: micro-indentation hardness in kg/mm^2	Miscellaneous
H: >chalcocite, <cuprite, goethite. VHN: 203-254 [159, 1715].	Forms sheaves of acicular crystals, concentric aggregates (earthy modification), fine-grained mosaics or pseudomorphs. Lamellar twinning not uncommon, esp. in coarser crystals. Replaces cuprite. Parag.: cuprite, "limonite", goethite, hematite, copper, delafossite, pyrite, marcasite, galena, tennantite, covellite, bornite, chalcocite. Ref.: 215, 332, 891, 1237, 1360, 1392, 1575, 1721.
H: no data. VHN: no data.	May form crystals with pseudo-cubic and pseudo-octahedral habit. Veins cuprite and tenorite. No cleavage. Twinning or zonal texture not observed. Parag.: goethite, tenorite, cuprite, copper. Ref.: 443, 1697.
H: >chalcopyrite, copper, tenorite, <goethite. VHN: 179-218 [159, 194, 1715]. VHN_{50}: 195-218 [810, 1715].	Occurs as idiomorphic crystals and in "earthy" form; the two kinds may be intimately associated. Cleavage $\parallel$ (111) rarely visible. No twinning observed. Replaces chalcocite, copper; may contain remnants of the replaced minerals. Included tiny blebs of copper, sometimes orientated, might have formed simultaneously and not be due to replacement; exsolution intergrowths with covellite have been observed. Occurs as inclusions in goethite. Cuprite is not stable with magnetite or hematite at any temperature. Parag.: goethite, "limonite", tenorite, chalcocite, bornite, covellite, delafossite, paratenorite, chalcopyrite, pyrite, marcasite, copper, silver. Ref.: 8, 321, 334, 676, 726, 1122, 1237, 1360, 1392, 1603, 1629, 1721.

Description of minerals from Table III

Name / Formula / Crystal System	1: Colour / 2: Bireflectance / 3: Anisotropy / 4: Internal reflections	Reflectance in % (Wavelength in nanometres)
STANNITE Cu_2SnFeS_4 Tetr.	For the description of these minerals see at Sn-sulphosalts, p.307–319.	
HOCARTITE Ag_2SnFeS_4 Tetr.		
STANNITE JAUNE $Cu_{2+x}Sn_{1-x}FeS_4$ Tetr. (?)		
STANNOIDITE $Cu_5(Fe,Zn)_2SnS_8$ Ortho.		
KESTERITE Cu_2SnZnS_4 Tetr.		
SAKURAIITE $Cu_2(In,Sn)ZnS_4$ Tetr.		
RHODOSTANNITE $Cu_2FeSn_3S_8$ Hex.		
"STANNITE (?) III"		
"STANNITE (?) IV"		

Description of minerals from Table III

Name Formula Crystal System	1: Colour 2: Bireflectance 3: Anisotropy 4: Internal reflections	Reflectance in % (Wavelength in nanometres)
NICKEL Ni Cubic	1. White. → heazlewoodite, more bluish white. 2. + 3. Isotropic. 4. Not present.	470 nm 58.9 546 63.2 589 64.2 650 67.0 [194] (on artificial material)
HEAZLEWOODITE Ni_3S_2 Hex.	1. Yellowish cream. → pyrite, slightly darker, more yellow. → pyrrhotite, lighter, yellow; pyrrhotite appears light brown in contrast. → pentlandite, lighter, yellow; pentlandite is pinkish in contrast. → awaruite, more yellowish. 2. Very weak, only visible on grain boundaries (oil). 3. Strong with vivid colours: lilac to greyish green. 4. Not present.	486 nm 48.6 546 52.5 589 56.5 650 56.5 Dispersion curve: [771].
MILLERITE β-NiS Hex.	1. Pure yellow. → chalcopyrite, lighter; chalcopyrite appears greenish yellow in contrast. → linnaeite, lighter, yellower, no pinkish tint. → pentlandite, yellower, no brownish tint. → pyrite, slightly yellower. 2. In oil: distinct, bright yellow to greyish yellow. 3. Strong: straw- or lemon-yellow to a pronounced iris-blue and violet. Straight but no complete extinction. Basal sections appear isotropic. 4. Not present.	R_o R'_e 470 nm 45.5 44.8 546 51.5 56.2 589 53.2 59.0 650 54.5 61.0 The dispersion curves [301a] cross at about 470 nm. See also [504].
HAUCHECORNITE $(Ni,Co)_9(Bi,Sb)_2S_8$ Tetr.	1. Depending on the surroundings varying from greyish brown to creamy yellow. → pyrrhotite, very similar. → millerite, faint olive tint. 2. Weak but distinct. 3. Distinct to strong: brown and blue-grey tints; parallel extinction. 4. Not present.	About 40 %.

Hardness H: polishing hardness VHN: micro-indentation hardness in kg/mm^2	Miscellaneous
H:<heazlewoodite. VHN$_{50}$: 186-210 [194].	Occurs as euhedral grains (cubes) in heazlewoodite. Also as anhedral "spider"-like irregular masses in the intergranular spaces of heazlewoodite aggregates. Twinned // (111). Ref.: 1240.
H: slightly > chalcopyrite, bornite, slightly < pentlandite, awaruite, pyrrhotite, <millerite. VHN$_{50}$: 231-321 [243]. VHN$_?$: 221-274 [771].	Always occurs in serpentinized rocks, in a granular or interlocking mosaic texture. Cleavage // (10$\bar{1}$1) may be visible. May show twin lamellae. As rounded aggregates and strings of grains in chalcopyrite and bornite. May contain awaruite or magnetite in a "cell" or "net" texture. Forms orientated intergrowths with shandite and pentlandite. Forms granular intergrowths with pentlandite. Occurs as inclusions in pentlandite and chalcocite. Contains flames of millerite and inclusions of sphalerite, nickel, bornite and chalcocite. Enclosed by perovskite and magnetite. Replaces magnetite. Replaced by millerite, magnetite, pentlandite, bornite, chalcocite. Alters to bravoite, vaesite, millerite, pentlandite and goethite. Other ass.: copper, chromite, cuprite, pyrrhotite, pyrite. Ref.: 243, 325, 771, 774, 904, 980, 1108, 1130, 1152, 1186, 1237, 1240, 1692, 1694.
H:>chalcopyrite, < sphalerite, pentlandite, linnaeite. VHN: 192-376 [159, 194, 1715].	Forms radiated or bundle-like aggregates of needle-shaped crystals; more rarely allotriomorphic granular in cross section. Cleavage // (10$\bar{1}$1) commonly visible. Twinning very common. Basal sections may show zonal texture. As orientated lamellae enclosed in or intergrown with linnaeite (commonly due to exsolution); as reaction rims between violarite and pyrrhotite. Replaces linnaeite, hauchecornite, pyrite, heazlewoodite, pentlandite. Replaced by violarite, bravoite, gersdorffite, chalcopyrite. May be enclosed in pyrrhotite and pentlandite (as lath-shaped crystals). Forms intergrowths with heazlewoodite, sphalerite, galena, linnaeite. Alters to bravoite and violarite. Parag.: pentlandite, violarite, pyrrhotite, sphalerite, pyrite, chalcopyrite, ullmannite, gersdorffite, hauchecornite, linnaeite, hematite, magnetite, gold, ilmenite, Pt-minerals, cinnabar, metacinnabarite, marcasite, bravoite, chalcocite, covellite. Ref.: 136, 227, 321, 337, 439, 504, 588, 626, 765, 774, 995, 1170, 1237, 1342, 1379, 1392, 1629, 1703.
H: slightly > millerite. VHN: no data.	Usually occurs as granular aggregates, seldom as idiomorphic tabular crystals. No cleavage. Forms intergrowths with millerite and bismuthinite. Occasionally replaces ullmannite. Other ass.: gersdorffite, sphalerite, galena, niccolite. Ref.: 1132, 1237, 1248, 1692.

Name Formula Crystal System	1: Colour 2: Bireflectance 3: Anisotropy 4: Internal reflections	Reflectance in % (Wavelength in nanometres)
ALABANDITE α-MnS Cubic	1. Grey. $\longrightarrow$ sphalerite, distinctly lighter. 2. Not present. 3. Isotropic, complete extinction. Sometimes a very weak anomalous anisotr. may be visible: light grey to grey-black. 4. Typical, dark green or brownish.	About 25 %.
GREENOCKITE β-CdS (forms a complete solid solution series with wurtzite). Hex.	1. Grey. $\longrightarrow$ sphalerite, bluish tint, lighter. 2. Not perceptible. 3. Not perceptible. 4. Very abundant, esp. in oil: light yellow, brownish red and blood-red.	About 20 %.

Hardness H: polishing hardness VHN: micro-indentation hardness in kg/mm^2	Miscellaneous
H:<sphalerite. VHN: 138-266 [159, 1715]. VHN$_{50}$: 129-254 [194, 1715].	Occurs as idiomorphic crystals and as coarse-grained aggregates. Cleavage may be visible in badly polished surfaces. Lamellar twinning very common; zonal texture occurs. May show inclusions of chalcopyrite (tiny blebs), pyrite, pyrrhotite (commonly small idiomorphic crystals ∥ (111). probably due to exsolution). May be replaced by pyrite, marcasite, pyrolusite. Parag.: pyrite, pyrrhotite, sphalerite, chalcopyrite, galena, stannite, jamesonite, tellurium, petzite and other tellurides, Mn-carbonates, enargite. Ref.: 557, 603, 604, 617, 648, 696, 726, 1174, 1237, 1255, 1392, 1603. Note: Fe-alabandite is a high-temperature solid solution compound with Mn:Fe∼1:1 from Bühl and from the Kaiserstuhl. Shows a tetrahedrite-like colour and refl., and a galena-like cleavage. May contain very small exsolution bodies of pyrrhotite. Ass.: pyrrhotite, chalcopyrite, sphalerite, arsenopyrite, ilmenite, rutile. Ref.: 1235.

H:<sphalerite.
VHN$_{10-20}$: 52-91 [810].
	Forms coatings upon sphalerite, pyrite, franckeite, canfieldite. Occurs as interstitial mineral; often forms merely a film on clastic grains. Cleavage not observed. No twinning.
	Commonly associated with wurtzite. Replaces sphalerite.
	Other ass.: marcasite, pyrrhotite, stannite, miargyrite, cassiterite, goethite, galena, pyrolusite.
	Ref.: 7, 15, 390, 514, 585, 646, 1237, 1615, 1682.
	
	Notes:
	1. Hawleyite, α-CdS, cubic.
	   Occurs as extremely fine-grained coatings on sphalerite and siderite, in vugs and along fractures. May be associated with greenockite.
	   Ass.: sphalerite, arsenopyrite, pyrite, pyrrhotite, marcasite, galena, chalcopyrite.
	   It seems probable that hawleyite is a more common mineral. Much material identified as greenockite may prove to be hawleyite when identified by X-ray diffraction.
	   Ref.: 234, 1563.
	2. Monteponite, CdO, cubic.
	   Occurs as octahedral crystals and fine-grained masses on Zn-minerals. Greyish black colour in polished section, isotropic; no cleavage.
	   Ref.: 384, 1702.

Description of minerals from Table III

Name / Formula / Crystal System	1: Colour / 2: Bireflectance / 3: Anisotropy / 4: Internal reflections	Reflectance in % (Wavelength in nanometres)
SPHALERITE ZnS (practically always contains Fe)* Cubic	1. Grey (in air); in oil very dark grey, sometimes with a brownish tint. The colour of sphalerite is not always related to the Fe-content. ⟶ alabandite, darker. ⟶ magnetite, darker. ⟶ wurtzite, very similar. 2. Not present. 3. Usually isotropic. Anisotropism, light grey to dark grey, mostly disturbed by the int. refl., may be due to tectonic deformation or to chemical admixtures: a high Fe-content causes distinct anisotropism. 4. Always visible, esp. in oil and in badly polished spec.; for Fe-rich spec. reddish or reddish brown; for Fe-poor spec. yellowish brown or yellowish white.	Range due to differences in Fe-content. 480 nm 17.88–20.02 540 17.28–19.60 580 17.15–19.50 660 16.85–18.51 Dispersion curves: [524a]; see also: [43, 300].
WURTZITE ZnS** (forms a complete solid solution series with greenockite). Hex.	1. In air: grey with a slight bluish tint. In oil: very dark grey. ⟶ sphalerite, very similar. 2. + 3. Not perceptible. 4. Very common, yellow to brown.	Similar to sphalerite.

* For minor, but occasionally important, elements in sphalerite, see Evrard (1945), Gabrielson (1945), Neumann (1944), Stoiber (1944), Warren and Thompson (1945b), Fleischer (1955) and Fryklund and Fletcher (1956).
** For minor elements in wurtzite, see Fleischer (1955).

Hardness H: polishing hardness VHN: micro-indentation hardness in kg/mm^2	Miscellaneous
H:>chalcopyrite, tetrahedrite, stannite, slightly >enargite,<pyrrhotite, magnetite, ilmenite. Int. refl. may prevent determination of H of other minerals. VHN: 128-276 [159, 194, 730, 1113, 1592, 1715]. VHN increases sharply with an Fe-content up to 4 %; further increasing Fe-content slowly diminishes VHN [1592, 1715]. The results of Henriques (1957) and Grafenauer et al. (1969) are not reliable.	Cleavage // (110) commonly visible in coarser-grained spec. Lamellar twinning and zonal texture very common. May contain exsolution bodies, sometimes orientated, of chalcopyrite, pyrrhotite, cubanite, stannite, tetrahedrite. May contain inclusions of bravoite and cattierite. Occurs as inclusions, sometimes due to exsolution, in stannite, bornite, tetrahedrite, chalcopyrite, pyrrhotite, cubanite, pyrite. Pseudomorphically replaces marcasite, pyrite, or intergrowths of these minerals; also replaces pyrrhotite, arsenopyrite, ilmenite, alabandite, uraninite. May be replaced by chalcopyrite, enargite, bornite, tetrahedrite, galena, chalcocite, covellite, argentite, pyrite, marcasite, electrum, silver, Ag-Sb-sulphosalts. Forms intergrowths, many of them orientated, with pyrite, quartz, chalcopyrite, stannite, tetrahedrite, cubanite, shandite, pyrrhotite, covellite, bournonite, enargite, millerite. Often intergrown with wurtzite; etching with HI(D 1.96) may show difference between wurtzite and primary sphalerite. Ref.: 185, 202, 262, 339, 347, 350, 412, 459, 474, 475, 588, 665, 726, 730, 800, 802, 805, 997, 1002, 1045, 1058, 1201, 1263, 1279, 1280, 1290, 1343, 1346, 1358, 1385, 1418, 1450, 1454, 1467, 1511, 1592, 1603, 1628, 1629, 1654, 1660, 1670, 1696, 1705, 1732.
H: ~ sphalerite. VHN: 146-264 [1715].	Cleavage // (0001) often distinct. Twinning not observed; if therefore, etching with HI(D 1.96) develops twinning, sphalerite is indicated; an untwinned aggregate with or without ice flower-texture indicates wurtzite or the primary wurtzite-texture, as wurtzite paramorphically alters to sphalerite (a thin section may be needed in the latter case). Concentric texture very common and often already shown with crossed nicols. Also idiomorphically, or sheaf-like developed or in radiating aggregates. Veined and replaced by pyrite, marcasite, galena. Replaces teallite, galena. Other ass.: tennantite, franckeite, cylindrite, cassiterite, chalcopyrite, pyrrhotite. Ref.: 7, 339, 350, 973, 1237, 1279, 1412, 1529, 1558, 1578, 1603. Note: "Voltzite" has been described as a probable oxysulphide of Zn [1100, 1237]. The material from Sterling Hill, however, has been shown to be a mixture of wurtzite and an organometallic compound containing Zn (Frondel, 1967).

Description of minerals from Table III

Name Formula Crystal System	1: Colour 2: Bireflectance 3: Anisotropy 4: Internal reflections	Reflectance in % (Wavelength in nanometres)
IRON α-Fe (usually contains some C and Ni). Cubic	1. White. ⟶ silver, light grey; silver is creamy in contrast. ⟶ pentlandite, much whiter. ⟶ cohenite, slightly bluish. ⟶ platinum, very similar, slightly darker. 2. + 3. Isotropic. 4. Not present.	About 65 %.
WAIRAUITE CoFe Cubic	1. ⟶ awaruite, very similar. 2. + 3. Isotropic. 4. Not present.	510 nm 54 % [242].
AWARUITE (Ni, Fe) Cubic	1. White, sometimes with a light green tinge, depending on comp. ⟶ heazlewoodite, lighter. ⟶ magnetite and chromite, bright white. ⟶ niccolite, lighter. ⟶ bismuth, darker. 2. + 3. Isotropic. 4. Not present.	Increases with increasing Ni-content. 530 nm 60-61 % [1187].

Hardness H: polishing hardness VHN: micro-indentation hardness in kg/mm^2	Miscellaneous
H:<magnetite, cohenite. VHN_{15-100}: 116-288 [194, 245, 810, 1715].	Usually occurs as drop-like segregations. Intimate exsolution intergrowths with fine-lamellar cohenite, orientated $/\!/$ (111) of iron, are called pearlite. If less than 0.75 % C is present, iron was the first component to be formed, followed by cristallization of pearlite, as groundmass; if more C is present, cohenite crystallized first, followed by pearlite. The texture of pearlite is often only visible with very high magnification (x 1000). By oxidation of the iron-component of pearlite, this eutectoid alters to oxidepearlite (greyish blue in air) or to hydroxidepearlite (when Fe-hydroxide is formed), the cohenite lamellae remaining unaltered in the new aggregate*. No cleavage, twinning or zoning were observed in natural terrestrial iron. Cohenite may be present in the iron as fringes and thin veinlets. Iron may also contain inclusions of wuestite. May be surrounded by pyrrhotite, cohenite, copper or magnetite. Surrounds mackinawite. Alters to colloform hematite, goethite and magnesioferrite. Wuestite alters to magnetite and iron. Other ass.: ilmenite, pentlandite, troilite, schreibersite, ulvöspinel. Ref.: 88, 97, 98, 245, 270, 845, 1230, 1237, 1369, 1437, 1603, 1639.
H: = awaruite. VHN_{15-100}: 185-329 [245]	Occurs as euhedral crystals with cubic or octahedral forms in serpentines. Closely associated with awaruite, with which it forms zoned grains, awaruite forming the core and wairauite the outer layers. Parag.: chromite, copper, magnetite. Diff.: the euhedral form, together with its slightly lower refl. may be used to distinguish wairauite from awaruite. Ref.: 242, 245, 1240.
H: slightly>heazlewoodite, ~pentlandite,<magnetite. VHN_{15-100}: 209-420 [245, 1007, 1715].	Always occurs in serpentinized rocks, commonly as coarse-granular aggregates. Also as veins in magnetite and silicates. Mostly occurs as rounded grains. May form skeletal crystals or elongated tabular grains, and reniform or botryoidal aggregates, sometimes with alternating layers of copper. Rarely observed as tiny needles or as almost perfect cubes a few microns in diameter. Forms pseudomorphs after magnetite and pentlandite. Occurs in a net-like intergrowth with heazlewoodite. As irregular rims around pentlandite. May contain replacement relicts of pentlandite, graphite, iridium. Together with heazlewoodite enclosed by magnetite. May form by oxidation of pentlandite, together with Fe-sulphides, or as the result of the breakdown of primary olivine and enstatite containing about 0.2 % Ni. May contain small granules of graphite and inclusions of copper. Parag.: heazlewoodite, magnetite, copper, pyrrhotite, chromite, hematite, pentlandite, cuprite, chalcocite, covellite, millerite, mackinawite, wairauite. Ref.: 164, 243, 245, 695, 766, 768, 1007, 1152, 1184, 1185, 1186, 1187, 1226, 1237, 1240, 1694.

* Von Schwarz's (1937) description is misleading as he, in citing Benedicks (1912), completely turns about Benedicks' explanation of the formation of oxidepearlite. Much of Von Schwarz's "ferrite" must be cohenite, perhaps even pearlite. The subscripts of some of his figures are accordingly erroneous. Van der Veen (1925) too mistakes "ferrite" for cohenite (cementite) when describing oxidepearlite.

Name Formula Crystal System	1: Colour 2: Bireflectance 3: Anisotropy 4: Internal reflections	Reflectance in % (Wavelength in nanometres)
GUDMUNDITE FeSbS Mono.	1. White, in oil with a pinkish tint. ⟶ arsenopyrite, slightly creamier. ⟶ galena, more yellowish. 2. Oil: distinct: ∥ c (∥ striation of the prism): pure white, perhaps with a faint green tint. ∥ a (darkest): faint pinkish, similar to pyrrhotite, but brighter. ∥ b pinkish white. 3. Strong, like arsenopyrite, with clear colours: white, grey, green, blue, black, purple, pink, brown. 4. Not present.	Values estimated from the dispersion curves. Range due to differences in chem. comp. 470 nm 36-50 to 45-54 546 41-54 to 50-57 589 41-55 to 50-57 650 39-56 to 49-58 [1606].
PENTLANDITE $(Fe,Ni)_9S_8$ (may contain Co in various amounts, see at cobaltpentlandite; also Ag up to 4 %). Cubic	1. Light creamy or yellowish. ⟶ pyrrhotite, much lighter; pyrrhotite is brown in contrast. ⟶ linnaeite, darker, without pinkish tint. ⟶ cobaltpentlandite, somewhat darker. Ag-containing pentlandite is fox-red. 2. + 3. Isotropic. 4. Not present.	470 nm 39.8 546 44.1 589 47.0 650 50.5 [194].

Hardness H: polishing hardness VHN: micro-indentation hardness in kg/mm^2	Miscellaneous
H: slightly $>$ sphalerite, pyrrhotite, $<$ breithauptite, $\ll$ arsenopyrite. VHN: 1094-1221 [194]. VHN: 402-588 [1606]. VHN$_{100-200}$: 588-683 [810]. VHN$_{50}$: 570 [1086].	Often forms idiomorphic, needle-shaped crystals or crystal aggregates. Twinning very common. Almost invariably associated with tetrahedrite, pyrrhotite, galena, sulphosalts. Forms regular or irregular intergrowths with pyrrhotite and Pb-Sb- or Cu-Sb-sulphosalts. Occurs as a decomposition product of tetrahedrite; in myrmekitic intergrowths with pyrrhotite as reaction rims between pyrrhotite and tetrahedrite. Replaces pyrrhotite, bournonite, galena, pyrite, sphalerite. May contain inclusions of sphalerite. Replaced by pyrrhotite, breithauptite, stibnite, berthierite, Pb-Sb-sulphosalts. May be enclosed by rims of bournonite, boulangerite, galena. Other ass.: arsenopyrite, cobaltite, marcasite, bravoite, chalcostibite, stannite, bismuth, antimony, loellingite, pyrargyrite. Ref.: 180, 183, 219, 473, 474, 889, 891, 1047, 1059, 1086, 1209, 1212, 1227, 1267, 1317, 1318, 1321, 1606.
H: $>$ chalcopyrite, slightly $<$ pyrrhotite. Co increases, Ag decreases the H. VHN: 202-231 [159, 1715]. VHN$_{50}$: 195-303 [194, 244, 810, 1715].	Occurs as coarse- or fine-grained idiomorphic or xenomorphic crystals. Cleavage $/\!/$ (111) commonly distinct and typical, esp. in larger grains. Twinning and zonal texture not observed. Usually found as flame-like segregations or as feather-like, finger-like or star-shaped bodies in pyrrhotite, sometimes orientated; partly formed by exsolution. The exsolution of pentlandite in pyrrhotite is accompanied by the conversion of some hex. pyrrhotite into mono. pyrrhotite. Also observed as exsolution bodies in chalcopyrite and cubanite. May contain exsolution bodies of pyrrhotite, chalcopyrite and mackinawite. Replaces pyrrhotite, niccolite. May be replaced by chalcopyrite, cubanite, pyrrhotite, magnetite, violarite, millerite, mackinawite, heazlewoodite. Rarely occurs as beautiful exsolution intergrowths with pyrrhotite, pentlandite forming a lamellar network. Also forms intergrowths with heazlewoodite, awaruite, chalcopyrite, cubanite and mackinawite. May contain minute inclusions of pyrrhotite, mackinawite, pyrite, marcasite. Occurs as inclusions in magnetite. Fills interstices between grains of pyrrhotite and magnetite. Alters to bravoite and violarite, the process starting along the cleavage planes and cracks; tiny blebs of pyrite and marcasite may be formed simultaneously. May also alter to goethite. Ag-containing pentlandite occurs as small grains and in orientated intergrowths with normal pentlandite. Other ass.: galena, ilmenite, cohenite, iron, copper, cuprite. Ref.: 38, 137, 302, 321, 339, 351, 353, 356, 367, 434, 588, 590, 845, 916, 967, 981, 1142, 1152, 1186, 1187, 1202, 1237, 1259, 1342, 1358, 1392, 1600, 1603, 1629.

132   Description of minerals from Table III

Name Formula Crystal System	1: Colour 2: Bireflectance 3: Anisotropy 4: Internal reflections	Reflectance in % (Wavelength in nanometres)
VIOLARITE $(Ni,Fe)_3S_4$ Cubic	1. White, usually with a violet tint. Sometimes yellowish or brownish. ⟶ pentlandite, distinct violet tint; pentlandite is yellow in contrast. ⟶ pyrrhotite, no violet tint is shown. ⟶ bravoite, usually more violet. ⟶ millerite, darker, pinkish violet. ⟶ chalcopyrite, pinkish cream. 2. + 3. Isotropic; a faint anomalous anisotr. has been reported [302]. 4. Not present.	470 nm 38.3 546 42.25 589 44.1 650 48.2 [558a].
BRAVOITE $(Fe,Ni,Co)S_2$ (complete solid solution series FeS_2-CoS_2-NiS_2). May contain up to 10 mol. % Cu. Cubic	1. Varying with composition: pinkish cream, pink, brown, dark brown or slightly violet. Ni-rich varieties closely resemble pyrrhotite in colour and commonly tarnish violet after some weeks. Fe-rich varieties are brownish violet and Co-rich varieties are reddish violet. Cu-bravoite shows a cream to pinkish cream colour. 2. + 3. Isotropic. 4. Not present.	Strongly dependent on chem. comp.: high R = Fe, low R = Ni, Co. 470 nm 32.0-44.5 546 30.9-52.2 589 31.0-53.9 650 31.3-54.6 Dispersion curves: [300, 301]; see also: [1597a].

Hardness H: polishing hardness VHN: micro-indentation hardness in kg/mm^2	Miscellaneous
H:>chalcopyrite, sphalerite, $\geq$ pentlandite, commonly slightly <bravoite, pyrrhotite, ≪arsenopyrite, pyrite. VHN: 241-373 [1715]. VHN$_{50}$: 458 [810]. VHN$_{25}$: 306-327 [194].	Cubic and octahedral cleavage commonly well developed. Twinning and zonal texture not observed. Usually occurs as an alteration product, often together with bravoite, of pentlandite or millerite. The alteration commonly starts along the cleavages. Replaces pentlandite, millerite, pyrrhotite, gersdorffite. May be replaced by chalcopyrite. May contain exsolution lamellae of millerite. Forms pseudomorphs after pentlandite and millerite. Other ass.: niccolite, magnetite, ilmenite, hematite. Difficult to distinguish from bravoite. Bravoite has a greater H and a less developed cleavage. Ref.: 66, 134, 137, 302, 319, 321, 367, 504, 588, 626, 700, 765, 909, 1170, 1237, 1323, 1392, 1692.
H: varying between sphalerite and pyrite. Usually slightly > pentlandite, ~pyrrhotite. VHN$_{100-200}$: 668-1535 [159, 194, 810, 1597a, 1715, 1718].	Twinning not observed. Zonal texture very common, due to variation in composition; the zones may show differences in colour as well as in H. Zones of high Ni-content coincide with the zones of low refl.; Co-contents can almost not be distinguished microscopically. May be formed by alteration of pentlandite, together with tiny spots of marcasite or pyrite, the cleavage system of pentlandite often remaining visible. Pentlandite-flames, -stringers etc. in pyrrhotite may thus completely be replaced by bravoite. The bravoite formed by alteration of pentlandite does not show zonal texture. Then difficult to distinguish from violarite. Bravoite is also formed by alteration of millerite. Replaced by goethite, marcasite, delafossite. Forms orientated intergrowths with skutterudite. Other ass.: pyrrhotite, chalcopyrite, sphalerite, galena, tetrahedrite, linnaeite, siegenite, villamaninite, parkerite, maucherite, safflorite, bismuth, bismuthinite, niccolite, langisite. Diff.: Violarite does not show zonal texture and commonly exhibits a distinct cubic cleavage. Ref.: 107, 300, 301, 325, 413, 626, 700, 725, 1155, 1237, 1256, 1329, 1342, 1432, 1597a, 1629, 1717, 1718.

Description of minerals from Table III

Name Formula Crystal System	1: Colour 2: Bireflectance 3: Anisotropy 4: Internal reflections	Reflectance in % (Wavelength in nanometres)

CATTIERITE
CoS_2 (may contain some Ni and Cu).
Cubic
	1. In air: pinkish.
	   In oil: pinkish violet.
	   $\longrightarrow$ siegenite, more pinkish.
	2. + 3. Isotropic.
	4. Not present.
		470 nm 33.5
		546 33.4
		589 33.9
		650 35.1
		Dispersion curve: [300].

VAESITE
NiS_2 (may contain some Co and Cu; up to 20 % Se).
Cubic
	1. In air: grey.
	   $\longrightarrow$ sphalerite, very similar, but somewhat lighter.
	   $\longrightarrow$ tennantite, darker.
	2. + 3. Isotropic.
	4. Not present.
		470 nm 31.8
		546 30.7
		589 30.7
		650 31.5
		Dispersion curve: [300].

Hardness H: polishing hardness VHN: micro-indentation hardness in kg/mm^2	Miscellaneous
H: slightly > sphalerite, > siegenite. VHN_{15}: (on synthetic material) 953-1113 [300].	Forms granular intergrowths with a linnaeite-group mineral. Larger crystals may show a perfect cubic cleavage; Co-end-member of the bravoite-series. May be replaced by pyrite and siegenite. Forms intergrowths with pyrite and galena. Never occurs together with vaesite in the same vein. Other ass.: chalcopyrite. Ref.: 300, 301, 303, 710, 1372, 1612.
H: no data VHN_{15}: (on synthetic material) 773-856 [300].	Ni-end-member of the bravoite-series. Occurs as octahedral and cubic crystals, and as aggregates. May show cubic cleavage almost as perfect as that of galena. Formed by alteration of Ni-skutterudite and other Ni-sulphides and -arsenides. Replaced by millerite, polydymite, siegenite and gold. Replaces pyrite. Never occurs together with cattierite in the same vein. cattierite in the same vein. Other ass.: linnaeite, chalcocite, bornite, chalcopyrite, uraninite, digenite, covellite, molybdenite. Ref.: 300, 301, 303, 538, 710, 734.

Name Formula Crystal System	1: Colour 2: Bireflectance 3: Anisotropy 4: Internal reflections	Reflectance in % (Wavelength in nanometres)
INDITE $FeIn_2S_4$ Cubic	1. White. 2. + 3. Isotropic. 4. Not present.	477 nm 29.3 545 29.0 589 27-28 658 26.8 [493].
DZHALINDITE $In(OH)_3$ Cubic	1. Dark grey. 2. + 3. Isotropic. 4. Not indicated.	589 nm 8.2 [493].
COBALTPENTLANDITE $(Co,Ni,Fe)_9S_8$ (contains up to 66.6 wt. % Co.) Cubic	1. Yellowish white. ⟶ pentlandite, slightly lighter. 2. + 3. Isotropic. 4. Not present.	470 nm 49.5 546 53.7 589 55.2 650 57.2 [1155].

Hardness H: polishing hardness VHN: micro-indentation hardness in kg/mm^2	Miscellaneous
H: no data. VHN$_{20-50}$: 293-325 [493].	Occurs in concentrically zoned cassiterite. Formed by the replacement of In-bearing cassiterite. Replaced along cracks by dzhalindite and Fe-hydroxides. Other ass.: arsenopyrite, pyrite, chalcopyrite. Ref.: 493, 1382.
H: no data. VHN: no data.	Originates as a secondary mineral under supergene conditions. Replaces indite along cracks. Parag.: cassiterite, arsenopyrite, chalcopyrite, pyrite, Fe-hydroxides. Ref.: 493, 1382.
H>pentlandite, slightly ≦pyrrhotite,<linnaeite. VHN$_{25-50}$: 245-363 [753, 1155]. VHN increases with increasing Co-contents.	Occurs as xenomorphic grains and as exsolution lamellae ∥ (0001) of pyrrhotite or ∥ (100) of linnaeite and siegenite. As veinlets in pyrite and marcasite. Alters to bravoite. Other ass.: parkerite, bismuth, bismuthinite, safflorite, maucherite, sphalerite, Cu-Fe-sulphides. Ref. 588, 753, 1155, 1319, 1483.

Name Formula Crystal System	1: Colour 2: Bireflectance 3: Anisotropy 4: Internal reflections	Reflectance in % (Wavelength in nanometres)

PYRRHOTITE
Stoichiometric FeS is called troilite (hex.).
$Fe_{1-x}S$ (48.1-47.5 at. % metals): Hex.
$Fe_{1-x}S$ (46.5 at. % metals): Mono.
May contain some Ni, Co, Mn*.
	1. Cream with a faint pinkish brown tint. Tarnishes slowly in air.
	⟶ pentlandite, much darker; pentlandite is white in contrast.
	⟶ cubanite, more pinkish.
	⟶ niccolite, much darker, without reddish tints.
	⟶ bismuth, greyish creamy brown.
	2. Very distinct: E brownish creamy, O reddish brown.
	3. Very strong: yellow-grey, greenish grey or greyish blue; intensity of anisotr. depends on orientation. System of orientation under the microscope has been described by Kanehira (1966).
	4. Not present.
		For hexagonal pyrrhotite (47.3 at. % metals):
		$\quad\quad R_o \quad R_e$
		470 nm 30.9 36.2
		546 34.0 39.2
		589 35.8 40.7
		650 38.6 42.5
		
		Dispersion curves: [486]
		see also: [43].
		
		546 nm R_o 36.3 (46.67 at. % metals)
		             34.7 (47.3 at. % metals)
		             34.1 (47.7 at. % metals) [305].

* For minor elements in pyrrhotite, see Fleischer (1955).
**Foslie (1946) reports an exceptional low Ḣ for a pyrrhotite spec., even<sphalerite; Ödman (1941a) also describes a pyrrhotite spec. with very low H, even<chalcopyrite. However, if these minerals really are pyrrhotite, such H must be regarded as exceptions.
***In iron from Ovifak, Disco Island, Greenland, pyrrhotite occurs equally distributed as minute, rod-like, orientated inclusions in cohenite, formed by exsolution; also as larger grains showing exsolution intergrowths with pentlandite. In the description [845] named troilite, but shown to be pyrrhotite by one of the authors (W.U.). Described by Ramdohr since 1950 in the different editions of his textbook as "chalcopyrrhotite".

Hardness H: polishing hardness VHN: micro-indentation hardness in kg/mm^2	Miscellaneous
H: $\gg$ chalcopyrite, $\sim$ pentlandite, niccolite, $\ll$ arsenopyrite, pyrite**. VHN: 230-390 [159, 194, 1113, 1715].	Forms allotriomorphic granular aggregates or compact masses, less common idiomorphic crystals. Cleavage in unaltered spec. only distinct in larger crystals; in altered spec. usually well visible. Distinctness of cleavage may also depend on degree of pentlandite exsolutions $/\!/$ (0001), a distinct cleavage indicating a high Ni-content. Twinning not uncommon. Zonal texture may occur. The three phases of pyrrhotite can occur as: - two-phase mixtures of hex. $Fe_{1-x}S$ and mono. $Fe_{1-x}S$: either phase can occur as lamellae in a matrix of the other, but in a majority of cases, the two phases co-exist as irregularly shaped grains. Distinction in polished section: structural etching with a saturated solution of CrO_3 in water: mono. $Fe_{1-x}S$ is etched more strongly (appears darker) than hex. $Fe_{1-x}S$. - two-phase mixtures of FeS and hex. $Fe_{1-x}S$: each phase generally occurs as regularly shaped lamellae in a matrix of the other. FeS appears to have a somewhat higher refl. and a somewhat lower H than hex. $Fe_{1-x}S$, but tarnishes more quickly in air. - single-phase hex. $Fe_{1-x}S$. - single-phase mono. $Fe_{1-x}S$. The intergrowths between the different pyrrhotite phases can probably be attributed to exsolution. Pyrrhotite occurs as fillings between cassiterite clusters. May be interlocked with chalcopyrite or pentlandite. Usually contains flames, spindles, blebs, lenses, granular aggregates and veins of exsolved pentlandite, commonly starting from cracks, fissures or grain boundaries or orientated $/\!/$ crystallographic directions. These pentlandite bodies may completely be replaced by violarite or bravoite. Pyrrhotite may also contain exsolution bodies of magnetite and chalcopyrite. Enclosed in chalcopyrite, sphalerite, pentlandite or alabandite as more or less orientated exsolution bodies of different, even worm-like, shape. Shows inclusions of chalcopyrite, cubanite, arsenopyrite, sphalerite, marcasite, magnetite, cassiterite, ilmenite, gold, niccolite. Forms orientated intergrowths with sphalerite, pentlandite, chalcopyrite, galena, pyrite, marcasite, arsenopyrite, magnetite, cubanite. Replaces pyrite, magnetite; replaced by pentlandite, pyrite, marcasite, chalcopyrite, tetrahedrite, tennantite, sphalerite, stannite, gudmundite, berthierite and other sulphosalts, galena, gold, magnetite, hematite, covellite, chalcocite. Alters to pyrite, marcasite, magnetite, hematite; alteration to pyrite and marcasite may produce so called "bird's eye" textures. Other ass.: bravoite, millerite, chromite, valleriite, mackinawite, awaruite, heazlewoodite, silver, copper, Co-Ni-Fe-arsenides, platinum, cohenite, iron***. Ref.: 45, 46, 56, 137, 177, 229, 302, 305, 321, 339, 351, 353, 367, 397, 411, 412, 413, 421, 423, 438, 474, 487, 539, 588, 590, 694, 756, 757, 845, 891, 997, 1047, 1058, 1237, 1342, 1358, 1392, 1444, 1450, 1462, 1464, 1576, 1600, 1603, 1629, 1662. Note: Jaipurite, γ-CoS, hex., is the Co-analogue of pyrrhotite. Doubtful as mineral. Ref.: 1237, 1475.

Description of minerals from Table III

Name / Formula / Crystal System	1: Colour / 2: Bireflectance / 3: Anisotropy / 4: Internal reflections	Reflectance in % (Wavelength in nanometres)
GREIGITE Fe_3S_4 (may contain some Ni). Cubic	1. Pale creamy white. ⟶ marcasite, less yellow and more pinkish. ⟶ pyrrhotite, distinctly whiter, less pinkish. ⟶ pyrite, pyrite is yellow in contrast. Ni-containing greigite shows a more pronounced pink colour. 2. + 3. Isotropic*. 4. Not present.	No data.
SMYTHITE Fe_3S_4 Hex.	1. Pinkish cream. ⟶ pyrrhotite, very similar. 2. Strong, greyish yellow to reddish brown; noticeably stronger than for pyrrhotite. 3. Strong: yellow to blue-grey. 4. Not present.	588 nm 34.5 [258].
MACKINAWITE FeS (15 at. % of Fe may be replaced by Ni, Co and/or Cu; may contain up to 9 wt. % Cr). Tetr.	1. Strongly varying, even in well polished sections. Usually pinkish or reddish grey. No bronze or brown colours as for valleriite. ⟶ pyrrhotite, very similar. Cr-bearing mackinawite shows a distinct brownish tint in contrast to "normal" mackinawite. 2. Moderate to strong, pinkish grey to grey. 3. Very strong. Completely crossed nicols: greyish white to dark grey or black. Not completely crossed nicols: bluish white to sienna brown. No bronze polarization colours as for valleriite. The anisotropy decreases markedly after a few weeks; polishing restores the original effects. 4. Not present.	450 nm 21.0-37.0 550 21.8-44.5 589 22.0-46.0 650 23.2-47.5 [1382a]. Dispersion curves: [1597b].

* Jedwab (1967) describes greigite without birefl., but with a distinct anisotr., creamy white to bluish grey.

Hardness H: polishing hardness VHN: micro-indentation hardness in kg/mm^2	Miscellaneous
H: slightly>pyrrhotite. VHN$_?$: 312 [1199]. VHN$_{15}$: (on loose aggregates) 35-63 [669].	Very similar to bravoite and to minerals of the linnaeite-group. Tends to form idiomorphic crystals. No cleavage. Occurs as fine-grained globular aggregates in stibnite. Fills fractures in stibnite. Replaced by stibnite. Other ass.: realgar, orpiment, marcasite, pyrrhotite. Ref.: 669, 1199, 1406, 1700.
H: no data. VHN: no data.	Occurs as clusters of very fine patches in radial, more rarely in parallel arrangement. Perfect basal cleavage. Parag.: pyrrhotite, pyrite, sphalerite. Diagn.: only distinguishable from pyrrhotite by X-ray diffraction. Ref.: 258, 364, 365.
H: similar to that of pyrrhotite, but varying with Co- and Ni-contents. VHN$_{50}$: 52-58 [244, 267]. VHN$_{25}$: (for Ni-mackinawite) 94-181 [1597b]	Occurrence: 1) Sometimes as idiomorphic crystals. Perfect basal cleavage; consequently mackinawite tends to flake somewhat like graphite. 2) By exsolution in pentlandite, pyrrhotite, chalcopyrite and cubanite as lamellae, sometimes orientated; also as patches, lenticular bodies and herringbone textures. 3) By alteration as irregular, subhedral to anhedral flakes and microveinlets in pentlandite, cobaltpentlandite, magnetite, chalcopyrite, pyrrhotite, galena, schapbachite, silver. 4) as reaction rims between chalcopyrite and pyrrhotite. Replaces pentlandite, cobaltpentlandite, chalcopyrite, cubanite, pyrrhotite. Other ass.: chromite, sphalerite, wurtzite, awaruite, iron, valleriite. The numerous references on valleriite are ambiguous and many could equally well apply to mackinawite; in fact, much of the valleriite mentioned in the literature has been proved to be mackinawite. It is practically impossible to distinguish between mackinawite and valleriite under the microscope; identification is only possible by X-ray diffraction or electron-microprobe techniques. It appears that mackinawite is a common and widespread mineral, while valleriite is only found in contact-metamorphic environments associated with carbonates and serpentine. Ref.: (only to mackinawite, see also at valleriite): 41, 108, 135, 186, 244, 266, 267, 270, 271, 353, 370, 372, 755, 757, 1254, 1259, 1338, 1342, 1426, 1597b.

Name Formula Crystal System	1: Colour 2: Bireflectance 3: Anisotropy 4: Internal reflections	Reflectance in % (Wavelength in nanometres)
OREGONITE Ni_2FeAs_2 Hex.	1. Pure metallic white. ⟶ awaruite, bluish. 2. Weak. 3. Weak, only visible at grain boundaries; straight extinction. 4. Not present.	470 nm 45.2-45.6 546 49.0-50.0 589 50.5-51.5 650 52.0-52.9 [194] (on synthetic material).
COHENITE* Fe_3C (some Ni and Co may be present). Ortho.	1. Creamy white. ⟶ pearlite, creamy or slightly brownish. ⟶ pyrrhotite, light creamy. ⟶ pentlandite, very similar. ⟶ iron, very similar. 2. Weak but distinct. 3. Weak but distinct. 4. Not present.	No data.
SCHREIBERSITE $(Fe,Ni,Co)_3P$ Tetr.	1. White (in air), in oil with a brownish pink tint. ⟶ cohenite, lighter (in air). ⟶ iron, in lightest position very similar. 2. In oil distinct: pinkish brown to yellowish. 3. Weak but distinct in oil. 4. Not present.	No data.

* For Fe_3C, whether or not containing some Ni or Co, two names have appeared in literature, cohenite and cementite, the first name being given by Weinschenk (1889, Ann. Mus. Wien, 4: 94) to the natural mineral, the last name by Osmond (1894, Compt. Rend., 118: 307; 119: 329; Z. Krist., 27: 5, 1897) to the artificial substance. The name cementite, however, has been used by many authors for natural material [98, 845, 1369, 1603]. Here the name cohenite is used, having priority.

Hardness H: polishing hardness VHN: micro-indentation hardness in kg/mm^2	Miscellaneous
H:> pyrrhotite VHN: (on synthetic material): 605-635 [194].	Occurs as aggregates of rounded to oval grains; as crusts around niccolite; between pyrrhotite and chalcopyrite; rimmed by maucherite. May contain inclusions of copper. Parag.: copper, bornite, chalcopyrite, pyrrhotite, maucherite, cuprite, molybdenite, chromite, niccolite, awaruite. Ref.: 41, 1156, 1240, 1247.
H:≫ pearlite, >iron, pyrrhotite, <schreibersite*. VHN: no data.	Terrestrial cohenite forms fine-lamellar intergrowths with iron, so called pearlite (see iron). If iron has been replaced by iron oxide or hydrous iron oxide, this intergrowth is called respectively oxide-pearlite or hydroxidepearlite**. Larger cohenite grains may occur in a groundmass of pearlite. These crystals may be stabilized by admixture of some Mn or S; otherwise cohenite is metastable and dissociates into graphite and iron. Minute orientated needles of pyrrhotite and schreibersite may be enclosed, due to exsolution (only visible at highest magnifications). Aggregates of cohenite and iron may be rimmed by fine fringes of copper. Other ass.: magnetite, pentlandite. Ref.: 98, 845, 1237, 1369, 1437, 1603.
H:>cohenite*,~iron. VHN: no data.	Occurs as orientated needle-like inclusions in cohenite in the iron of Ovifak (only known terrestrial occurrence). Other ass.: pyrrhotite, pentlandite. Ref.: 845, 1237.

---

* Only if Ni is present; some Ni decreases the hardness of cohenite, but increases that of schreibersite [845].
** See footnote on page 129.

Name Formula Crystal System	1: Colour 2: Bireflectance 3: Anisotropy 4: Internal reflections	Reflectance in % (Wavelength in nanometres)
WULFENITE $PbMoO_4$ Tetr.	1. Grey-white. 2. In air weak, more distinct in oil. 3. Obscured by the int. refl. 4. Strong and abundant: colourless, yellow, and orange-yellow.	$\quad\quad\quad R_0 \quad R_e$ 450 nm 18.75 17.00 530 $\quad$ 17.70 16.50 600 $\quad$ 17.25 16.05 [525]
DESCLOIZITE $Pb(Zn,Cu)[OH/VO_4]$ Ortho.	1. See at birefl. 2. Strong in air and in oil: grey to greyish blue. 3. Very strong. 4. Rare, colourless to brownish.	$\quad\quad\quad // a \quad // c$ 450 nm 19.25 16.35 530 $\quad$ 17.60 15.35 600 $\quad$ 17.25 14.95 [525]
VILLAMANINITE Idiomorphic villamaninite: $Cu_{0.60}Ni_{0.14}Co_{0.03}Fe_{0.23}S_2$. Nodular villamaninite: $Cu_{0.28}Ni_{0.35}Co_{0.11}Fe_{0.26}S_2$. Always contains some Se. Cubic	1. Idiomorphic villamaninite shows a reddish violet tinge, similar to that of ilmenite, while nodular villamaninite is more light bluish grey to violet-grey. 2. Not present. 3. Isotropic; a very weak anomalous anisotr. may be visible in idiomorphic villamaninite. 4. Not present.	About 25-30 %.

Hardness H: polishing hardness VHN: micro-indentation hardness in kg/mm^2	Miscellaneous
H:<descloizite. VHN_{25}: 211-333 [525].	Occurs as idiomorphic crystals and as aggregates in the oxidation zone, associated with molybdenite, goethite and oxidation minerals. Ref.: 525.
H:>wulfenite. VHN_{25}: 310-491 [525].	See at wulfenite. Ref.: 525.
H:≫chalcopyrite, >penroseite, ≧linnaeite, <bravoite, pyrite. VHN_{200}: (on idiomorphic vill.): 440-520 [1718]. VHN_{200}: (on nodular vill.): 535-710 [1718].	Villamaninite as described by Schoeller and Powell (1920) was shown to consist of three compounds (Ypma et al., 1968): Cu-rich villamaninite, relatively Cu-poor villamaninite and Cu-bravoite (up to 10 mol. % Cu.). The Cu-rich villamaninite, because of its occurrence, is called idiomorphic villamaninite. It is surrounded by, or forms the cores of, Cu-poor villamaninite, which occurs as spherical aggregates and, therefore, is called nodular villamaninite. Between these two phases a rim of Cu-bravoite may occur. Zonal banding between Cu-rich villamaninite and Cu-bravoite has been observed. Idiomorphic villamaninite may show cleavage ∥ (100) and twin lamellae. Zonal texture is very common in individual crystals. Villamaninite alters to, and is replaced by, "normal" bravoite, linnaeite, bornite, chalcopyrite, marcasite and pyrite. Ramdohr (1937b) described two minerals of the linnaeite-group occurring in villamaninite: an unusual white phase, shown to be normal linnaeite; the second phase being probably Cu-bravoite (Ypma et al., 1968). Other ass.: clausthalite. Ref.: 621, 1210, 1341, 1377, 1717, 1718.

146   Description of minerals from Table III

Name Formula Crystal System	1: Colour 2: Bireflectance 3: Anisotropy 4: Internal reflections	Reflectance in % (Wavelength in nanometres)
LINNAEITE Co_3S_4 Cubic	1. White with a creamy tint. ⟶ skutterudite, greyish white. ⟶ ullmannite, gersdorffite, creamy or yellowish; both minerals are white in contrast. ⟶ millerite, much darker, very light pinkish brown; millerite is light yellow in contrast. ⟶ chalcopyrite, distinctly pinkish. 2. + 3. Isotropic. 4. Not present.	About 45-50 %.
SIEGENITE $(Ni,Co)_3S_4$ (may contain up to 11 % Se). Cubic	1. Creamy white, sometimes with a slight pinkish tinge. ⟶ cattierite, less pinkish. 2. + 3. Isotropic. 4. Not present.	470 nm 42.5-43.1 546 44.9-45.4 589 46.3-46.9 650 47.6-49.0 [1155].
POLYDYMITE Ni_3S_4 Cubic	1. Light brass-yellow. 2. + 3. Isotropic. 4. Not present.	470 nm 42.9 546 46.0 589 47.3 650 51.7 [194].
CARROLLITE Co_2CuS_4 (Cu- and Co-contents vary strongly; may contain some Ni). Cubic	1. Creamy white, similar to linnaeite, but somewhat lighter. Some spec. have a pinkish tint. 2. + 3. Isotropic. 4. Not present.	546 nm 42.6 [194].

Hardness H: polishing hardness VHN: micro-indentation hardness in kg/mm^2	Miscellaneous
H: $\gg$ chalcopyrite, $>$ sphalerite, pyrrhotite, $\sim$ skutterudite, $<$ gersdorffite, $\ll$ arsenopyrite, pyrite. VHN: 351-566 [159, 194, 1715]. VHN$_{200}$: 460-485 [1718].	Occurs as idiomorphic crystals and as anhedral grains. Cleavage or parting $/\!/$ (100) commonly visible. Twinning and zonal texture not observed. Forms regular exsolution intergrowths with millerite, chalcopyrite, cobaltpentlandite, pentlandite, these minerals occurring as a network in linnaeite. Found as exsolution lamellae in pyrrhotite. Forms intergrowths, sometimes orientated, with bornite, bravoite, bismuth, Occurs as inclusions in chalcopyrite, pyrrhotite. Replaces skutterudite; may be replaced by chalcocite, covellite, chalcopyrite, pyrite, bismuth, wittichenite. Other ass.: safflorite, rammelsbergite, niccolite, hematite, magnetite, tennantite, cubanite, digenite, villamaninite. Ref.: 215, 333, 658, 801, 995, 1165, 1214, 1237, 1318, 1379, 1383, 1392, 1483, 1711, 1718.
H: $\sim$ linnaeite, $<$ cattierite. VHN: 336-579 [159, 1155, 1643, 1715].	Occurs as idiomorphic crystals and as irregular grains. Cubic cleavage not always visible. May be replaced by Cu-sulphides. Forms intergrowths with cobaltpentlandite. Replaces bismuthinite, pyrite, vaesite, cattierite, chalcopyrite, uraninite. Other ass.: bornite, digenite, covellite, tennantite, parkerite, bismuth, safflorite, maucherite, marcasite, bravoite, niccolite, langisite, sphalerite. Very similar to linnaeite in polished sections. Tests on Ni and Co may be decisive. Ref.: 303, 480, 1155, 1237, 1392, 1617, 1620, 1643.
H: $\sim$ other minerals of the linnaeite-group. VHN: 437-444 [194]. VHN$_{50}$: 362-449 [810].	Very similar to other members of this group. Twinning or zonal texture not observed. Perhaps rarer than linnaeite, violarite and siegenite. Parag.: chalcopyrite, pyrrhotite, pyrite, millerite, galena, sphalerite, bismuth, gersdorffite, ullmannite. Reported from a pitchblende association, together with Co-Ni-arsenides and sulphides of Fe, Zn, Pb, Cu and Ag. Ref.: 718, 774, 1237, 1392.
H: $\sim$ linnaeite-group members. VHN: 351-566 [159, 194].	Occurs as euhedral, subhedral and anhedral grains. Replaced by Cu-minerals, such as chalcocite, bornite, chalcopyrite, digenite, covellite, and by pyrite and pyrrhotite. Occurs as inclusions in chalcopyrite. May show inclusions of sphalerite, chalcopyrite, other Cu-sulphides. Very similar to linnaeite in polished section. Ref.: 293, 538, 1065, 1149, 1307, 1392, 1617, 1620.

Description of minerals from Table III

Name Formula Crystal System	1: Colour 2: Bireflectance 3: Anisotropy 4: Internal reflections	Reflectance in % (Wavelength in nanometres)
ORCELITE $Ni_{\lesssim 5}As_2$ Hex.	1. Pinkish bronze. → niccolite, browner. 2. Weaker than for niccolite or breithauptite. 3. Distinct to strong, green and violet tints. 4. Not present.	530 nm 44-48 600 52-54 650 56-58 [216].

BREITHAUPTITE
NiSb
Hex.
	1. Pink with a violet tint.
	→ niccolite, slightly darker with violet tint.
	2. Strong, more distinct than for niccolite:
	  O: light pinkish,
	  E: bright pinkish violet.
	3. Very strong, slightly higher than for niccolite: bluish green, bluish grey, violet-red.
	4. Not present.
		R_o R'_e
		470 nm 45.5 37.4
		546 48.2 36.9
		589 53.0 43.7
		650 58.0 51.0
		              [194].

LANGISITE
(Co,Ni)As
Hex.
	1. Pinkish buff.
	2. Very weak.
	3. Moderate, bluish grey to light brown.
	4. Not present.
		470 nm 44.9-46.0 to
		        46.0-46.8
		546 46.1-46.8 to
		        46.7-47.4
		589 47.1-47.7 to
		        47.9-48.7
		650 49.4-50.3 to
		        50.3-51.3
		Dispersion curves: [1155].

MAUCHERITE $Ni_{\lesssim 3}As_2$ Tetr.	1. When massive and isolated from other minerals: plain white. → niccolite, chalcopyrite, greyish with a mauve tint. → cobaltite, very similar. → loellingite, distinctly brownish grey. → breithauptite, bluish grey. 2. Not observable. 3. Weak to distinct, only visible in oil: light grey to dark grey. 4. Not present.	470 nm 46.3 546 48.0 589 50.6 650 54.2 [194].

Hardness H: polishing hardness VHN: micro-indentation hardness in kg/mm^2	Miscellaneous
H: no data. VHN: no data.	Shows a lamellar structure. Only observed in serpentinized rocks; occurs as inclusions in awaruite; may contain inclusions of pentlandite. Other ass.: pyrrhotite, magnetite. Ref.: 216, 1240, 1475.
H: slightly<niccolite, <rammelsbergite, skutterudite, safflorite, ≪cobaltite, arsenopyrite, glaucodot. VHN_{50-100}: 412-584 [159, 194, 810, 1715].	Commonly allotriomorphic granular; also idiomorphically developed. Cleavage not observed, no twinning. Zonal texture very common. Replaced by niccolite, silver, Ag-minerals, safflorite. Replaces safflorite. Occurs as inclusions in chromite, pentlandite, galena, safflorite. Forms intergrowths with pyrrhotite. Parag.: niccolite and other Ni-minerals, silver and Ag-minerals, Co-minerals. Ref.: 227, 619, 1059, 1164, 1202, 1237, 1392, 1488, 1603.
H: no data. VHN_{50}: 780-857 [1155].	Occurs as irregular grains and lamellae in safflorite. Parag.: cobaltpentlandite, siegenite, parkerite, bismuth, bismuthinite, safflorite, maucherite, pyrite, marcasite, bravoite, sphalerite, chalcopyrite. Ref.: 1155.
H:>chalcopyrite, sphalerite,≧niccolite,≳ minerals of the skutterudite series, <rammelsbergite, safflorite, loellingite. VHN: 602-724 [159, 194, 1715].	Commonly occurs as allotriomorphic aggregates of elongated crystals; also as idiomorphic tabular or long-columnar crystals. Cleavage not observed; twinning occurs. May be intergrown with or enclosed by niccolite (texture of such intergrowths may be brought out by etching with $FeCl_3$, maucherite turning black). Forms pseudo-eutectic intergrowths with niccolite by replacement of gersdorffite, and with pyrrhotite by breakdown of ferroan gersdorffite. Replaces niccolite and gersdorffite. May be replaced by niccolite, chalcopyrite, cubanite. Sometimes contains inclusions of gold and siegenite. Occasionally encloses euhedral and subhedral crystals of gersdorffite. Diff.: niccolite is strongly anisotr. and shows strong birefl.; safflorite and rammelsbergite are whiter and anisotropic; cobaltite is much harder; skutterudite shows other texture when etched. Parag.: about as for niccolite; also: pyrrhotite, pentlandite, chalcopyrite, parkerite, langisite. Ref.: 290, 588, 591, 594, 765, 916, 1059, 1123, 1133, 1155, 1237, 1379, 1392.

Description of minerals from Table III

Name Formula Crystal System	1: Colour 2: Bireflectance 3: Anisotropy 4: Internal reflections	Reflectance in % (Wavelength in nanometres)	
NICCOLITE NiAs Hex.	1. + 2. Strong and typical birefl.: yellowish pink or light pink (O) to brownish pink (E). ⟶ maucherite, more pinkish, lighter; maucherite is bluish violet-grey in contrast. ⟶ breithauptite, paler pinkish yellow. ⟶ skutterudite, distinct pinkish. ⟶ bismuth, more pinkish. ⟶ arsenic, pinkish. ⟶ pyrrhotite, lighter and pinkish, no brownish tint. ⟶ bornite, much lighter. 3. Very strong: yellowish, greyish green, violet-blue, bluish grey. Straight, but no complete extinction. Basal sections appear isotropic. 4. Not present.	470 nm 546 589 650 Dispersion curves: [1155].	38.5-39.2 to 46.2-46.8 48.9-49.1 to 52.6-52.9 54.4-54.8 to 56.6-56.9 59.6-61.6 to 60.3-62.4

Hardness H: polishing hardness VHN: micro-indentation hardness in kg/mm^2	Miscellaneous
H: $\gg$ silver, dyscrasite, $>$ chalcopyrite, $\sim$ breithauptite, pyrrhotite, $\leq$ maucherite, $<$ rammelsbergite, skutterudite, $\ll$ loellingite, pyrite. VHN: 308-533 [159, 194, 1113, 1715].	Cleavage commonly not visible. A chevron-like texture of alternating isotropic and anisotropic lamellae may occur, probably due to twinning. Usually forms concentric aggregates, often with radiated texture, associated with safflorite and rammelsbergite, or subparallel aggregates. Idiomorphic crystals and cross-shaped twins also occur. Forms pseudo-eutectic intergrowths with pyrrhotite, chalcopyrite and maucherite (in the last case, etching with $FeCl_3$, which blackens maucherite, brings out texture). Intergrowths with breithauptite, pyrrhotite, maucherite, safflorite, loellingite, galena, pitchblende, and many others also occur. Replaces skutterudite, pyrrhotite, pentlandite, chalcopyrite, silver, bismuth, maucherite, galena, sphalerite. May be replaced by cobaltite, gersdorffite, rammelsbergite, pararammelsbergite, maucherite, skutterudite, chalcopyrite, silver, gold, pentlandite, safflorite. Mutual replacement of and by chalcocite results in a myrmekitic intergrowth of niccolite with chalcocite. Occurs enclosed in skutterudite, rammelsbergite forming a reaction rim. Also occurs as inclusions in pentlandite, galena, chalcopyrite, safflorite. May contain inclusions of gold, pitchblende and coffinite. Forms coatings around silver. Decomposition of niccolite yields a compound closely related to rammelsbergite, and vaesite. Other ass.: linnaeite, millerite, glaucodot, arsenopyrite, bismuth, arsenic, dyscrasite, allargentum, pyrargyrite, polybasite. Ref.: 227, 290, 378, 379, 426, 458, 588, 591, 619, 632, 707, 708, 765, 791, 916, 997, 1059, 1154, 1155, 1164, 1224, 1237, 1342, 1356, 1379, 1392, 1488, 1603, 1732.

Notes:
1. There exists a cubic mineral with the name dienerite, Ni_3As. Occurs as cubic crystals. No optical data available. The only crystals ever found have disappeared in Vienna.
   Ref.: 559, 1247.
2. Modderite, CoAs, ortho. No optical data available. The mineral has been found associated with niccolite in Au-concentrates in S.Africa.
   Ref.: 282.
3. A new Ni-As-mineral with a Ni/As ratio of about 5:4 has been described by Frenzel and Ottemann (1968) from Příbram. It replaces niccolite. Strong birefl.: (O) very similar to niccolite; (E) much darker, grey-black. Enormous anisotropy.

Description of minerals from Table III

Name Formula Crystal System	1: Colour 2: Bireflectance 3: Anisotropy 4: Internal reflections	Reflectance in % (Wavelength in nanometres)
SKUTTERUDITE-series a: skutterudite (cobaltian) b: nickelian skutterudite c: ferrian skutterudite $(Co,Ni,Fe)As_{3-x}$ Cubic	1. Differing with composition; zones with a creamy-white colour (a) may alternate with more greyish white or even bluish white zones. ⟶ rammelsbergite and safflorite, slightly more yellowish. ⟶ ullmannite and gersdorffite, less yellow. ⟶ niccolite, bluish grey. ⟶ cobaltite, white; cobaltite is distinctly pinkish. ⟶ maucherite, white; maucherite is distinctly pinkish. ⟶ bismuth, whiter. ⟶ silver, dull bluish grey. ⟶ galena, yellowish white. ⟶ pyrite, pure white. 2. + 3. Isotropic; sometimes a weak anomalous anisotropy may be visible in markedly zoned crystals. 4. Not present.	For (a): 470 nm 53.8 546 53.4 589 53.0 650 52.9 [194].

Hardness H: polishing hardness VHN: micro-indentation hardness in kg/mm^2	Miscellaneous
H: c>a>b; a ~ rammelsbergite, safflorite, >linnaeite, niccolite; b commonly slightly <rammelsbergite, safflorite,≧niccolite, linnaeite; for all members: >maucherite,<loellingite, ullmannite, gersdorffite, ≪arsenopyrite, pyrite, cobaltite. VHN$_{50-200}$: 268-974 [159, 194, 810, 1715].	Cleavage may be distinct; in cobaltian skutterudite rare. Twinning not observed. Differences in H may often already be sufficient to reveal texture in composite crystals. Pure skutterudite commonly forms homogeneous idiomorphic crystals; zonal crystals may occur. All other members form coarse crystals with very fine zonal texture (developed esp. by etching), or zonal intergrowths with other members of the series or with rammelsbergite, safflorite. Later generations, replacing younger ones, may be finer grained or even cryptocrystalline. Crystal aggregates not uncommon. Intergrowths with niccolite or bismuth may occur. Replaces pyrite. May be replaced by rammelsbergite, safflorite, loellingite, niccolite, bismuth, pararammelsbergite (sometimes resulting in myrmekitic intergrowths), silver, argentite, other Ag-minerals, linnaeite, stannite, chalcopyrite, arsenic. Veined by loellingite, arsenic, bismuth. Forms coatings on silver (esp. Ni-rich members). Shows inclusions of bismuth, gold, silver, niccolite, valleriite, pyrrhotite, pentlandite, cobaltite, arsenopyrite, loellingite. Occurs as inclusions in pitchblende, cobaltite, gersdorffite, arsenopyrite. Other ass.: uraninite, molybdenite, tetrahedrite, proustite, Ag-Sb-minerals, sulphides. The members higher in Ni prefer association with Ag to Bi, Co-rich members prefer Bi to Ag. Ref.: 290, 475, 632, 638, 707, 708, 718, 763, 866, 906, 960, 995, 997, 1076, 1133, 1154, 1165, 1189, 1202, 1224, 1237, 1286, 1322, 1392, 1447, 1537, 1603.

Note:
The names "chloanthite" and "smaltite" have been applied to respectively isometric Ni- and Co-diarsenides. These names should definitely be dropped because a series of isometric Ni- and Co-diarsenides does not exist (Holmes, 1947, and Roseboom, 1962a).

Name Formula Crystal System	1: Colour 2: Bireflectance 3: Anisotropy 4: Internal reflections	Reflectance in % (Wavelength in nanometres)
RAMMELSBERGITE $(Ni,Co,Fe)As_2$ including the end-member $NiAs_2$ Ortho.	1. White; considerably whiter than other Ni-Co-Fe-arsenides. ⟶ maucherite, no pinkish tint, pure white. ⟶ safflorite, yellowish. ⟶ niccolite, pure white. ⟶ bismuth, white. ⟶ arsenopyrite, white; arsenopyrite shows a creamy tint in contrast. ⟶ galena, white; galena appears light grey in contrast. 2. In air very weak, weaker than for safflorite; in oil more distinct: yellowish white to bluish white. 3. Strong, esp. in oil (not as strong as for safflorite or loellingite), showing pinkish, brownish, greenish and blue colours. 4. Not present.	About 60 %.
PARARAMMELSBERGITE $NiAs_2$ Ortho.	1. Pure white, smoother and whiter than associated white arsenides. ⟶ rammelsbergite, whiter. 2. Very weak, distinct in oil: ∥ elongation: yellowish white, ⊥ elongation: bluish white. 3. Strong, but not as strong as for rammelsbergite, and usually without bluish colours*. 4. Not present.	470 nm 55.1-57.6 546 56.2-57.8 589 56.0-58.6 650 55.4-60.3 [558a].

* Paděra (1962) reported pararammelsbergite, identified by X-ray diffraction, with blue-green polarization colours.

Hardness H: polishing hardness VHN: micro-indentation hardness in kg/mm^2	Miscellaneous
H:~ minerals of the skutterudite-series, niccolite, slightly <safflorite, loellingite, ≪arsenopyrite, cobaltite, glaucodot. VHN: 459-830 [159, 194, 810, 1113, 1715].	Forms compact or fine-grained aggregates or mosaics of interlocking grains. Zonal spherulitic and radiated textures very common, in the latter case with fibrous or bladed crystals. Skeletal crystals also occur. Cleavage ∥ (110) rarely visible. Simple and lamellar twinning very common. Shows intergrowths with niccolite (concentric) and with minerals of the skutterudite-series. Forms coatings around dendritic or skeleton-shaped crystals of silver and bismuth. Replaces niccolite, pitchblende, Ni-skutterudite; as reaction rims between niccolite and Ni-skutterudite, and between niccolite and safflorite; veins silver. May be replaced by niccolite, safflorite, arsenic, proustite, gersdorffite. Associated with Ag rather than with Bi. Other ass.: other Co-Ni-Fe-arsenides, breithauptite, cobaltite, glaucodot, arsenopyrite, linnaeite, sulphides of Fe, Cu, Pb; stephanite, polybasite, pyrargyrite. Diff.: pararammelsbergite tends to form tabular crystals which very rarely show twinning; blue polarization colours not typical. It may be very difficult to distinguish between rammelsbergite and safflorite in polished section. Ref.: 290, 458, 475, 632, 638, 689, 707, 708, 718, 807, 938, 960, 997, 1050, 1076, 1135, 1138, 1154, 1164, 1165, 1188, 1237, 1288, 1322, 1379, 1392, 1514, 1603, 1719, 1732.
H:>niccolite,<Ni-skutterudite,<loellingite. VHN: 673-824 [159, 194, 1715].	Usually forms tabular crystals with rectangular transverse sections; occasionally radial subrectangular crystals or mosaics of interlocking grains. Cleavage may be visible. Twinning rare and not lamellar. Zonal texture occurs. Replaces niccolite and Ni-skutterudite, sometimes resulting in myrmekitic intergrowths. Shows inclusions of silver. Forms rims around rammelsbergite. Other ass.: gersdorffite, cobaltite, chalcopyrite, pyrite, proustite. Diff.: rammelsbergite almost invariably shows lamellar twinning, other crystal habits and differing polarization colours. Ref.: 91, 1092, 1094, 1135, 1138, 1154, 1188, 1224, 1237, 1288.

Description of minerals from Table III

Name Formula Crystal System	1: Colour 2: Bireflectance 3: Anisotropy 4: Internal reflections	Reflectance in % (Wavelength in nanometres)
SAFFLORITE* $(Co,Fe,Ni)As_2$: solid solution series $CoAs_2$–$FeAs_2$; usually contains less than 20 mol. % $NiAs_2$. Ortho; only Co-rich safflorite shows mono. symmetry.	1. White, often with a bluish tint. ⟶ antimony, slightly grey. ⟶ bismuth, bluish. ⟶ silver, greyish white. ⟶ rammelsbergite, white; rammelsbergite is yellowish. 2. Very weak, less distinct than for loellingite; bluish to greyish; more distinct at higher Fe-contents. 3. Strong; in zonal textures differing for different zones; higher than for glaucodot. 4. Not present.	About 55–60 %.
LOELLINGITE* $FeAs_2$ (may contain some S and Sb). Ortho.	1. White, often with a yellowish tint. ⟶ arsenopyrite, less yellow. ⟶ rammelsbergite and safflorite, very similar. 2. Weak but distinct in oil: white or bluish white to yellowish white; weaker than for arsenopyrite, but stronger than for safflorite. 3. Very strong: bright orange-yellow, reddish brown, pale brown, blue, pale slaty blue, green. 4. Not present.	About 55 %.

* After the classification given by R.J. Holmes (1947): loellingite, $(Fe,Co,Ni)As_2$, with $Fe > 85$ %, including the end-member $FeAs_2$; cobaltian loellingite, $(Fe,Co,Ni)As_2$, with $Fe < 85$ %, and $Co > Ni$; nickelian loellingite, $(Fe,Ni,Co)As_2$, with $Fe < 85$ %, and $Ni > Co$; the different members only being distinguishable by quantitative chemical analysis. After Radcliffe (1968b) loellingite is $FeAs_2$ with less than 3 mol. % metal impurity, whereas safflorite applies to the $(Co,Fe,Ni)As_2$-series.

157

Hardness H: polishing hardness VHN: micro-indentation hardness in kg/mm^2	Miscellaneous
H:>skutterudite-series, ≧niccolite,<loellingite, ≪arsenopyrite, cobaltite, glaucodot. VHN$_{100-200}$: 430-988 [194, 810, 1715].	Usually shows concentric or radiated textures forming alternating zones with other minerals (such as niccolite, glaucodot, skutterudite, etc.); also tubercle-like textures. Occurs as aggregates of idiomorphic crystals. Cleavage not observed. Twinning very common and typical (star-shaped or compound star-shaped trillings). Replaces minerals of the skutterudite-series, rammelsbergite, silver, pitchblende, magnetite, chalcopyrite, sphalerite, galena, niccolite, breithauptite, cobaltite. May be replaced by cobaltite, arsenopyrite, Ag-sulphosalts, galena, sphalerite, chalcopyrite, marcasite. May contain inclusions of niccolite, breithauptite, langisite. Occurs as inclusions in pyrite. Forms intergrowths with loellingite, arsenopyrite, gold. Associated with bismuth rather than with silver. Parag.: other Co-Ni-Fe-arsenides, niccolite, breithauptite, linnaeite, siegenite, bismuthinite, parkerite, bismuth, loellingite, arsenopyrite, arsenic, sulphides of Fe, Cu, Zn, Pb. Diff.: rammelsbergite usually shows fine compound twinning. Ref.: 290, 292, 458, 475, 638, 707, 708, 718, 938, 997, 1050, 1076, 1154, 1155, 1164, 1165, 1190, 1191, 1237, 1288, 1310, 1379, 1488, 1549, 1603, 1629, 1732.
H: varying with composition;≫chalcopyrite, sphalerite, slightly >pyrrhotite, safflorite, rammelsbergite,≧gersdorffite,slightly<magnetite,≪arsenopyrite, pyrite. VHN$_{100-200}$: 368-1048 [159, 194, 810, 1715].	Zoned coarse crystals may occur; not as common as for rammelsbergite or safflorite. Usually forms idiomorphic crystals, radiated aggregates or fine-grained massive aggregates; also as skeleton-shaped aggregates, as thin crusts (e.g. enclosing arsenic spheroids) or as interlocking mosaics with safflorite. Cleavage not observed. Twinning, simple or compound, very common. Shows intergrowths with niccolite, safflorite, arsenopyrite. Inclusions of dyscrasite, arsenic, sphalerite, chalcopyrite, galena. Occurs as inclusions in arsenopyrite, skutterudite, maucherite. Replaces sphalerite, uraninite, pitchblende, arsenopyrite. May be replaced by arsenopyrite, cobaltite, sphalerite, chalcopyrite, galena, stibarsenic, antimony. Veins niccolite, rammelsbergite, skutterudite. Other ass.: pyrite, pyrrhotite, argentite, electrum, gold. Ref.: 179, 180, 265, 290, 322, 339, 458, 475, 638, 707, 802, 866, 972, 1059, 1076, 1154, 1190, 1191, 1237, 1288, 1379, 1392, 1447, 1458, 1549, 1629.

158   Description of minerals from Table III

Name Formula Crystal System	1: Colour 2: Bireflectance 3: Anisotropy 4: Internal reflections	Reflectance in % (Wavelength in nanometres)
GERSDORFFITE (Ni,Co,Fe)AsS Cubic*	1. White with a yellowish or pinkish cream tint. Colour changes with increasing Sb-content gradually to the colour of ullmannite. ⟶ skutterudite, more yellowish. ⟶ linnaeite, much less pink. ⟶ pyrite, less yellow, pinkish or cream, darker. ⟶ chalcopyrite, pinkish cream. ⟶ galena, lighter, pinkish cream. ⟶ arsenopyrite, white. ⟶ ullmannite, very similar, perhaps slightly more yellowish. ⟶ niccolite, bluish. 2. Not present. 3. Isotropic, complete extinction. Anomalous anisotropism is related to the amount of distortion in the structure. 4. Not present.	Range due to differences in chem. comp. 470 nm 45.4-52.6 546 46.7-53.8 589 46.3-53.8 650 46.6-53.4 [558a].
ULLMANNITE NiSbS (some Co, Fe, Bi or As may be present). Cubic	1. White, in oil with a faint bluish grey tint. ⟶ gersdorffite, very similar; slightly less yellow. ⟶ skutterudite, more yellowish, lighter. ⟶ linnaeite, white; linnaeite is pinkish. 2. + 3. Isotropic; anomalous anisotropism may occur. 4. Not present.	470 nm 46.7 546 44.6 589 44.5 650 46.1 [194].

* Gersdorffite may show three possible structures related to temperature conditions: high temp., pyrite space group; intermediate temp., ullmannite space group; low temp., triclinic space group, geometrically cubic.

Hardness H: polishing hardness VHN: micro-indentation hardness in kg/mm^2	Miscellaneous
H: >linnaeite,~skuterudite, ullmannite,≤loellingite, ≪ pyrite, arsenopyrite. VHN: 520-907 [159, 194, 810, 1715].	Commonly forms idiomorphic crystals often showing zonal texture (developed by etching with HNO_3 1:1 or by difference in H or colour); also as skeletal crystals, or as fine grains with irregular boundaries. Cleavage // (100) nearly always distinct and typical; commonly more distinct than for ullmannite. Triangular pits not uncommon. Twin lamellae have been observed. Shows inclusions of silver, niccolite, pyrite, skutterudite, rammelsbergite, bismuth, cobaltite, uraninite, brannerite. Replaces pyrite, niccolite, millerite. May be replaced by tetrahedrite, violarite, rammelsbergite, niccolite, pyrargyrite. Replacement or breakdown of gersdorffite may lead to pseudo-eutectic intergrowths between niccolite or maucherite with pyrrhotite or chalcopyrite. Occurs as reaction rim between pyrrhotite and niccolite. Other ass.: pararammelsbergite, cobaltite, linnaeite, arsenopyrite, marcasite, sphalerite, gold, galena, jamesonite, stephanite. Diff.: Ni-skutterudite does not show cubic cleavage. Ref.: 93, 94, 104, 288, 290, 322, 355, 457, 458, 480, 504, 552, 588, 594, 746, 765, 802, 807, 965, 1133, 1136, 1154, 1170, 1224, 1237, 1322, 1356, 1392, 1455, 1690, 1692. Notes: 1. Dunn (1937b) described a β-modification of gersdorffite which is Co-rich and tarnishes readily in a moist atmosphere. Dunn explained it as a submicroscopic intergrowth of gersdorffite and "chloanthite". Lyakhnitskaya and Shumskaya (1966) described a similar modification associated with "chloanthite", with optical properties close to these of "chloanthite" and a refl. 5 % higher than for normal gersdorffite. 2. Corynite, Ni(Sb,As)S, is an intermediate member of the probable gersdorffite-ullmannite series.
H: >linnaeite, ≧gersdorffite, ≪pyrite. VHN: 460-560 [159, 194, 1113, 1715].	Commonly idiomorphically developed. Cleavage // (100) often visible; triangular pits may occur, but more rarely than in gersdorffite. Zonal texture occurs. Replaced by chalcopyrite and breithauptite. Replaces breithauptite. Forms intergrowths with chalcopyrite and galena. May show inclusions of gold. Parag.: about same as for linnaeite and gersdorffite, but relatively rare. Diff.: gersdorffite commonly shows a more distinct cleavage. Ref.: 76, 94, 95, 1237, 1392, 1629. Note: (Co,Ni)SbS with Co:Ni = 1:1 has been called willyamite; kallilite is Ni(Sb,Bi)S. Corynite is an intermediate member of a probable gersdorffite-ullmannite series.

Name Formula Crystal System	1: Colour 2: Bireflectance 3: Anisotropy 4: Internal reflections	Reflectance in % (Wavelength in nanometres)

LEPIDOCROCITE
γ-FeOOH
Ortho.
	1. Greyish white.
	    ⟶ goethite, lighter and whiter.
	    ⟶ hematite, dull greenish tint.
	    ⟶ rutile, the latter is pinkish blue.
	2. Weak to distinct (in oil); more distinct than for goethite.
	3. Strong, shades of grey.
	4. Reddish, often visible; not so abundant as in goethite.
		About 15 to 25 %.

GOETHITE
α-FeOOH
Ortho.
	1. Grey, varying from dull grey to bright grey; often with a bluish tint. Colour largely depends on texture and orientation; crystalline material is brighter than cryptocrystalline material.
	    ⟶ sphalerite, distinctly bluish.
	    ⟶ hematite, much darker.
	    ⟶ lepidocrocite, darker; lepidocrocite has a brighter colour.
	2. Weak; in oil more distinct but often masked by the int. refl.
	3. Distinct: grey-blue, grey-yellow, brownish, greenish grey; in crystalline goethite stronger than in cryptocrystalline material; in oil masked by the int. refl.
	4. Brownish yellow or reddish brown (commonly lighter than for lepidocrocite). Most abundant in cryptocrystalline material.
		About 15 to 20 %.

Hardness H: polishing hardness VHN: micro-indentation hardness in kg/mm^2	Miscellaneous
H: <goethite. VHN: 147-782 [159, 194, 1715].	Usually occurs as thin plates or tabular crystals in goethite; often associated with magnetite and hematite; also in intergrowths called "limonite". Alteration product of pyrite and arsenopyrite. Forms pseudomorphs after pyrite. Lepidocrocite is much less common than goethite and even relatively rare compared with goethite. Diff.: rutile has a different colour and int. refl. Ref.: 321, 324, 1202a, 1237. Note: Akaganeite is β-FeOOH, tetr. Occurs in "limonite", sometimes resulting from the alteration of pyrrhotite. Ass.: pyrrhotite, goethite, lepidocrocite, hematite. Ref.: 856, 1590.
H: (for crystalline material)≥lepidocrocite, <magnetite, maghemite, ilmenite, hematite. VHN: 525-1010 [159, 1715].	Colloform texture and spherolitic aggregates very common; banding or intimate intergrowths with hematite and ilvaite occur. In zonal textures the central parts commonly show the highest refl. Forms pseudomorphs after pyrite, siderite. Replaces pyrite, huebnerite, magnetite, ilmenite, hematite, Fe-sulphides, chalcopyrite. Forms the principal constituent of most "limonites". Goethite has rarely been observed as a primary mineral. Other ass.: maghemite, rutile, lepidocrocite, Mn-oxides, Cu-oxides, sphalerite, galena, boulangerite, wurtzite, pyrrhotite, tetrahedrite. Diff.: rutile has a higher refl., greater H and much lighter int. refl. Ref.: 154, 169, 321, 324, 344, 345, 462, 824, 1237, 1379, 1392, 1629.

162   Description of minerals from Table III

Name Formula Crystal System	1: Colour 2: Bireflectance 3: Anisotropy 4: Internal reflections	Reflectance in % (Wavelength in nanometres)

HETEROGENITE
$CoOOH(H_2O)_{0\ to\ 1}$ (the phase CoOOH occurs in nature in different degrees of stoichiometry and crystallization)
Hex.
	1. + 2. Very strong and characteristic birefl.
	  In air: O creamy white.
	  $\longrightarrow$ hematite, yellower.
	  $\longrightarrow$ magnetite, very similar.
	  $\longrightarrow$ goethite, brighter.
	  E much darker, greyish brown.
	  In oil: O creamy white, slightly darker than in air.
	  E dark brown.
	3. Very strong, brown to grey; basal sections appear isotropic; complete extinction except for very fine-grained spec. or parts of concentric spec.
	4. Not observed.
		About 10 to 25 %.

ZINCITE ZnO (the reddish colour is due to the MnO-content). Hex.	1. Pinkish brown. 2. + 3. Masked by the int. refl. 4. Abundant, yellowish or red.	About 10 %.
HAUERITE MnS_2 Cubic	1. Air: greyish white with a light brown tinge. Oil: darker, brownish grey. 2. + 3. Isotropic. 4. Always present; commonly red as in cuprite, occasionally brownish red as in manganite; esp. well visible in oil.	About 25 %.

PLATTNERITE
PbO_2
Tetr.
	1. Grey.
	  $\longrightarrow$ sphalerite, in air slightly darker; in oil similar.
	2. Weak, but distinct on grain boundaries: grey to grey with a bluish tint.
	3. Distinct in crystalline material: blue to greenish.
	4. Abundant, reddish brown.
		470 nm 18.3-19.5
		546 17.2-17.9
		589 16.1-16.8
		650 14.8-15.5
		                [194].

Hardness H: polishing hardness VHN: micro-indentation hardness in kg/mm^2	Miscellaneous
H: no data. VHN: no data.	Cleavage or parting often visible. Basal sections show hexagonal symmetry and often a fine concentric texture. Spherulites and radiated aggregates very common; central parts usually very fine-grained, the grain size increasing towards the outer parts of the spec. May be accompanied by a cryptocrystalline or colloform modification which appears to be isotropic, slightly softer and dull white. Parag.: Fe-oxides, hydrous Fe-oxides, Cu-oxides, hydrous Mn-oxides, carbonates, some ill-defined hydrous Co-oxides. Ref.: 281, 622, 816, 1075, 1237, 1349.
H: $\ll$ franklinite. VHN: 150-318 [159, 194, 1715].	Occurs as rounded and xenomorphic grains. Cleavage $\parallel$ (0001) may be distinct in not well polished sections. Forms orientated intergrowths with hausmannite and manganosite. Ass.: franklinite. Ref.: 441, 1237.
H: no data. VHN: 485-508 [1715].	Generally occurs as idiomorphic octahedral crystals. Cubic cleavage usually distinct; triangular pits may occur. Besides pyrite no associated minerals. Only found in clays or in weathered igneous rocks rich in sulphur. Ref.: 1237, 1392.
H: $\gg$ all associated minerals. VHN: 490-642 [194].	Occurs as short needle-like crystals. Parag.: cerussite, hemimorphite, pyromorphite, malachite, azurite, and other minerals of the oxidation zone. Ref.: 72, 199, 1237.

Description of minerals from Table III

Name Formula Crystal System	1: Colour 2: Bireflectance 3: Anisotropy 4: Internal reflections	Reflectance in % (Wavelength in nanometres)
LUDWIGITE* $(Mg,Fe^{2+})_2(Fe^{3+},Al)BO_5$ Ortho.	1. Air: grey and blue. ⟶ magnetite, darker grey. ⟶ "grey" direction of lepidocrocite, R_p is darker and more blue, R_m is darker and very slightly bluish, R_g is very similar. ⟶ "white" direction of lepidocrocite, R_p is much darker and more blue. Oil: much darker, blue colour more intense, but contrast between R_p and R_m reduced. R_g looks slightly pinkish grey. Mg-rich ludwigite shows no blue colour! 2. Mg-rich: air: strong, different shades of grey to brownish grey; oil: darker, the brownish tint is intensified. Mg-poor: air: very strong, bluish (R_p), grey (R_m), brownish grey (R_g); oil: darker, but colours seem to be intensified. 3. Mg-rich: air: strong, grey to black; 45°: very dark grey; oil: moderate to strong, greatly reduced. Mg-poor: air: very strong, dark blue to bright pinkish grey under medium-power objective and grey to brownish under high-power objective; 45°: greyish orange or pinkish grey; oil: slightly reduced, colour differences are less pronounced; 45°: grey. 4. Mg-rich: faint reddish brown, only visible at grain boundaries; Mg-poor: not present.	About 8-10 %.
VONSENITE* $(Fe^{2+},Mg)_2(Fe^{3+},Al)BO_5$ Ortho.	1. Air: light grey to brown or blue. ⟶ magnetite, R_p looks very bluish, R_m slightly bluish, R_g approaches the colour of magnetite, but is darker and slightly more pinkish at low magnif. and more greyish at high magnif. ⟶ hematite, R_g of vonsenite is darker and considerably browner. Oil: darker, but the colours are intensified. 2. Air: very strong, $R_p = R_b$ = blue or blue-grey; $R_m = R_a$ = grey, brown relative to b, but blue relative to c; $R_g = R_c$ = brown or pinkish brown. Oil: slightly intensified, but darker colours. 3. Air: very strong, blue, pink, reddish brown; 45°: fiery orange. Oil: slightly reduced; 45°: pinkish grey. 4. Not present, except where Fe^{2+}-content has greatly been reduced.	About 10-15 %.

* Brovkin et al. (1963) proposed to name the isomorphous series Fe_2FeBO_5-Mg_2FeBO_5 "ferro-vonsenite"-"magnesioludwigite", with vonsenite and ludwigite as intermediate members.

Hardness H: polishing hardness VHN: micro-indentation hardness in kg/mm^2	Miscellaneous
H: $>$ lepidocrocite, $\sim$ magnetite. VHN$_{50}$: 537-1486 [810, 820].	Occurs as single prisms, clusters of prisms and radiating aggregates of minute fibers. Lamellar twinning is rarely visible in sections $/\!/$ c. Replaces magnetite, sphalerite, pyrite. Replaced by franklinite, magnetite, hematite, lepidocrocite. May contain inclusions of franklinite. Forms intimate intergrowths with lepidocrocite. Pseudomorphosed by magnetite. Ref.: 24, 251, 740, 820, 823, 1331.
H: $\sim$ magnetite, $<$ hematite. VHN: 707-1003 [820].	Occurs as aggregates of polygonal to subrounded grains. Sometimes as lath-, wedge-, and diamond-shaped crystals, commonly with rounded corners. Locally broad twin lamellae may be visible. Replaced by goethite, hematite, chalcopyrite. Other ass.: magnetite, pyrrhotite, sphalerite. Diff.: very similar to ilvaite, but this mineral is much harder. Ref.: 24, 820, 821, 822, 823, 1504.

Name Formula Crystal System	1: Colour 2: Bireflectance 3: Anisotropy 4: Internal reflections	Reflectance in % (Wavelength in nanometres)
AVICENNITE Tl_2O_3 Cubic	1. Grey. 2. + 3. Isotropic. 4. Not present.	No data.
WUESTITE FeO Cubic	1. In contrast to each other, wüstite is grey with a greenish tint, and magnetite brownish to pinkish grey, especially in oil. $\longrightarrow$ hematite, darker. $\longrightarrow$ goethite, lighter. 2. + 3. Isotropic. 4. Not present.	Somewhat lower than for magnetite.
MURDOCHITE Cu_6PbO_8 (may contain some Br and Cl) Cubic	1. Grey with a distinct yellowish brown tint. $\longrightarrow$ magnetite, more distinctly brownish. 2. + 3. Isotropic. 4. Not present.	470 nm 16.3 546 17.0 589 17.4 650 17.6 [194].

Hardness H: polishing hardness VHN: micro-indentation hardness in kg/mm^2	Miscellaneous
H: no data. VHN: no data.	Occurs as small euhedral cubic crystals. No cleavage or twinning. Tentatively identified as inclusions in chalcothallite (Semenov et al., 1967). Other ass.: hematite, Fe-hydroxides, pyrite, stibnite, U-minerals. Ref.: 698, 748, 1305.
H: ~magnetite. VHN: no data.	Occurs with magnetite, hematite, goethite and iron as curved bodies in a tuff-breccia. Also with magnetite and hematite in a natural coke. Replaced by magnetite wherein it may occur as replacement-relicts. Decomposes to an intergrowth of magnetite and native iron. May contain very small inclusions of native iron. Ref.: 246, 1639.
H: slightly > plattnerite. VHN: 519-657 [194].	Occurs as euhedral crystals. Twinning not uncommon. Zonation always present. Octahedral cleavage has been observed. Ass.: plattnerite, wulfenite, hematite. Ref.: 73, 194, 254, 283.

Description of minerals from Table III

Name Formula Crystal System	1: Colour 2: Bireflectance 3: Anisotropy 4: Internal reflections	Reflectance in % (Wavelength in nanometres)
MAGNETITE $FeFe_2O_4$ (Fe may be replaced by some Mg, Mn, Zn, Ti, Al, V, or Cr). Cubic	1. Grey, commonly with a brownish tint; Ti may cause a brownish pink tint, Mn a more yellowish green tint, Al and Mg a yellowish tint, Cr a darker grey tint. ⟶ hematite, much darker, distinctly brown. ⟶ coulsonite, distinctly brown. ⟶ ilmenite, lighter or paler, less pink; ilmenite is distinctly pink with a violet tinge in contrast. ⟶ sphalerite, much lighter. ⟶ maghemite, brownish; maghemite is bluish. ⟶ psilomelane and cryptomelane, slightly darker and duller. 2. Not present. 3. Isotropic; a slight anomalous anisotropism, due to internal tension by zoning or tectonic deformation may be present; distinct anisotropism, caused by twinning, was reported by Steinike (1959). 4. Not present, except in spec. with high Mn-content.	470 nm 20.6 546 20.8 589 21.0 650 21.1 Dispersion curve: [300].
TITANOMAGNETITE $Fe^{2+}_{1+x}Fe^{3+}_{2-2x}Ti^{4+}_{x}O_4$ Cubic	1. Air: white with a brownish tint. Oil: distinctly brown (intermediate between magnetite and ilmenite). 2. Not present. 3. Usually isotropic; sometimes slightly anisotropic. 4. Not present.	480 nm 16.65 540 16.95 580 17.2 640 16.05 Dispersion curve: [236].

Hardness H: polishing hardness VHN: micro-indentation hardness in kg/mm^2	Miscellaneous
H: varying with composition; ≫ pyrrhotite, commonly < ilmenite, < braunite, ≪ hematite. VHN: 440-1100 [159, 194, 300, 810, 1113, 1306, 1401, 1715].	Usually occurs as euhedral (octahedral, dodecahedral) crystals; tabular crystals may be pseudomorphs after hematite; also forms irregular strings and granular aggregates, rarely colloform aggregates. Cleavage commonly not distinct. Lamellar twinning // (111) and zonal texture not uncommon. Replaces hematite, pyrrhotite, chromite, pentlandite, heazlewoodite; may be replaced by hematite, maghemite, lepidocrocite, pyrolusite, pyrite, pyrrhotite, chalcopyrite, pentlandite, cubanite, sphalerite, bornite, stannite, galena, gold. The alteration of magnetite to hematite (martitization) may be regular or irregular. In the first case the octahedral planes are followed, producing a triangular network of hematite lamellae. This texture, however, is not as regular as that produced by exsolution (with hematite lamellae of very uniform width). Rarely martitization may follow growth zones. Magnetite alters to maghemite and Fe-hydroxides. Occurs as inclusions in sulphides; may show inclusions of hematite, ilmenite, awaruite, sulphides, minerals of the spinel-group. Forms solid solutions with ilmenite and ulvöspinel at higher temperatures. Rapid cooling produces titanomagnetite, slow cooling may produce fine intergrowths with blades of ilmenite // octahedral planes in magnetite, or, the final product, granular intergrowths of both minerals as the blades of ilmenite tend to coalesce into irregular or rounded grains. The exsolution of ilmenite may be accompanied by a cloth-like texture due to exsolution of ulvöspinel // (100) of magnetite. Magnetite may also contain exsolution bodies of pyrophanite, geikielite, hercynite and other minerals of the spinel-group, and of corundum. Magnetite may occur as exsolution bodies in ilmenite. Other ass.: rutile, marcasite, bismuthinite. Diff.: sphalerite and chromite are much darker and often show int. refl.; braunite and ilmenite are anisotropic; jacobsite shows other colours and a red int. refl. Ref.: 26, 38, 81, 137, 155, 221, 226, 283, 296, 302, 321, 324, 325, 331, 332, 333, 339, 340, 362, 382, 411, 413, 474, 554, 588, 627, 640, 641, 726, 808, 828, 880, 899, 931, 1044, 1152, 1187, 1215, 1222, 1237, 1306, 1358, 1368, 1392, 1401, 1440, 1448, 1482, 1600, 1603, 1629, 1721.
H: slightly > magnetite. VHN: 715-734 [194].	Homogeneous titanomagnetite is only formed by rapid cooling of a high temperature solid solution of magnetite with ilmenite or ulvöspinel. May contain exsolution lamellae of ilmenite, ulvöspinel and rutile. Replaced by oxidation by titanomaghemite, pseudobrookite, hematite. Much material described as magnetite probably is titanomagnetite. Ref.: 82, 419, 420, 1177, 1350, 1689.

Description of minerals from Table III

Name Formula Crystal System	1: Colour 2: Bireflectance 3: Anisotropy 4: Internal reflections	Reflectance in % (Wavelength in nanometres)
MAGHEMITE $\gamma\text{-}Fe_2O_3$* Cubic	1. Bluish grey. ⟶ goethite, grey, lighter. ⟶ hematite, bluish grey, darker. ⟶ magnetite, distinctly bluish, magnetite is brownish. 2. Not present. 3. Isotropic; occasionally slightly anisotropic (transition to hematite). 4. Very rare, brownish red, more yellow than in hematite, and more red than in goethite.	Value estimated from the dispersion curve: 550 nm 26 % [1177].
ULVÖSPINEL** Fe_2TiO_4 Cubic	1. Brown to reddish brown. ⟶ ∥ E of ilmenite, very similar. ⟶ magnetite, darker brown. 2. + 3. Isotropic. 4. Not observed.	No data.
MAGNESIOFERRITE $MgFe_2O_4$ (may contain Fe^{2+}, Mn, Al, Ti). Cubic	1. Air: grey with a brownish tint. Oil: distinctly brown. ⟶ hematite, brown tint very distinct. ⟶ lightest position of ilmenite, similar. 2. + 3. Isotropic. 4. Not present.	470 nm 18.8 546 17.5 589 16.8 650 16.0 Dispersion curve: [300]; see also: [1181].

* Unstable form of Fe_2O_3; polymorphous, monotropic (no transition point); the stable form, $\alpha\text{-}Fe_2O_3$ is hematite [880].
** Some authors prefer the name ulvite for Fe_2TiO_4 (Strunz, 1966).

Hardness H: polishing hardness VHN: micro-indentation hardness in kg/mm^2	Miscellaneous
H: slightly $>$ magnetite*, $\ll$ hematite. VHN$_{50-100}$: 357-988 [159, 194, 810].	Only formed by oxidation of magnetite; all stages in the oxidation of magnetite into maghemite may occur in nature, which may explain the differences in VHN. Commonly irregularly spread in the magnetite grains, as fine networks, or as cloud- or sponge-like formations, occasionally using crystallographic directions or fissures. Ilmenite lamellae present in magnetite appear unaltered in maghemite. Alters gradually to hematite. Other ass.: goethite, lepidocrocite, spinel. Ref.: 19, 81, 306, 419, 627, 880, 1044, 1121, 1177, 1237. Note: Titanomaghemite is a member of the series γ-FeTiO$_3$ to γ-Fe$_2$O$_3$, cubic. Homogeneous mineral with varying composition, formed by oxidation of titanomagnetite. Occurs in the same way as maghemite in magnetite. Bluish grey colour, somewhat lighter than titanomagnetite. Does not alter to hematite (in contrast to maghemite). Ref.: 81, 241, 704, 1003, 1177.
H: somewhat $>$ magnetite. VHN: no data.	Rarely occurs as idiomorphic octahedral crystals. Usually found as very fine exsolution bodies in Ti-rich magnetite, $/\!/$ (100) and (110) of the magnetite. These intergrowths have the appearance of a closely-woven piece of cloth. These magnetites often also contain exsolution lamellae of spinel and ilmenite $/\!/$ (111). In rare cases, when the original solid solution had a very high Ti-content, ulvöspinel serves as the groundmass for exsolved orientated cubes of magnetite. Ilmenite as very fine lamellae in magnetite may have been produced by the alteration of the original ulvöspinel. Ulvöspinel breaks down to magnetite and ilmenite, and sometimes also to iron. Ref.: 221, 631, 931, 1006, 1230, 1231, 1613, 1614.
H: $>$ magnetite, $\ll$ hematite. VHN: 627-925 [194, 300, 1401]. VHN increases with increasing MgO-content.	Forms euhedral crystals. Usually found in fumaroles or in volcanic rocks; sometimes in complex intergrowths with hematite. Observed as reniform alteration bodes on native iron. Also found in metasomatic deposits as a reaction product between hematite and dolomite. Other ass.: pyrochlore. Ref.: 100, 578, 642, 784, 1181, 1230, 1249, 1401.

---

* According to Pavlov (1957), H $<$ magnetite.

Name Formula Crystal System	1: Colour 2: Bireflectance 3: Anisotropy 4: Internal reflections	Reflectance in % (Wavelength in nanometres)
TREVORITE $NiFe_2O_4$ Cubic	1. Dull grey. $\longrightarrow$ magnetite, no pinkish tinge, clear grey. 2. + 3. Isotropic. 4. Not present.	On Fe-trevorite: 548 nm 21.0 589 20.8 [1632a].
COULSONITE $Fe^{2+}V_2^{3+}O_4$ Cubic	1. Bluish grey. $\longrightarrow$ magnetite, bluish; magnetite is brownish. 2. + 3. Isotropic; locally slightly anisotropic. 4. Not present.	About 25 %.
DONATHITE $(Fe,Mg,Zn)(Cr,Fe)_2O_4$ Tetr.	1. Grey. $\longrightarrow$ chromite, very similar, but somewhat more brownish. 2. Weak: O brownish, darker, E bluish, lighter. 3. Distinct, yellow-brown. 4. Not present.	Somewhat higher than for chromite.

Hardness H: polishing hardness VHN: micro-indentation hardness in kg/mm^2	Miscellaneous
H: no data. VHN: (on Fe-trevorite): 773 [1632a].	Often shows octahedral outlines, but usually occurs as rounded grains. Replaced by magnetite. May contain inclusions of millerite. Other ass.: violarite, goethite. Ref.: 1114, 1632a, 1640.
H: slightly <magnetite. VHN: no data.	May form by exsolution of vanadium along the octahedral planes of magnetite. Also found separated from magnetite, not related to exsolution. Occurs as patches in magnetite; forms segregations within magnetite. Often appears to be related to coarser intergrowths of ilmenite. Replaced by hematite. If not in crystallographic continuity, boundaries between magnetite and coulsonite may be quite distinct; in oil, with high-power magnif., in some spec. a noticeable but rapid graduation from normal brownish magnetite to the blue-grey of coulsonite may be observed. Other ass.: rutile, Fe-hydroxides. Ref.: 321, 324, 1193, 1195.
H:<chromite. VHN: no data.	Usually occurs as xenomorphic and hypidiomorphic grains, seldom idiomorphic. Twinning abundant. Shows intricate intergrowths of individual grains. Contains inclusions of chromite and a magnetite-like mineral. Ref.: 1375.

Description of minerals from Table III

Name Formula Crystal System	1: Colour 2: Bireflectance 3: Anisotropy 4: Internal reflections	Reflectance in % (Wavelength in nanometres)
CHROMITE $(Fe,Mg)(Cr,Al,Fe)_2O_4$ (may contain up to 12 % ZnO, 8 % V, and some Mn). Cubic	1. Shows differences in colour shades and refl. with varying chemical composition. Dark grey to brownish grey. ⟶ magnetite, much darker. ⟶ sphalerite, similar, slightly darker. ⟶ ilmenite, much darker; ilmenite is brown-red in contrast. 2. Not present. 3. Isotropic. Many spec. show weak to distinct anisotropism, especially zoned crystals. All Zn-containing chromite is anisotropic. Tectonic deformation may also cause anisotropism. 4. Red-brown, very common, esp. in Mg- and Al-rich spec. Chromite with high Fe-content has no int. refl.	Strongly dependent on chem comp.: High R = Fe, Cr; low R = Mg, Al. 470 nm 11.9-13.8 546 11.5-13.3 589 11.5-13.0 650 11.2-12.7 Dispersion curves: [300]. See also: [221, 911].
DAVIDITE $(Fe,U,Ce,La)_2$ $(Ti,Fe,Cr,V)_5O_{12}$ Hex.	1. Grey with a brownish tint. ⟶ ilmenite, slightly darker. ⟶ sphalerite, very similar. 2. Not present. 3. Usually metamict; sometimes weakly anisotropic. 4. Rare, deep-brown.	About 15-20 %.

Hardness H: polishing hardness VHN: micro-indentation hardness in kg/mm^2	Miscellaneous
H:>magnetite,<hematite. VHN$_{100-200}$: 1036-2000 [159, 194, 221, 300, 810, 911, 1311, 1715].	Usually forms homogeneous rounded idiomorphic crystals or coarsely crystalline aggregates. Cataclastic texture very common. Cleavage commonly not visible; in some spec. distinct. Twinning not observed. Zonal texture not uncommon. This is due to differences in chem. comp. Usually the marginal zones have a higher Fe- and Cr-content and show a higher refl. than the core which is richer in Mg and Al. Sometimes the outer zones are darker than the core; then the darker zones have a higher Zn-content than the core. A cloud-like appearance has been observed in some spec. (in serpentinites), different shades of grey merging gradually into each other; sharp boundaries are exceptions and only occur in a very fine feather-like or fine-lamellar form; the lighter substances contain more Fe. May show exsolution intergrowths, sometimes orientated, with hematite, ilmenite, magnetite, rutile, ulvöspinel. May be replaced by magnetite. "Magnetic chromite" is normal chromite veined by magnetite. Radioactive inclusions have been observed, causing haloes. Alters to higher reflecting material resembling magnetite. The final product of alteration may be ilmenite or magnetite. Other ass.: platinum, pentlandite, millerite, pyrrhotite, chalcopyrite, heazlewoodite, pyrolusite, cryptomelane, eskolaite, graphite. Ref.: 38, 49, 97a, 174, 221, 240, 325, 411, 434, 670, 726, 911, 1105, 1204, 1216, 1237, 1245, 1323, 1392, 1519, 1520, 1603, 1668. Note: The pure end-member $MgCr_2O_4$ is called magnesiochromite. VHN_{500}: 1200-1385 [956].
H:<ilmenite, ≪hematite, rutile. VHN: 693-890 [159, 194, 810, 1715].	Occurs as rounded grains or as tabular crystals. May form colloform textures. Cleavage ∥ (0001). Forms complex intergrowths with ilmenite, hematite, rutile, magnetite, pyrite, chalcopyrite, and molybdenite. May contain very small particles of rutile, hematite, ilmenite. Contains exsolution bodies of rutile and hematite. Replaces ilmenite and other Fe-Ti-oxides, sometimes causing radial fractures in the replaced minerals. May be replaced by, or altered to, rutile and ilmenite. Develops radiogenetic haloes in ilmenite. At the contact between davidite and ilmenite, an intermediate zone may occur which shows a dull bluish grey anisotropism contrasting strongly with the normal greyish white of ilmenite and the isotropism of davidite. Parag.: rutile, molybdenite, hematite, magnetite, chalcopyrite, pyrite, ilmenite, galena, gold, pseudobrookite. Ref.: 50, 205, 279, 298, 597, 711, 808, 999, 1090, 1112, 1242, 1473, 1584, 1642, 1674, 1683. Note: The following minerals belong to the same group as davidite; optical data are not available. Crichtonite, $(Fe^{2+},Fe^{3+})_2(Ti,Fe^{3+})_5O_{12}$, hexagonal. Formerly the name was used for pure $FeTiO_3$. Ref.: 67, 624a. Senaite, $Pb(Ti,Fe,Mn,Mg)_{24}O_{38}$, hexagonal. Found as rough crystals and rounded fragments in diamond-bearing sands in Brazil. Ref.: 1100, 1291. Mohsite, $(Fe,Ca,Mg)_2Ti_5O_{12}$, hexagonal. Ref.: 1475.

Name Formula Crystal System	1: Colour 2: Bireflectance 3: Anisotropy 4: Internal reflections	Reflectance in % (Wavelength in nanometres)
ILMENITE $FeTiO_3$ (often contains Fe^{3+}, Mn^{2+}, and Mg). Hex.	1. Light to dark brown, sometimes with a faint pinkish or violet tint. ⟶ magnetite, darker, esp. for E; brown. ⟶ sphalerite, brighter, distinctly brown. ⟶ hematite, much darker. ⟶ chromite, distinctly lighter, more reddish brown. Homogeneous Fe_2O_3-rich ilmenites are much lighter. 2. Distinct: O light pinkish brown, E dark brown. 3. Strong: light greenish grey, brownish grey. 4. Dark brown, rare; in Mg-rich spec. much more common.	Depending on the MgO-content; pure ilmenite: R_o 480 nm 19.6-20.2 540 19.5-20.1 580 19.6-20.0 640 19.3-19.5 R'_e 480 nm 16.8-17.4 540 16.5-17.4 580 16.9-17.5 640 16.6-17.5 Dispersion curves: [236]; see also: [486].
GEIKIELITE $MgTiO_3$ (forms a complete solid solution series with ilmenite). Hex.	1. Somewhat darker than for ilmenite. 2. Distinct. 3. Strong, colours more vivid than for ilmenite. 4. More frequent than for ilmenite, orange-red.	Depending on the MgO-content; for geikielite containing 29.4 wt. % MgO: R_o R'_e 480 nm 14.85 12.2 540 14.5 11.85 580 14.35 11.65 640 13.95 11.55 Dispersion curves: [236].

Hardness H: polishing hardness VHN: micro-indentation hardness in kg/mm^2	Miscellaneous
H: commonly > magnetite, slightly < hematite. VHN$_{100-300}$: 501-752 [159, 194, 236, 406, 810, 1715].	Cleavage not observed; parting $\parallel$ (0001) may occur. Lamellar twinning usually well developed; lamellae of equal width. Forms exsolution intergrowths with hematite ($\parallel$ 0001), corundum, rutile, magnetite, tantalite. Orientated intergrowths with hematite, rutile or magnetite are not necessarily due to exsolution. Regular intergrowths with sulphides (pyrrhotite, sphalerite, galena) less common, probably due to replacement. Replaced by rutile, magnetite, hematite, sphalerite, pyrite, pentlandite, davidite. Davidite may develop radiogenetic haloes in ilmenite. Ilmenite alters to material that resembles rutile, with Fe-oxide present in a colloidal state. The final product of weathering is a mixture of Fe- and Ti-oxide. Other ass.: maghemite, coulsonite, pseudobrookite, chromite, Fe-hydroxides. Ref.: 92, 142, 165, 221, 236, 283, 321, 324, 332, 335, 339, 382, 406, 474, 588, 641, 808, 1215, 1216, 1222, 1237, 1358, 1365, 1366, 1392, 1413, 1454, 1515. Notes: 1. Crichtonite, formerly used for pure $FeTiO_3$, is a distinct mineral species; see at davidite. 2. Melanostibite, $Mn(Sb,Fe)O_3$, hexagonal. Occurs as aggregates of small crystals in parallel growth. Ref.: 939, 941, 942. 3. Pyrophanite, $MnTiO_3$, is the Mn-analogue of ilmenite. Resembles ilmenite, but shows lower refl., less brown colour, less strong birefl. and anisotr., and often beautiful reddish int. refl. Ref.: 332, 815, 998. 4. "White ilmenite": see at hematite.
H: slightly > ilmenite. VHN: dependant on the Mg-content: 0 % MgO: 560 25.7 % MgO: 930 [406].	Occurs as irregular to rounded grains. May contain exsolution lamellae of magnetite. Alters to brucite and periclase. Other ass.: spinel, rutile, pyrite, pyrrhotite. Ref.: 123, 236, 247, 971, 1701.

Description of minerals from Table III

Name Formula Crystal System	1: Colour 2: Bireflectance 3: Anisotropy 4: Internal reflections	Reflectance in % (Wavelength in nanometres)
ANATASE TiO_2 Tetr.	1. Grey. 　⟶ sphalerite, very similar. 　⟶ rutile, very similar. 2. Very weak (may be visible in oil on close examination). 3. Masked by the int. refl. 4. Strong and abundant, white to bluish and blue-grey (in contrast to sphalerite).	About 20 %.
PSEUDOBROOKITE $Fe_2^{3+}TiO_5$ Ortho.	1. Different shades of grey depending on chem. comp. 　⟶ rutile, very similar. 2. Weak. 3. Distinct. 4. Reddish yellow (in contrast to rutile which shows more abundant light yellow int. refl.).	About 15 %.

Hardness H: polishing hardness VHN: micro-indentation hardness in kg/mm^2	Miscellaneous
H:<rutile. VHN: 576-623 [1715]. VHN$_{50}$: 601 [560].	Occurs as idiomorphic crystals and as rounded grains. Replaces ilmenite. May contain exsolution spindles of ilmenite and vice-versa. Forms orientated intergrowths with rutile and pseudobrookite. Other ass.: hematite, magnetite. Diff.: very similar to sphalerite (if no int. refl. are visible); can most easily be mistaken for rutile. Ref.: 215, 560, 827, 1237, 1629.
H:<rutile. VHN: no data.	Sometimes idiomorphically developed. Occurs intergrown with hematite formed by oxidation of ilmenite and titanomagnetite, or commonly associated with hematite, ilmenite, rutile, magnetite, bixbyite. Occurs as inclusions in rutile, titanomagnetite, goethite. May contain inclusions (sometimes due to exsolution) of rutile and hematite. Decomposes to an aggregate of, and is pseudomorphosed by rutile and hematite. Ref.: 419, 420, 1116, 1215, 1237, 1411a.

Notes:
1. Kennedyite, $Fe_2^{3+}MgTi_3O_{10}$, isostructural with pseudobrookite. Occurs as laths in feldspars of olivine-rich rocks. No birefl., weakly anisotropic, steel-grey to purplish brown. Contains exsolution lamellae of rutile and patches of secondary hematite. Ref.: 735.
2. Landauite, $(Zn,Mn,Fe)_{0.92}(Ti,Fe)_3O_7$, monoclinic.

    The mineral is analogous to kennedyite, but has another space-group, and other interplanar spacings. Occurs in albite veinlets as microgranular aggregates. Ref.: 1173.
3. Derbylite, $Fe_3^{3+}Ti_3SbO_{11}OH$, monoclinic. Occurs as prismatic crystals in cinnabar-bearing gravels in Brazil. Ass.: rutile, hematite. Ref.: 1101, 1475.

Description of minerals from Table III

Name / Formula / Crystal System	1: Colour / 2: Bireflectance / 3: Anisotropy / 4: Internal reflections	Reflectance in % (Wavelength in nanometres)
RUTILE TiO_2 (may contain some Fe, Nb, Ta, V, Sn, Cr in solid solution). Tetr.	1. Grey, sometimes with a faint bluish tint. → magnetite and chromite, no great difference. → ilmenite, no brownish tint. → hematite and goethite, pinkish blue. → cassiterite, much lighter. 2. Distinct (in oil): ⊥ c (O) lightest. 3. Strong; colours usually masked by the int. refl. Nb-containing rutile is weakly anisotr. 4. Strong and abundant: white, yellowish, brown, reddish brown, brownish violet, green, depending on the content and the nature of the admixed elements.	R_o 470 nm 21.0-21.1 546 19.8-20.0 589 19.4-19.6 650 18.8-19.1 R'_e 470 nm 24.0-24.6 546 23.0-23.5 589 22.3-23.0 650 22.0-22.6 [194].
Nb-RUTILE and Ta-RUTILE $(Fe \gtrless Mn)^{2+}_{x+y}(Nb \gtrless Ta)^{5+}_{2x}$ $Ti^{4+}_{2-(3x+y)}O_{4-y}$ Tetr. (monorutile structure)	1. Grey, darker than pure rutile; sometimes with a pinkish tinge. The optical properties for Nb-rutile and Ta-rutile are intermediate between those of rutile and tapiolite. 2. Hardly visible, only on grain boundaries. 3. Rather strong. 4. Much less abundant and usually darker than for rutile.	For Nb-rutile: 470 nm 20.3-21.4 546 19.3-20.2 589 19.0-19.9 650 18.7-19.6 [194]. For Ta-rutile: 470 nm 19.2-20.6 546 18.1-19.9 589 17.8-19.7 650 17.3-19.5 [194].

Hardness H: polishing hardness VHN: micro-indentation hardness in kg/mm^2	Miscellaneous
H: >ilmenite, <hematite, cassiterite. VHN: 933-1280 [159, 194, 1311, 1715].	Often idiomorphically developed. In minute needle-shaped crystals arranged $/\!/$ crystallographic directions of original host minerals. Cleavage often distinct. Lamellar twinning very common. Intergrown with Ti-hematite, Ti-magnetite, Fe_2O_3-rich ilmenite, tantalite, anatase and pseudobrookite, or in intergrowths of ilmenite and Ti-hematite, the development of rutile often preceeding the exsolution hematite-ilmenite. Replaces ilmenite; may enclose ilmenite crystals of differing size or hematiteflakes. May contain exsolution bodies of ilmenite and hematite. The orientated intergrowths of rutile with ilmenite, hematite and magnetite are not necessarily due to exsolution, but may be formed by alteration of ilmenite and Ti-magnetite. Other ass.: pyrite, maghemite, coulsonite, goethite, lepidocrocite. Diff.: habit, int. refl., and twinning in combination with parag. and H may suffice for identification. Ref.: 321, 324, 332, 339, 688, 775, 809, 827, 1081, 1212, 1215, 1237, 1333, 1392. Notes: 1. Redledgeite, $(Mg,Ca,OH,H_2O)_{\leq 2}[(Ti,Cr,Si)_8O_{16}]$, tetragonal, formerly called "chromrutile", is a distinct species. Colour and refl. very similar to rutile. Ass.: chromite. Ref.: 515, 1472, 1474. 2. Priderite, $(K,Ba)_{1.33}[(Ti,Fe)_8O_{16}]$, tetragonal. Similar to rutile in its optical properties. Occurs as elongate acicular crystals. Ass.: anatase, ilmenite. Ref.: 228, 1013, 1604. 3. Cafetite, $(Ca,Mg)(Fe,Al)_2Ti_4O_{12} \cdot 4H_2O$, monoclinic, pseudo-orthorhombic. Forms columnar and acicular crystals. No reflected light data available. Ass.: Ti-magnetite, ilmenite, anatase, perovskite. Ref. 770.
H: somewhat>rutile. VHN: for Nb-rutile: 803-1180 [194, 810, 838, 1715] for Ta-rutile: 911-1239 [194, 810, 838, 1403, 1715].	In view of their crystal-chemical properties "ilmenorutile" and "strueverite" should be included in the rutile-group as varieties of rutile, and be named Nb-rutile (with atomic ratio Nb/Ta>1, and Ta-rutile (with atomic ratio Nb/Ta<1) (Černy et al., 1964). In the Nb-rutile - Ta-rutile series, homogeneous samples have an atomic content of Nb subordinate to that of Ta; inhomogeneous mixtures on the contrary contain Nb in excess over Ta. To distinguish between the two types in polished section: all samples containing columbite exsolutions should be called Nb-rutile; homogeneous minerals are probably Ta-rutile, but chemical investigation is always necessary. The optical properties of the groundmass of the inhomogeneous samples are the same as for Ta-rutile. This groundmass should be called Nb-rutile. The inclusions in Nb-rutile are always columbite, sometimes accompanied by rutile or ilmenite. Twinning lamellae are very common. Ass.: columbite, fergusonite, pyrochlore, ilmenite, molybdenite, cassiterite. Ref.: 233, 235, 335, 408, 647, 838, 1403, 1615.

Name Formula Crystal System	1: Colour 2: Bireflectance 3: Anisotropy 4: Internal reflections	Reflectance in % (Wavelength in nanometres)
FREUDENBERGITE $(Na,K)_2(Ti,Nb)_6(Fe,Si)_2(O,OH)_{18}$ [427] or $Na_2Fe_2Ti_6O_{16}$ [1376] Mono. [427] or Hex. [1376]	1. $\longrightarrow$ rutile, very similar. 2. Very weak. 3. Weak. 4. Abundant, brown-yellow.	About 15-20 %.
PEROVSKITE group ABO_3 A=Ca, Na, rare b. earths. B=chiefly Ti, Nb Ortho., pseudocubic (for naturally occurring compounds).	1. Varying with chemical composition, shades of dark grey (in oil). Perovskite shows a distinct blue tint in oil. $\longrightarrow$ magnetite and ilmenite, darker, dark blue tint more distinct. 2. + 3. Not observable. 4. Strong, and always present; colour varying from white to brown.	About 15 %.
HÖGBOMITE $(Fe, Mg)_6(Al,Fe)_{16}TiO_{32}$ Hex.	1. Dark grey. $\longrightarrow$ spinel, slightly lighter. $\longrightarrow$ sphalerite, much darker. 2. Weak; in oil distinct on twin boundaries, O > E. 3. Air: distinct. Oil: masked by the int. refl. 4. Abundant and strong: yellowish brown.	About 10 %.

Hardness H: polishing hardness VHN: micro-indentation hardness in kg/mm^2	Miscellaneous
H:<hematite. VHN: no data.	Occurs as xenomorphic, rarely tabular crystals. Rock-forming mineral in apatite-rich alkali-syenite. Forms basis-parallel intergrowths with hematite. Ass.: högbomite. Ref.: 427, 428, 895, 1376.
H:>or~magnetite. VHN: perovskite: 925-1131 [810, 1715], loparite: 648-895 [810, 1404], lueshite: 490-760 [787, 1601].	Note on the nomenclature: the classification and nomenclature of the perovskite-group is based on the valency of the A cation and on the predominance of Ti or Nb as B cation (valid mineral species in CAPITALS): - monovalent A (Na): LUESHITE, NaNbO$_3$. - divalent A (Ca): PEROVSKITE, CaTiO$_3$ (knopite and uhligite = perovskite, "dysanalyte" = niobian perovskite). If Nb > Ti in perovskite, the mineral should be named LATRAPPITE. - trivalent A (Ce): LOPARITE, (Na,Ce,Ca)TiO$_3$ ("irinite" = loparite, "nioboloparite" = niobian loparite, since Ti>Nb). For all varieties: occur usually in alkaline rocks or carbonatites as idiomorphic cubic or octahedral crystals, sometimes rounded or with carious or allotriomorphic texture. Cleavage in fine-grained material not observable. Complicated lamellar twinning practically always present, may be visible by the int. refl. Very difficult to identify in polished section. Perovskite may replace ilmenite and form intergrowths with ilmenite and magnetite. May contain inclusions of heazlewoodite, hematite, ilmenite. Other ass.: spinel, sulphides of Fe and Cu. Ref.: 706, 968, 1044, 1186, 1210, 1233. Latrappite is replaced by pyrochlore. Ass.: magnetite, pyrrhotite. Ref.: 1008, 1009. Lueshite. Ref.: 58, 787, 1111, 1313. Loparite. Ref.: 693, 1404. Niobian loparite. Ref.: 1548.
H: slightly > spinel, ilmenite, magnetite. VHN: 1048-1214 [194].	Forms small idiomorphic crystals, basal sections showing hexagonal symmetry; also occurs as tabular, prismatic and barrel-shaped crystals and as irregular grains. Twinning very common. Replacement product of spinel. Parag.: magnetite, ilmenite, pyrrhotite, minerals of the spinel-group. Ref.: 255, 435, 992, 1062, 1237, 1376, 1661. Note: Nel (1949) and Friedman (1952) both described two varieties of högbomite: högbomite (A) has a lower Fe-content than högbomite (B), and seems to be transitional between spinel (pleonast) and högbomite (B). Both authors give data for transmitted light only.

Description of minerals from Table III

Name Formula Crystal System	1: Colour 2: Bireflectance 3: Anisotropy 4: Internal reflections	Reflectance in % (Wavelength in nanometres)
PLUMBOFERRITE $PbO.2Fe_2O_3$ (?) Hex.	1. Grey. ⟶ magnetite, very similar. 2. Not observable. 3. Weak, even in oil. 4. Not present.	About 15–20 %.
MAGNETOPLUMBITE $Pb(Fe,Mn,Al,Ti)_{12}O_{19}$ Hex.	1. In oil: grey. ⟶ magnetite, no pinkish tint. 2. Very weak. 3. Distinct only in oil. 4. Very rare, even in oil.	$\quad\quad R_o \quad R'_e$ 470 nm 24.9 22.8 546 $\quad$ 24.0 22.0 589 $\quad$ 23.0 21.0 650 $\quad$ 22.4 20.7 [194].
SCHEELITE $CaWO_4$ (Mo may substitute part of W). Tetr.	1. In air: grey-white. ⟶ gangue minerals, practically the same refl. ⟶ chalcopyrite, dark grey with a faint violet tint. ⟶ arsenopyrite, very dark grey; arsenopyrite appears white. In oil: very dark grey, much darker than in air. ⟶ gangue minerals, much lighter grey ⟶ chalcopyrite, dark grey. ⟶ arsenopyrite, dark grey; arsenopyrite is slightly pinkish. 2. Not observable. 3. Distinct, but highly masked by the int. refl., which already are distinct with one nicol. 4. Abundant, white.	$\quad\quad R_o \quad R'_e$ 470 nm 10.2 10.5 546 $\quad$ 10.0 10.3 589 $\quad$ 9.9 10.1 650 $\quad$ 9.85 10.05 [194].

Hardness H: polishing hardness VHN: micro-indentation hardness in kg/mm^2	Miscellaneous
H: no data. VHN: no data.	Occurs as rounded grains. Two cleavages visible: prismatic and $/\!/$ (0001). Twinning rare. Ref.: 1237.
H: no data. VHN: 841-868 [194].	Occurs as idiomorphic barrel-shaped crystals. Replaced by hematite. Ref.: 115, 1237.
H: < wolframite. VHN: 285-464 [159, 194, 1715].	Replaces wolframite or is interstitial to this mineral; the replacing scheelite may show a pale yellow int. refl. in contrast to the white of interstitial scheelite. Scheelite and wolframite are mutually replacive, giving raise to alternating zones of these minerals. May also be replaced by ferberite, huebnerite, Fe-hydroxides. May contain inclusions of molybdenite (sometimes along the cleavage planes of scheelite), arsenopyrite, wolframite, chalcopyrite, pyrite. Occurs as inclusions in chalcopyrite, pyrrhotite, arsenopyrite, pyrite. Other ass.: cassiterite, magnetite, hematite, bismuth, bismuthinite, bursaite, sphalerite, gold. Ref.: 7, 13, 47, 323, 339, 349, 507, 511, 540, 546, 824, 828, 862, 1223, 1315, 1320, 1489, 1693, 1736.

Description of minerals from Table III

Name Formula Crystal System	1: Colour 2: Bireflectance 3: Anisotropy 4: Internal reflections	Reflectance in % (Wavelength in nanometres)
FERBERITE $FeWO_4$ (contains max. 20 % $MnWO_4$). Mono.	1. In air: greyish white; in oil creamy or yellowish grey. ⟶ goethite, lepidocrocite, distinctly yellowish, no bluish tint. 2. Not observable. 3. Distinct: greenish yellow to dark grey. 4. Not present.	Somewhat higher than for wolframite.
WOLFRAMITE $(Fe,Mn)WO_4$ Mono.	1. In air: grey or greyish white. In oil: grey with a faint brownish or yellow tint. ⟶ sphalerite, in air very similar, in oil darker. ⟶ magnetite, similar tint, but slightly darker. ⟶ cassiterite, lighter. ⟶ chalcopyrite, brownish grey. 2. Weak to distinct at twin boundaries, esp. in sections ∥ c: grey to a more brownish grey. 3. Weak to distinct; in air better than in oil: yellow to dark grey, sometimes with a violet or green tint. Oblique extinction. 4. Deep-red; in oil distinct, esp. in Mn-rich members.	470 nm 15.8-18.5 546 16.0-18.7 589 15.7-18.4 650 15.4-18.0 [194].
HUEBNERITE $MnWO_4$ (contains max. 20% $FeWO_4$). Mono.	1. Grey, similar to sphalerite. 2. Distinct. 3. Strong. 4. Red, lighter than in wolframite.	Somewhat lower than for wolframite.

Hardness H: polishing hardness VHN: micro-indentation hardness in kg/mm^2	Miscellaneous
H: ∼wolframite. VHN: 387-418 [194].	May occur in large crystals or in concentric aggregates with radial texture. Cleavage distinct in larger crystals. Replaces tellurides, scheelite; may be replaced by scheelite, sphalerite, stibnite, goethite (along cleavage planes). Other ass.: huebnerite, pyrite. Ref.: 12, 13, 586, 847, 1392, 1446, 1580, 1688.
H:>magnetite, scheelite, ≪pyrite, arsenopyrite, cassiterite. VHN: for wolframite-series: 258-657 [126, 159, 194, 1715].	Forms tabular idiomorphic crystals; irregular grains also occur. Zoning occurs (growing zones), sometimes with scheelite. Cleavage often distinct in two directions. Twinning very common. May contain inclusions of idiomorphic arsenopyrite crystals, and of chalcopyrite, molybdenite, bismuth, bismuthinite, gold, columbite, microlite. Occurs as central cores in cassiterite crystals. Partly altered to or replaced by scheelite; occasionally replaces scheelite. The fact that wolframite and scheelite are mutually replacive causes alternate zoning of these minerals. Also replaced by arsenopyrite, pyrrhotite, pyrite, chalcopyrite, galena and bismuth. Forms intergrowths with cassiterite, hematite and scheelite. May show tendency to marginal alteration. Other ass.: other W-minerals, tantalite, sphalerite, stannite. Diff.: magnetite is isotropic, does not show int. refl.; ilmenite is more brownish and more strongly anisotropic; cassiterite shows lighter and stronger int. refl. Ref.: 6, 7, 11, 126, 172, 315, 323, 339, 344, 546, 687, 1223, 1237, 1320, 1392, 1462, 1489, 1608, 1629, 1691, 1693, 1736, 1737. Notes: 1. Saari et al. (1968) report a niobian wolframite from Mozambique, containing 20.25 wt. % Nb_2O_5, considered to be an intermediate member between wolframite and an Fe-Mn niobate of the type $FeNbO_4$. 2. Sanmartinite, $(Zn,Fe,Ca)WO_4$, monoclinic. Occurs as fine granular aggregates, in part as microscopic tabular crystals and as reticular aggregates. Parag.: willemite, scheelite. Alteration product of scheelite. Ref.: 39.
H: ∼wolframite. VHN: no data, see at wolframite.	Occurs as coarse, separate crystals or as finely crystalline aggregates. Zonal crystals occur due to differences in Fe/Mn ratio. Replaces most of the accompanying minerals. Replaced by scheelite, Fe-hydroxides. Alters to tungstite and pyrolusite. Parag.: other W-minerals, pyrite, sphalerite, molybdenite, stibnite, other sulphides and sulphosalts. Ref.: 12, 13, 34, 377, 498, 709, 824, 1392.

Description of minerals from Table III

Name Formula Crystal System	1: Colour 2: Bireflectance 3: Anisotropy 4: Internal reflections	Reflectance in % (Wavelength in nanometres)
PYROCHLORE-series $A_2B_2X_7$ A = Ca, Na, U, Ce-earths, Y-earths, Fe, Th, Mn, Mg, Sr, K, Ba, Pb, Bi, Cs, Sb. B = Nb, Ta, Ti, Fe, Sb. X = O, OH, F, perhaps Cl, Na. Cubic	1. In air very similar to sphalerite, in oil much darker. Colour and refl. strongly varying, even within a single crystal. 2. + 3. Isotropic. 4. Always present in different colours: from colourless to dark brown.	About 12-16 %.
CALZIRTITE $Ca(Ca,Zr)_2Zr_4(Ti,Fe)_2O_{16}$ Tetr.	1. Light grey. 2. Weak to distinct. 3. Distinct only in oil. 4. Strong, reddish brown and yellowish brown.	About 15 %.

Hardness H: polishing hardness VHN: micro-indentation hardness in kg/mm^2	Miscellaneous
H:<columbite. VHN: range due to meta-mictization, alteration and chem. comp. pyrochlore: 255-826 betafite: 173-815 microlite: 341-916 obruchevite: 317-412 pandaite: 353-550 rijkeboerite: 485-498 [1601].	Note on nomenclature: the pyrochlore-series can be divided into three groups, based on the B-ions: - pyrochlore-group, B-ions mainly Nb(Ta,Ti); valid mineral species: pyrochlore, betafite, obruchevite, pandaite, koppite. - titanopyrochlore-group, B-ions mainly Ti(Nb,Ta); valid mineral species: titanopyrochlore, titanobetafite, titano-obruchevite. - microlite-group, B-ions mainly Ta(Nb,Ti); valid mineral species: microlite, tantalobetafite, tantalo-obruchevite, rijkeboerite, plumbian microlite, westgrenite. Usually occurs as single, idiomorphic crystals. Rarely as aggregates. Members of the pyrochlore-microlite series are fairly often partially or completely metamict. Zoning very common. Twinning rare: spinel law $/\!/$ (111). Replaced and associated with many minerals, especially Nb-Ta-Ti-minerals, and Fe-minerals, cassiterite. Very difficult to identify as a group, and impossible to distinguish between the different members under the microscope. Ref.: Van der Veen (1963) gives complete literature on the whole series.
H:~zircon. VHN?: 626-1035 [1729].	Occurs as single grains, as tabular concretions composed of complex trillings, and as irregular grains and aggregates. Twinning very common. Forms intergrowths with niobian perovskite and hematite. Other ass.: magnetite, ilmenite, rutile, anatase, goethite, Mn-oxides, pyrochlore, pandaite, perovskite, pyrrhotite, chalcopyrite, pyrite. Ref.: 63, 190, 1180, 1602, 1729, 1731. Note: Zirconolite, $CaZrTi_2O_7$, cubic, is related to calzirtite and to the pyrochlore-group. Occurs as irregular masses and as imperfect octahedra. May contain some Nb. Ass.: perovskite. Ref.: 151, 152.

Name Formula Crystal System	1: Colour 2: Bireflectance 3: Anisotropy 4: Internal reflections	Reflectance in % (Wavelength in nanometres)
GLAUCODOT $(Co,Fe)AsS$ (with $Co > 9$ wt. %)* Ortho.	1. White to light cream. ⟶ arsenopyrite, similar but more bluish white instead of creamy white. 2. Weak, weaker than for arsenopyrite. 3. Distinct, not as strong as for arsenopyrite. 4. Not present.	About 45-50 %.
ARSENOPYRITE $FeAs_{0.9}S_{1.1}$ to $FeAs_{1.1}S_{0.9}$ (may contain Co up to 9 wt. %*, some Au or Sb**). As-rich: tricl., pseudo-ortho. As-poor: tricl., pseudo-mono.	1. White with a faint creamy or pinkish tint. ⟶ pyrite, almost white; pyrite is creamy yellow with a brown tint. ⟶ loellingite, safflorite, creamy. ⟶ antimony, greyish white with a suggestion of violet. ⟶ galena, sphalerite, very pale yellow. ⟶ silver, much darker, white. ⟶ cobaltite, lighter and whiter. 2. Weak, but noticeable. 3. Strong: blue, green, reddish brown-yellow (nicols not completely crossed). 4. Not present.	470 nm 49.3-51.7 546 51.0-51.5 589 51.4-52.0 650 50.9-52.6 [194].

* Arsenopyrite with up to 9 wt. % Co is called cobaltian arsenopyrite (the name "danaite" for this material should be dropped). Material with Co-content > 9 wt. % is glaucodot (Gammon, 1966).
** For minor elements in arsenopyrite, see Fleischer (1955).

Hardness H: polishing hardness VHN: micro-indentation hardness in kg/mm^2	Miscellaneous
H: slightly <arsenopyrite, <cobaltite. VHN$_{100-200}$: 841-1277 [159, 810, 1715].	Commonly idiomorphically developed, often showing inclusions (e.g. galena). Cleavage in two directions may be distinct, ∥ (001) and (110). Occurs as inclusions in skeletal crystals of cobaltite. Forms coatings around pyrite and cobaltite. Veined by cosalite and cobaltite. Occurs as exsolution lamellae in cobaltian arsenopyrite. Other ass.: rammelsbergite, safflorite, skutterudite, niccolite, pitchblende, gold, sulphides of Fe, Zn, Cu, Pb and Ag. Ref.: 391, 458, 468, 718, 1171, 1237, 1322, 1379, 1388, 1392, 1536. Results published by Çagatay (1968) are doubtful.
H:≫ skutterudite, loellingite, magnetite, slightly >glaucodot, <cobaltite, pyrite. VHN$_{100-200}$: 715-1354 [159, 194, 468, 810, 1113, 1715].	Usually idiomorphically developed with characteristic rhomb-shaped sections; also skeleton-shaped, hypidiomorphic granular, elongated (in spherulites), fine-grained or extremely fine-grained. Cleavage commonly absent. Lamellar twinning very common; two systems may occur, resulting from the pseudo-orthorhombic or pseudomonoclinic symmetry of the lattice. Zonal texture not uncommon. May be intergrown with pyrite, loellingite, tetrahedrite, pyrrhotite, galena, safflorite, cobaltite, skutterudite. May contain exsolution lamellae of glaucodot. Occurs enclosed in pyrite, pyrrhotite, tetrahedrite. May be replaced by sphalerite, chalcopyrite, tetrahedrite, stannite, boulangerite, Ag-minerals, loellingite, magnetite, ilmenite. Core replacement by chalcopyrite, galena, pyrrhotite, sphalerite, loellingite; this is due to the zonal texture of arsenopyrite. Replaces loellingite, cobaltite, pyrite, molybdenite, wolframite, galena. Occasionally shows inclusions of gold and molybdenite. Diff.: polishing hardness and colour may prevent arsenopyrite to be mistaken for glaucodot, loellingite, gudmundite, or other similar anisotropic minerals. Other ass.: cassiterite, scheelite, canfieldite, franckeite, silver, dyscrasite, argentite, pyrargyrite, bismuth, bismuthinite, Bi-tellurides, marcasite, gudmundite, niccolite, cubanite, valleriite, mackinawite, covellite, chalcocite, uraninite. Ref.: 7, 44, 178, 179, 180, 265, 280, 322, 339, 468, 474, 667, 777, 802, 852, 891, 945, 947, 1046, 1047, 1049, 1212, 1237, 1328, 1392, 1447, 1450, 1454, 1458, 1462, 1465, 1608, 1629, 1732.

Description of minerals from Table III

Name Formula Crystal System	1: Colour 2: Bireflectance 3: Anisotropy 4: Internal reflections	Reflectance in % (Wavelength in nanometres)
COBALTITE (Co,Fe)AsS (may contain Ni). Ortho. (ordered structure), or, above about 800°C, Cubic (disordered structure).	1. White with a distinct pinkish, violet or brown tint. ⟶ arsenopyrite, distinctly pinkish. ⟶ pyrite, whiter, no yellow tint. ⟶ chalcopyrite, distinctly pinkish, lighter. 2. Weak, white to pinkish white. 3. Weak to distinct (in oil): blue-grey and brown colour shades. 4. Not present.	550 nm 56.2 [468].
NOLANITE $(V,Fe,Al,Ti)_9O_{18}$ with V_2O_5/V_2O_4 about 1.50 Hex.	1. See at birefl. 2. ∥ c-axis: strong, medium brown to blue-grey. ⊥ c-axis: not present, medium brown. 3. Sections containing a- and c-axis: strong, deep-red to midnight-blue; parallel extinction. Basal sections are nearly isotropic. 4. Not present.	No data.

Hardness H: polishing hardness VHN: micro-indentation hardness in kg/mm^2	Miscellaneous
H: very high; different in different sections; $\gg$skutterudite, loellingite, $>$arsenopyrite,$<$pyrite. $VHN_{100-200}$: 948-1367 [159, 194, 468, 810, 1113, 1715].	Commonly occurs as idiomorphic crystals; also skeletons or allotriomorphic aggregates. Cleavage in coarser grains distinct. Twinning lamellae often visible; zoning occurs. Shows inclusions of cobaltian loellingite, gold, silver, niccolite, chalcopyrite. Replaces cobaltian loellingite, safflorite, pitchblende. Replaced by pyrrhotite, cobaltian arsenopyrite, chalcopyrite, arsenopyrite, pyrite, skutterudite, sphalerite, galena, tennantite, safflorite, niccolite, silver. Core replacement by bornite and silver. Forms intergrowths with pyrite, chalcopyrite, gersdorffite, safflorite, skutterudite. Other ass.: cubanite, bismuth, molybdenite, graphite, glaucodot. Ref.: 32, 94, 95, 290, 458, 468, 501, 726, 917, 1063, 1133, 1136, 1237, 1260, 1270, 1387, 1392, 1394, 1447, 1463, 1629. Note: Klemm and Weiser (1965) reported a high-temperature solid solution member between cobaltite and gersdorffite: (Co,Ni)AsS. Its colour is white with a weak bluish tint, and its refl. and H are somewhat higher than for pyrite. No birefl. is present, but the anisotropy is distinct: reddish brown to grey with a bluish green tinge.
H:$\gg$chalcopyrite, slightly $<$hematite. VHN_{50}: 717-766 [194].	Occurs as subhedral to euhedral tabular prismatic crystals with hexagonal outlines. Fills cracks in pyrite. Replaced by pyrite, chalcopyrite, pitchblende. Replaces hematite. As inclusions, together with hematite and ilmenite, in chalcopyrite. Fractured grains may be cemented with coloradoite and gold. Occurs intergrown with another Fe-V-oxide, which shows a more bluish colour. Other ass.: galena, tellurides, karelianite. Ref.: 570, 846, 1275, 1277, 1510.

Description of minerals from Table III

Name / Formula / Crystal System	1: Colour / 2: Bireflectance / 3: Anisotropy / 4: Internal reflections	Reflectance in % (Wavelength in nanometres)
PITCHBLENDE* Varying mixtures of UO_2 and UO_3 Cubic	1. Different shades of grey. Refl. decreases with increasing UO_3-content. 2. + 3. Isotropic. 4. Dark brown and yellow-brown, not uncommon.	About 10-15 %.
URANINITE UO_2 (may contain some Th, Pb, and small amounts of rare earths) Cubic	1. Grey with a brownish tint, in oil much darker than in air. ⟶ magnetite, similar or slightly darker, less pinkish. ⟶ sphalerite, distinct brownish tint. ⟶ pitchblende, lighter. 2. Not present. 3. Isotropic; some spec. show a weak anomalous anisotropy. 4. Very dark brown or reddish brown, not uncommon.	515 nm 17.1 [644].

* The name uraninite should only be used for crystalline material; pitchblende for the massive substance of colloidal origin.

Hardness H: polishing hardness VHN: micro-indentation hardness in kg/mm^2	Miscellaneous
H: varying. VHN: 314-803 [159, 1715].	Kidd and Haycock (1935) describe two kinds of pitchblende: I: high in UO_2; takes a good polish, smooth surface; homogeneous; lighter grey; comparatively hard. II: higher in UO_3; oxidation product of I; takes a bad polish, rough surface; appears to be porous; comparatively softer. Pitchblende occurs in a great variety of forms, showing all gradations, such as: botryoidal, colloform, cellular, dendritic, spherulitic, oolitic, ring-like, brecciated or vein forms. Concentric or circular cross sections commonly show radiating cracks. May enclose minerals of the skutterudite-series (often containing bismuth crystals), bismuth-dendrites niccolite, "cross-shaped skeleton" galena structures, pyrite, chalcocite, proustite. Replaces galena, pyrite, magnetite, ilmenite, hematite; occasionally fills central parts of skutterudite. Veined or replaced by safflorite, rammelsbergite, loellingite, other Co-Ni-Fe-arsenides, chalcopyrite, bornite, digenite, chalcocite, covellite, tetrahedrite, sphalerite, galena, silver, pearceite, bismuth, bismuthinite, hematite, pyrite, marcasite, gold, ilmenite, magnetite, rutile. Preferential replacement of colloform pitchblende gives raise to a type of cockarde texture. Forms intergrowths with gold, hematite, ilmenite, magnetite. May contain galena, often of radiogenetic origin. Other ass.: arsenopyrite, glaucodot, linnaeite, pentlandite, pyrrhotite, argentite, molybdenite, montroseite, Fe-hydroxides, jordisite, coffinite, tiemannite, clausthalite, cinnabar, emplectite, stromeyerite, klockmannite, U-hydroxides, U-Pb-minerals, U-Cu- and U-Ca-minerals. Ref.: 4, 50, 84, 308, 432, 445, 458, 506, 544, 545, 550, 632, 644, 707, 708, 717, 718, 783, 928, 960, 1182, 1237, 1270, 1282, 1537, 1538, 1543, 1584, 1671, 1672, 1705.
H: very high, > magnetite, ≦ pyrite. VHN_{50-100}: 625-929 [159, 194, 253, 1080].	Forms well-developed crystals. Cleavage ∥ (100) and (111) may be visible. Twinning ∥ (111) often occurs. Zonal texture very common; may be evident by orientated inclusions. Reported in a two-phase zoning polycrystal with pyrite. May contain lamellae of gold along the octahedral and cubic cleavage directions. Occurs as exsolution bodies in columbite. Replaces pyrite; may be replaced or veined by siegenite, pyrite, vaesite, sphalerite, pitchblende, gold. Alters to brannerite. Contains inclusions of galena, often of radiogenetic origin. Ass.: alteration products; magnetite, pyrite, uranothorite, chalcopyrite, molybdenite, pyrrhotite, hematite, Co-Ni-Fe-arsenides, arsenopyrite, marcasite, bornite, melonite, Fe-hydroxides. Ref.: 50, 253, 357, 445, 644, 719, 726, 761, 1080, 1182, 1237, 1278, 1282, 1335, 1392, 1447.

Description of minerals from Table III

Name Formula Crystal System	1: Colour 2: Bireflectance 3: Anisotropy 4: Internal reflections	Reflectance in % (Wavelength in nanometres)
THORIANITE $(Th,U,Ce)O_2$ Cubic	1. Grey. Refl. increases with increasing U-content. $\longrightarrow$ uraninite, very similar but darker. 2. + 3. Isotropic. 4. Red-brown and yellow-brown, often visible in oil (on close examination).	About 15 %.
BRANNERITE $(U, Th,Ca)[(Ti,Fe)_2O_6]$ Mono. (synthetic material seems to be tricl.).	1. Grey. 2. + 3. Usually metamict. 4. Coarse crystals: brownish grey or greyish white. Fine-grained material: blue-grey to bluish white, dark brown, yellowish (strong) partially due to anatase.	About 17 %.
COFFINITE $U(SiO_4)_{1-x}(OH)_x$ Tetr.	1. Grey; submicroscopic aggregates show a very low refl. (in oil like gangue minerals); idiomorphic crystals have a somewhat higher refl., but lower than uraninite. 2. Very weak, O>E. 3. Appears isotropic; sometimes very weakly anisotr. 4. Air: rare and weak. Oil: more pronounced, several brownish tints.	About 7-10 %.

Hardness H: polishing hardness VHN: micro-indentation hardness in kg/mm^2	Miscellaneous
H: >uraninite. VHN: 920-1235 [159, 1715]	Usually occurs as cubic crystals; as rounded grains in alluvial deposits. Cleavage $\parallel$ (100) and (111) may be visible. Zonal texture very common. Ass.: magnetite, ilmenite, cassiterite, uraninite, sulphides. Ref.: 445, 1237, 1278, 1308.
H: no data. VHN: 387-907 [133, 159, 519, 1715].	Occurs as idiomorphic columnar, prismatic, or needle-shaped crystals, or as aggregates of these. Also as rounded or irregularly shaped grains. Twinning or cleavage not observed. Brannerite of thermal origin is formed by replacement of uraninite and rutile. May form pseudomorphs after these minerals. Practically always contains very small laths of pyrrhotite and exsolution or alteration bodies of anatase. May show a characteristic dusting with minute inclusions of radiogenetic galena. Rutile clusters may be included in brannerite grains. Sometimes contains inclusions of pitchblende. Parag.: rutile, uraninite, pitchblende, pyrite, marcasite, davidite, coffinite, gersdorffite, gold, galena, sphalerite, tetrahedrite, aikinite, uranothorite, chromite, cassiterite, chalcopyrite, arsenopyrite, pyrrhotite, anatase, magnetite, molybdenite, bismuthinite. Ref.: 133, 355, 445, 478, 519, 618, 690, 929, 1085, 1089, 1115, 1236, 1237, 1336, 1377, 1512.
H: $\leqq$ pitchblende. VHN: 236-333 [159].	Occurs as idiomorphic crystals with tetragonal outlines, or as sub-microscopic aggregates, in colloform textures. May form veinlets, surrounded by carbonaceous material. Forms botryoidal encrustations. Occurs as intergranular films between quartz and fluorite; as inclusions in niccolite. Corrodes, and is corroded by pitchblende. Replaced by sphalerite, uraninite, pyrite, rammelsbergite, loellingite. Replaces bismuth. Parag.: uraninite, pitchblende, montroseite, paramontroseite, pyrite, wurtzite, galena, cobaltite, bornite, wittichenite, niccolite, skutterudite, rammelsbergite, safflorite, rutile, chalcopyrite, marcasite, sphalerite, bismuth, arsenic, tennantite, bismuthinite, hematite, aikinite. Ref.: 50, 394, 432, 445, 544, 731, 930, 985, 1055, 1237, 1238, 1396, 1449, 1477, 1478. Note: Uranothorite, (Th,U)SiO$_4$, tetr., forms prismatic crystals. Ref.: 1088.

198 Description of minerals from Table III

Name Formula Crystal System	1: Colour 2: Bireflectance 3: Anisotropy 4: Internal reflections	Reflectance in % (Wavelength in nanometres)
ILVAITE $CaFe_2^{2+}Fe^{3+}[OH/O/Si_2O_7]$ Both ortho. and mono. varieties are known.	1. Air: grey and blue. ⟶ sphalerite, R_b and R_a are much darker. ⟶ graphite, R_b and R_a are lighter than R_e of graphite. ⟶ magnetite, dark bluish grey. ⟶ pyrrhotite, bluish grey. 2. Air: very strong: R_c blue, R_b and R_a grey. Oil: extremely strong: R_c dark red or reddish purple, R_a and R_b grey or slightly bluish grey. 3. Very strong; air: blue, pink, reddish orange; 45°: fiery orange; oil: dark blue, purplish red; 45°: bluish orange. 4. Air: very rare, orange to brown and red. Oil: less rare, same colours.	470 nm 8.6-9.4 546 7.8-9.2 589 7.5-9.0 650 6.9-8.3 [1398].

HEMATITE*
Fe_2O_3 (may contain some $FeTiO_3$ or $MgTiO_3$ in solid solution; up to 17.64 wt. % Mn has been reported).
Hex.
	1. Grey-white with a bluish tint.
	⟶ ilmenite, white.
	⟶ magnetite, white.
	⟶ goethite and lepidocrocite, brighter white.
	⟶ pyrite, bluish grey.
	⟶ chalcocite, slightly brownish.
	⟶ pyrolusite, hollandite, bluish; no yellow tint.
	⟶ cuprite, white.
	Ti-rich hematite is slightly darker and more grey-white.
	⟶ ilmenite, more steely.
	2. Weak.
	3. Very distinct, esp. on twin boundaries: greyish blue, greyish yellow; sharp extinction.
	4. Deep-red (also on borders against gangue); very common in not well-polished spec.
		For hematite;
		         R_o R_e
		470 nm 31.5 28.0
		546 30.0 26.2
		589 28.8 25.1
		650 26.5 22.8
		
		Dispersion curves: [486]; see also: [300].
		
		For Ti-hematite:
		         R_o R_e
		470 nm 29.8 26.1
		546 28.3 24.6
		589 27.3 23.7
		650 25.2 22.0
		
		Dispersion curves: [486].

---

* Dunn (1937a) describes two varieties of hematite, α and β. α-Hematite replaces magnetite, shows a higher refl. and a definite birefl. in oil; β-hematite is younger, occurs only as interstitial material to α or as rims round α; shows more commonly a deep-red int. refl. except in coarser grains. Colour and anisotr. about the same for both types. An explanation of the varieties is not given.

Hardness H: polishing hardness VHN: micro-indentation hardness in kg/mm^2	Miscellaneous
H: $\gg$ graphite, $>$ sphalerite, $\sim$ hematite, magnetite, $<$ pyrite. VHN: 703-1055 [194, 1398, 1715].	Occurs as idiomorphic crystals or as aggregates. Also as hypidiomorphic grains. Twinning rare, lamellae following two systems. Alters to goethite; forms intergrowths with ilmenite, goethite and hematite. Replaces magnetite. Enclosed by chalcopyrite and magnetite. Other ass.: ulvöspinel, pentlandite, pyrrhotite, sphalerite, pyrite, graphite. Ref.: 61, 78, 304, 512, 759, 760a, 823, 855, 994, 1059, 1629.
H: very high, but variable, microcrystalline material showing lowest H. Commonly $\gg$ goethite, lepidocrocite, magnetite, maghemite, coulsonite, $>$ ilmenite, $\geq$ rutile, $\leq$ pyrite, $<$ cassiterite. VHN$_{100-200}$: 739-1114 [159, 194, 300, 810, 1113, 1715].	Usually idiomorphically developed, tabular or thin-tabular; also laminated, fibrous or as radiated aggregates of needle-shaped crystals (then probably pseudomorphically). Two pseudo-cleavages often distinct. Lamellar twinning very common, often caused by strain; lamellae may occur in different systems and show transverse twinning; they may be curved, tapering off, showing alternating wedges. "Micro-scale twinning" should occur frequently. Replaces magnetite, often pseudomorphous ("martite"), coulsonite, maghemite. May contain numerous remnants of replaced magnetite; occasionally inclusions of pyrite, bismuth, gold (introduced into magnetite before martitization). Replacement by magnetite, pyrite, or Fe-hydroxides not uncommon. May occur as inclusions in chromite. Forms exsolution intergrowths with ilmenite, magnetite, rutile, cassiterite, franklinite. The orientated intergrowths with ilmenite, magnetite and rutile are not necessarily due to exsolution. "Titanhematite" or "white ilmenite" is hematite which contains max. 10 % FeTiO$_3$ in solid solution; shows lower refl., stronger birefl. and anisotr., always lamellar twinning and no red int. refl.; more resistant to HF or HF + H$_2$SO$_4$ than ilmenite. If FeTiO$_3$-content exceeds 10 %, orientated intergrowths of ilmenite and Ti-hematite may be formed. Small bodies of Ti-hematite may occur in ilmenite (if Fe$_2$O$_3$-content in ilmenite exceeds 6 %); in both intergrowths inclusions are orientated // (0001) of the host mineral. May show inclusions of rutile, coarse-grained or as numerous orientated minute needle-shaped crystals. Other ass.: Mn-oxides, sulphides, tellurides, selenides. Ref.: 60, 154, 158, 165, 226, 283, 321, 324, 331, 332, 333, 339, 362, 451, 462, 554, 555, 1204, 1215, 1237, 1358, 1461, 1721.

Name Formula Crystal System	1: Colour 2: Bireflectance 3: Anisotropy 4: Internal reflections	Reflectance in % (Wavelength in nanometres)
IXIOLITE $(Ta,Fe,Sn,Nb,Mn)_4O_8$ Ortho.	1. Grey. 2. + 3. Usually metamict. 4. Yellow to dark red.	About 13-14 %.
WODGINITE $(Ta,Nb,Fe,Mn,Sn,Zr)_{16}O_{32}$ Mono.	1. Grey. ⟶ columbite-tantalite, very similar. ⟶ tapiolite, grey, darker. 2. Very weak, greyish tints. 3. Distinct, greyish tints. 4. Abundant, red to orange-red.	470 nm 14.7-15.8 546 14.3-15.4 589 14.2-15.4 650 13.9-15.4 [196].
TIN-TANTALITE* $Mn(Ta,Nb,Sn)_2O_6$ Mono. [886] or Ortho. [1010, 1011].	1. Grey, darker than manganotantalite, and much darker than stibiotantalite. 2. Distinct. 3. Strong. 4. Yellow-brown.	590 nm 14.8-15.9 [886].
THOREAULITE $Sn(Ta,Nb)_2O_7$ Mono.	1. Air: light grey; oil: grey to brownish grey. ⟶ tantalite, somewhat lighter. 2. Distinct (in oil): grey to brownish grey. 3. Strong (in oil): grey to dark brown. 4. Fairly abundant, yellow to yellowish brown.	470 nm 17.3-19.4 546 16.1-17.8 589 15.8-17.4 650 15.8-17.4 [1398].
STIBIOTANTALITE $Sb(Ta,Nb)O_4$ Ortho.	1. Grey. ⟶ tantalite, lighter. 2. Weak. 3. Weak to distinct, not as strong as for tantalite. 4. Strong, whitish and yellowish.	About 20 %.
BISMUTOTANTALITE $(Bi,Sb)(Ta,Nb)O_4$ Ortho.	1. Similar to stibiotantalite. 2. Weak. 3. Weak to distinct. 4. Strong, yellow-brown (in contrast to stibiotantalite).	About 20 %.

* Has also been referred to as "stannotantalite" and "olovotantalite".

Hardness H: polishing hardness VHN: micro-indentation hardness in kg/mm^2	Miscellaneous
H: no data. VHN: 860-947 [715].	Forms elongated tabular crystals. Also occurs as isometric and as circular grains. Polysynthetic twinning and one cleavage are discernable. Parag.: tapiolite, tantalite, wodginite. Ref.: 715, 1010.
H: ~tapiolite. VHN: 766-1080 [196, 716].	Grains are prismatic, spheroidal or anhedral in shape. Occasionally twinned ∥ (100). Cleavage not observed. Closely associated with tapiolite and microlite. Forms orientated intergrowths with cassiterite and tapiolite. May contain inclusions of sukulaite and tapiolite. Other ass.: staringite. Ref.: 157, 195, 196, 716, 853, 944, 1011, 1579, 1625.
H: no data. VHN$_?$: 660 [886].	Intermediate member between manganotantalite and thoreaulite. Occurs as well-shaped crystals. Two cleavages may be distinct. Shows simple and polysynthetic twinning. Alters to tapiolite and thoreaulite. Shows inclusions of thoreaulite, cassiterite, tapiolite, tantalite. Other ass.: stibiotantalite, simpsonite. Ref.: 886.
H: slightly < tantalite. VHN: 473-797 [1398, 1715].	Occurs as prismatic crystals. Cleavage ∥ (100) distinct. Contains inclusions of cassiterite and tantalite. Replaces tantalite, tin-tantalite, tapiolite. Ref.: 59, 206, 858, 886, 905, 1398, 1544.
H: no data. VHN: 441-607 [1424, 1715].	Forms prismatic crystals and irregular grains. Perfect cleavage ∥ (010). May be replaced by or intergrown with microlite, stibnite, antimony and senarmontite, which may produce a serpentine-like structure. Other ass.: tantalite, cassiterite, columbite, manganotantalite, bismuth. Ref.: 252, 335, 600, 647, 736, 1048, 1424. Note: Stibiocolumbite ("stibioniobite") is Sb(Nb,Ta)O$_4$, ortho.
H: no data. VHN$_?$: 764-824 [1615].	Forms prismatic crystals and rounded grains. Parag.: bismuth, cassiterite, Fe-hydroxides, Nb-Ta-minerals. Ref.: 131, 422, 645, 647, 1615.

Description of minerals from Table III

Name Formula Crystal System	1: Colour 2: Bireflectance 3: Anisotropy 4: Internal reflections	Reflectance in % (Wavelength in nanometres)
COLUMBITE*-TANTALITE MANGANOCOLUMBITE-* MANGANOTANTALITE $(Fe \leq Mn)(Nb \leq Ta)_2O_6$ Ortho.	1. Grey-white with a brown tint. $\longrightarrow$ magnetite, very similar, brownish tint not so distinct. 2. Weak, only visible at grain boundaries. 3. Distinct, straight extinction. 4. Distinct, probably related to Mn- and Fe-content: for manganotantalite yellow-brown, for ferro-tantalite deep-red.	470 nm 15.9-18.3 546 15.3-17.7 589 15.3-17.3 650 15.4-17.1 [194].
TAPIOLITE-MOSSITE** $(Fe,Mn)(Ta,Nb)_2O_6$ Tetr.	1. Grey-white to bluish grey. 2. Distinct to strong: O ~ columbite. E much lighter than columbite. 3. Strong, slaty grey to red-brown or dark blue to greenish grey. 4. Red or red-brown, not abundant.	R_o R'_e 470 nm 16.0 17.5 546 15.5 17.8 589 15.1 17.7 650 14.9 17.5 [194].
FERGUSONITE α-$YNbO_4$ Tetr. β-$YNbO_4$ Mono.	1. Greyish white to creamy grey. 2. Always absent. 3. Metamict, some spec. are anisotropic. 4. Some spec. are completely opaque; others have golden-brown int. refl. Strong yellowish white int. refl. are produced by thin films of alteration products along fractures.	About 11-14 %.
FORMANITE $YTaO_4$ Isostructural with fergusonite: tetr. and mono. upon heating all spec. are mono.	1. Greyish white. 2. Always absent. 3. Slight. 4. Different shades of brown.	Higher than for fergusonite.

* Also named niobite and manganoniobite respectively.
** Natural members of the series with Nb > Ta (mossite) have not yet been found; all data refer to tapiolite.

Hardness H: polishing hardness VHN: micro-indentation hardness in kg/mm^2	Miscellaneous
H: very high, $\leq$ tapiolite. VHN$_{100-400}$ (for columbite-tantalite series): 240-1021 [127, 159, 194, 810, 1601, 1715]. No relationship exists between Nb/Ta ratio and VHN.	Usually occurs as euhedral crystals and as granular aggregates. Cleavage $/\!/$ (100) may be distinct; $/\!/$ (010) less distinct. Twinning rare; zonal texture may occur. May show inclusions of cassiterite, galena, hematite, ilmenite, rutile, tapiolite, uraninite, wolframite. Contains exsolution bodies of cassiterite, ilmenite, rutile, uraninite. Occurs as exsolution bodies in cassiterite; in this case cassiterite contains some Ta_2O_5 or Nb_2O_5 in solid solution. Forms orientated intergrowths with uraninite (epitaxis). Replaced by stibiotantalite, microlite, tapiolite. Ref.: 50, 127, 335, 339, 601, 647, 903, 944, 977, 1148, 1237, 1471, 1480. Note: Magnesiocolumbite ("magnesioniobite"), $(Mg,Fe,Mn)(Nb,Ta,Ti)_2O_6$, orthorhombic. Occurs as acicular and tabular crystals, sometimes showing distinct zoning. Forms intergrowths with Nb-rutile. Ass.: cassiterite. Ref.: 887, 991.
H: $\geq$ columbite. VHN: 796-1132 [127, 194, 810].	Twin lamellae $/\!/$ (011) always present. Occurs as reaction rims in columbite; as exsolution bodies in the form of fine needles in cassiterite. Forms orientated intergrowths with cassiterite and wodginite. Other ass.: bismuthinite, tantalite, staringite. Ref.: 77, 127, 195, 196, 647, 903, 944, 1148; 1237, 1615, 1629.
H: > columbite. VHN$_{100-200}$: 683-897 [810, 1715].	Occurs as irregular masses of tabular crystals. Parag.: columbite, tantalite, uranothorite, magnetite, hematite, goethite, ilmenite, rutile, cassiterite, spinel, pyrochlore, gold. Ref.: 204, 287, 294, 392, 517, 743, 744, 924, 977, 1143, 1615.
H: probably > columbite. VHN: 772-870 [1715].	Forms tabular crystals. Not observed associated with other ore minerals. Ref.: 204, 392, 750, 1615.

Description of minerals from Table III

Name Formula Crystal System	1: Colour 2: Bireflectance 3: Anisotropy 4: Internal reflections	Reflectance in % (Wavelength in nanometres)

MARCASITE
FeS_2*
Ortho.
	1. Yellowish white with a slight pinkish or greenish yellow tint.
	   ⟶ pyrite, whiter.
	   ⟶ arsenopyrite, greenish yellow tint.
	   ⟶ silver, grey with a bluish green tint.
	2. Strong.
	   ∥ a: white with a brownish tint; very similar to pyrite.
	   ∥ b and ∥ c: yellow with a greenish tint; between b and c no great difference.
	3. Strong, especially in sections ∥ (100) or (010): blue, green-yellow, purplish, violet-grey.
	4. Not present.
		480 nm 44.8-50.5
		540 47.6-53.2
		580 48.0-52.1
		640 48.0-50.9 [194].

* For minor elements in marcasite, see Fleischer (1955).

Hardness H: polishing hardness VHN: micro-indentation hardness in kg/mm^2	Miscellaneous
H: >pyrrhotite, commonly <pyrite, but >gel-pyrite. VHN$_{100-200}$: 762-1561 [159, 194, 810, 1113, 1715]	Commonly shows colloform texture; forms alternating layers with pyrite or crusts on gel-pyrite; idiomorphic, lath-shaped crystals less common; radiated aggregates may occur. Cleavage may be distinct. Twinning (both coarse- and fine-lamellar) and zonal texture very common. Formed, often together with pyrite, by alteration of pyrrhotite ("bird's eye" texture), stannite, sternbergite, frieseite; fills cracks in pyrrhotite. Replaces pyrrhotite (orientated or irregularly), sphalerite; may be replaced by freibergite, galena, silver. Occurs intergrown with arsenopyrite, pyrite, chalcopyrite, covellite, proustite. Very fine-grained material may be mistaken for pyrite. Other ass.: bravoite, pentlandite. Ref.: 7, 273, 278, 321, 322, 339, 588, 671, 996, 997, 1058, 1237, 1379, 1392, 1576, 1629, 1659, 1732. Note: At the Maubach and Mechernich deposits in Germany, 10-40 microns thick rims occur on the borders between pure marcasite and bravoite. They show a reddish brown colour, similar to some bravoite varieties, but are strongly anisotropic with simultaneous extinction with the pure marcasite. These rims consist of marcasite with a Ni-content of 0.8 % and a Co-content of 17.4 % (Schachner-Korn and Springer, 1967).

Name Formula Crystal System	1: Colour 2: Bireflectance 3: Anisotropy 4: Internal reflections	Reflectance in % (Wavelength in nanometres)
PYRITE FeS_2* (may contain Co, Ni, As and perhaps Au in solid solution; Cu up to 10 wt. %). Cubic	1. Yellowish white. ⟶ marcasite, yellower. ⟶ arsenopyrite, creamy yellow with brown tint. ⟶ galena, pale yellow. ⟶ silver, greyer, slightly greenish. ⟶ chalcopyrite, much lighter, less yellow. ⟶ sperrylite, darker and yellower. Pyrite with higher Co-, Ni-, or Cu-contents shows pinkish, reddish or violet tints. 2. Not present. 3. Often weakly to distinctly anisotropic; blue-green to orange-red. This may be due to polishing methods, and/or lattice distortion caused by impurities (Stanton, 1957, Klemm, 1962b, Gibbons, 1967, Saager and Mihálik, 1967, Bayliss, 1969). 4. Not present.	470 nm 46.0 546 53.6 589 55.0 650 55.9 Dispersion curves: [300]; see also: [43].

* For minor elements in pyrite, see Auger (1941), Sztrókay (1944), Bjørlykke (1945 and 1947) and Fleischer (1955).

Hardness H: polishing hardness VHN: micro-indentation hardness in kg/mm^2	Miscellaneous
H: very high, varying with orientation; >arsenopyrite, cobaltite, slightly >marcasite, hematite, <sperrylite, laurite, cassiterite. Ni-, Co-, and As-containing pyrites are somewhat softer. $VHN_{100-200}$: 913-2056 [159, 194, 810, 1113, 1597a, 1715]. VHN_{150}: (depending on the temperature of formation): 913-1524 [558].	Generally idiomorphically developed, coarser grains showing rectangular or square outlines. Many other modes of occurrence have been described: coarse-grained aggregates of crystal fragments; coarse-grained sphere-like aggregates; fine-grained idiomorphic; very fine-grained skeleton-shaped; very fine-grained spherical, often with concentric texture. Most of the material named "gel-pyrites" consists for the greater part of cryptocrystalline pyrite, often containing some As. Cleavages $/\!/$ (100), $/\!/$ (311), $/\!/$ (111) may be visible. Twinning lamellae very rarely observed. Zoning not uncommon; may already be revealed by slight differences in colour or hardness, or by orientated inclusions. Being commonly one of the first minerals to crystallize, it may be replaced by many other minerals, such as: pyrrhotite, marcasite, chalcopyrite, sphalerite, tetrahedrite, bornite, enargite, chalcocite, galena, Fe-hydroxides, arsenopyrite, covellite, Ag-minerals, uraninite, hematite, glaucodot, pentlandite. Replaces chalcopyrite, safflorite, uraninite, Fe-Ti-oxides, sphalerite, alabandite, magnetite, chromite, rutile. May show inclusions of pyrrhotite, arsenopyrite, chalcopyrite, cinnabar, replacement relics of magnetite and sphalerite. May be formed by alteration of pyrrhotite, often together with marcasite. Pyrite may contain some submicroscopic Au in solid solution, or in colloidal form. Forms intergrowths, many of them orientated, with sphalerite, tetrahedrite, pyrrhotite, sternbergite, covellite, nagyagite, galena, gersdorffite, magnetite, cobaltite. Occurs with uraninite as alternate layers around a pyrite-nucleus (syntaxis). Perhaps most widely spread of all ore minerals; may be associated with many of them, such as: sulphides and sulphosalts of Fe, Cu, Pb, Zn; bravoite, gold, Au-Ag-tellurides, ilmenite, rutile, hematite, magnetite, cassiterite, graphite. "Melnicovite-pyrite" still consists partly of amorphous substance but commonly the greater part is cryptocrystalline; often contains some As (max. 8 %). Polishes better than pyrite; usually shows a more distinct brownish tint, but may be very similar in colour to pyrite; refl. in most cases lower than pyrite, sometimes about the same; H varies within wide limits, occasionally as low as for galena; a weak anisotr. may occur as for pyrite; less inert to reagents which may develop texture beautifully. Ref.: 48, 54, 91, 94, 136, 137, 215, 278, 297, 321, 322, 333, 334, 339, 352, 430, 431, 474, 538, 543, 558, 588, 649, 665, 719, 803, 891, 892, 996, 997, 1237, 1268, 1311, 1322, 1363, 1379, 1392, 1464, 1465, 1494, 1576, 1603, 1629, 1660, 1705.

Description of minerals from Table III

Name / Formula / Crystal System	1: Colour / 2: Bireflectance / 3: Anisotropy / 4: Internal reflections	Reflectance in % (Wavelength in nanometres)
CASSITERITE SnO_2 * Tetr.	1. Brownish grey, but colour may change from deep red-brown to almost colourless. Darker cassiterite has higher Fe- and W-contents. Refl. increases with increasing Ta-content. ⟶ stannite, greyish brown. ⟶ wolframite, grey to honey-brown. ⟶ ilmenite, rutile and magnetite, brownish grey. ⟶ tapiolite and tantalite, brownish grey. ⟶ cylindrite and franckeite, grey. 2. Moderately distinct, light grey to brownish grey; blotchy appearance not uncommon. 3. Very distinct, grey to dark grey, in oil usually masked by the int. refl. 4. Abundant, light yellow or whitish to yellow-brown.	R_o 470 nm 11.3-11.7 546 10.9-11.1 589 10.7-11.1 650 10.5-10.8 R'_e 470 nm 12.8-13.1 546 12.2-12.6 589 12.1-12.6 650 11.9-12.5 [194].
STARINGITE $(Fe,Mn)_{0.5}(Sn,Ti)_{4.5}$ $(Ta,Nb)_{1.0}O_{12}$ Tetr.	1. Dark grey (in oil). ⟶ cassiterite, very similar. 2. Moderately strong, lighter to darker grey. 3. Very distinct, different shades of grey. 4. Abundant (in oil), yellowish orange to brownish red.	R_o R'_e 470 nm 12.8 14.5 546 12.1 13.9 589 12.0 13.7 650 11.8 13.6 Dispersion curves: [195].
SUKULAITE $(Ta,Nb)_2Sn_2O_7$ Cubic	1. Light grey with a reddish or lilac tint. 2. Not present. 3. Apparently isotropic, but masked by the int. refl. 4. Strong, reddish brown (might be due to the surrounding cassiterite).	Somewhat higher than for cassiterite.

* For minor elements in cassiterite, see Noll (1948).

Hardness H: polishing hardness VHN: micro-indentation hardness in kg/mm^2	Miscellaneous
H: very high. VHN$_{100-200}$: 811-1532 [159, 194, 810, 1398, 1715].	Occurs: in coarse-grained crystals, intergrown crystals or irregular grains, often beautifully zoned; in very fine-grained aggregates; acicularly developed, sometimes with radiated texture (pinwheel-like clusters); banded; as colloform aggregates. Prismatic cleavage occasionally visible. Twinning very common. Replaces pyrite, arsenopyrite, sphalerite, stannite, frankeite, teallite, cylindrite, bismuthinite; may be replaced by pyrite, arsenopyrite, pyrrhotite, sphalerite, chalcopyrite, stannite. Occasionally found as inclusions in sulphides (sphalerite, galena, arsenopyrite); then commonly slightly corroded. May contain many inclusions, sometimes due to exsolution, of columbite, tantalite, rutile, tapiolite, wolframite, magnetite, hematite, ilmenite, sukulaite, wodginite, bismuth; some of these inclusions may cause magnetism in cassiterite. Forms exsolution intergrowths with Nb-Ta-minerals; then contains some Nb_2O_5 or Ta_2O_5 in solid solution. Forms graphic-like intergrowths with galena, boulangerite and franckeite. Other ass.: scheelite, wurtzite, Zn-teallite, herzenbergite, tetrahedrite, tennantite, Pb-Sb-sulphosalts, andorite, marcasite, bournonite, niccolite, cobaltite, skutterudite, stibnite. Diff.: sphalerite is isotropic and shows higher refl.; scheelite shows still lower refl., abundant white int. refl. and other habit. Ref.: 5, 7, 11, 91, 172, 224, 225, 239, 323, 335, 339, 667, 795, 828, 841, 1056, 1120, 1192, 1207, 1237, 1239, 1320, 1379, 1392, 1398, 1399, 1462, 1576, 1578, 1608, 1618, 1625, 1678, 1691, 1693, 1732, 1735.
H:>tapiolite and wodginite VHN: 1033-1187 [195].	Occurs as exsolution bodies in tapiolite, sometimes at the contact between tapiolite and wodginite. Can only be distinguished from cassiterite by X-ray diffraction. Ref.: 195.
H:>cassiterite and wodginite. VHN: no data.	Occurs as narrow rims or aggregates in inclusions of wodginite in tantalian cassiterite. Replaces wodginite. Ref.: 1625.

Name Formula Crystal System	1: Colour 2: Bireflectance 3: Anisotropy 4: Internal reflections	Reflectance in % (Wavelength in nanometres)
SPINEL-series (Spinel, hercynite, galaxite, gahnite, respectively $MgAl_2O_4$, $FeAl_2O_4$, $MnAl_2O_4$ and $ZnAl_2O_4$). Cubic	1. Grey, with very low refl., about as for gangue. 2. + 3. Isotropic. 4. Abundant and intense, colourless, various shades of green and brown.	For spinel: 470 nm 6.9 546 6.8 589 6.8 650 6.7 Dispersion curve: [300]. For hercynite: 470 nm 8.2 546 8.0 589 8.0 650 7.9 Dispersion curve: [300].
NIGERITE $(Sn,Mg,Zn,Fe)(Al,Fe)_4(O,OH)_8$ Hex.	1. In oil pinkish brown. ⟶ cassiterite, grey. 2. Weak to moderately strong (in oil): creamy brown to greyish brown. 3. Moderately strong (in oil): dark brown to greyish brown. 4. Not indicated.	470 nm 7.9-8.2 546 7.3-7.5 589 7.3-7.4 650 7.2-7.4 [1398].

Hardness H: polishing hardness VHN: micro-indentation hardness in kg/mm^2	Miscellaneous
H: very high, varying with comp. VHN: spinel: 861-1650 [300, 810, 1715]. gahnite: 1910-2420 [574]. hercynite: 1402-1561 [300]	For all members: usually occur as idiomorphic crystals; also massive, coarse-granular to compact and as irregular or rounded embedded grains. Twinning lamellae $\parallel$ (111) may be visible. Ref.: 636, 1100, 1237. Spinel: occurs as exsolution bodies and as inclusions in magnetite. May contain exsolution lamellae of magnetite. Other ass.: galena, sphalerite, chalcopyrite, pyrite, graphite, rutile, hematite. Hercynite: may contain exsolution lamellae of magnetite. Other ass.: cassiterite, högbomite. Ref.: 221. Galaxite: colour is dark brownish grey, reddish brown int. refl. Has been observed as groundmass containing exsolution lamellae of hausmannite in a vredenburgite-type intergrowth. Ref.: 1657. Gahnite: green int. refl. Occurs as exsolution lamellae and bodies in franklinite. May contain inclusions of pyrite. Replaced by sphalerite. Other ass.: cassiterite, pyrrhotite, arsenopyrite, cobaltite, chalcopyrite, rutile, ilmenite. Ref.: 407, 451, 574, 1619.
H: no data. VHN: 1206-1561 [503, 1398, 1715].	Forms hexagonal platelets, or aggregates of these. Occurs orientatedly overgrown on gahnite. Zoning may be visible. Forms intergrowths with cassiterite and columbite. Ref.: 70, 503, 660, 1141, 1398.

Description of minerals from Table III

Name Formula Crystal System	1: Colour 2: Bireflectance 3: Anisotropy 4: Internal reflections	Reflectance in % (Wavelength in nanometres)
KARELIANITE V_2O_3 Hex.	1. Brownish olive-grey. 2. Weak. 3. Strong, reddish brown to grey. 4. Not present.	About 18 %.
ESKOLAITE Cr_2O_3 Hex.	1. Air: grey; oil: more bluish grey. 2. Air: not observable; oil: distinct at grain boundaries and twin lamellae. 3. Strong; air: grey-blue and greenish grey; oil: grey to brownish grey. 4. Emerald-green.	R_o R'_e 470 nm 22.0 19.8 546 20.6 18.6 589 19.8 18.2 650 19.5 17.8 Dispersion curves: [300].

Hardness H: polishing hardness VHN: micro-indentation hardness in kg/mm^2	Miscellaneous
H: very high. VHN$_{50}$: 1790 [846].	Forms prismatic crystals. Contains inclusions of pyrrhotite, chalcopyrite, pyrite. Replaced by montroseite. Replaces pyrite. Other ass.: graphite, cobaltpentlandite, nolanite, and possibly coulsonite. Ref.: 479, 846.
H:>chromite. VHN: 2077-3200 [194, 300, 754, 1715].	Usually occurs as euhedral, tabular or prismatic crystals. Weak zoning may be visible on the darker outer parts of the crystals. Parag.: pyrrhotite, pentlandite, chalcopyrite, pyrite, chromite. Has been observed as a major component of "merumite", a mixture of eskolaite, bracewellite (CrOOH), grimaldiite (CrOOH), guyanaite (CrOOH) and mcconnellite (CrOOCu). "Merumite" is only known as placer material. Ref.: 411, 754, 913, 914, 1237.

SELENIDES

SELENIDES, isotropic or weakly anisotropic

Name Formula Crystal System	1: Colour 2: Bireflectance 3: Anisotropy 4: Internal reflections	Reflectance in % (Wavelength in nanometres)
ACHAVALITE FeSe Hex. (?)	1. Dark grey (?). 2. Not indicated. 3. Isotropic (?). 4. Not indicated.	No data.
CADMOSELITE β-CdSe Hex.	1. Grey, in oil slightly brownish. $\longrightarrow$ sphalerite, very similar, somewhat lighter. 2. Not observed. 3. Weak; straight extinction relative to the cleavage. 4. Brown, noticeable at medium magnification; sharply intensified (to light brown) in oil.	About 20 %.
BERZELIANITE $Cu_{2-x}Se$ Cubic (at normal temp.)	1. Air: bluish grey, rapidly tarnishing to dark blue. Oil: much darker, distinctly blue tint. $\longrightarrow$ chalcocite, air and oil, more distinct blue. $\longrightarrow$ digenite, air and oil, more distinct blue. $\longrightarrow$ clausthalite, much darker. 2. + 3. Isotropic. Sometimes slightly anisotropic. 4. Not present.	480 nm 31.4 540 28.5 580 25.6 650 23.5 Dispersion curve: [1066].
STILLEITE ZnSe Cubic	1. $\longrightarrow$ tetrahedrite, very similar, but without olive-brown and bluish green tints. 2. + 3. Isotropic. Sometimes slightly anisotropic, esp. in oil, due to tectonic deformation: greyish blue to greyish pink. 4. Air: rare. Oil: more frequent, deep-grey.	About 30 %.
TIEMANNITE HgSe Cubic	1. Air: white-grey with a slight brown tint. Oil: much darker, distinctly brown tint. $\longrightarrow$ galena, brownish, darker. $\longrightarrow$ clausthalite, much darker. $\longrightarrow$ chalcocite, much darker, no bluish tint. 2. + 3. Isotropic. Sometimes very weakly anisotropic only visible on twin lamellae and on grain boundaries, dark grey to grey-black. 4. Not present.	About 30 %.

217

Hardness H: polishing hardness VHN: micro-indentation hardness in kg/mm^2	Miscellaneous
H:~galena. VHN: no data.	Occurs intergrown with clausthalite. Ass.: other selenides. Ref.: 1061, 1397 Note: Data partly contradictory and highly insufficient.
H: no data. VHN$_{20-50}$: 203-222 [810].	Oxidation causes peripheric replacement of cadmoselite by powdery native selenium. Parag.: ferroselite, clausthalite, selenium, sphalerite, pyrite, greenockite. Ref.: 201, 1397, 1615.
H:~clausthalite and umangite VHN$_{15}$: 22-28 [194, 1066]. VHN$_{10-20}$: 79-99 [810].	May contain orientated inclusions of klockmannite. Parag.: other Cu-selenides, aguilarite, gold, pyrite, marcasite, stromeyerite, polybasite, pearceite, uraninite, Se-linnaeite. Ref.: 62, 99, 330, 1066, 1237, 1397, 1552, 1603, 1629.
H:>tetrahedrite. VHN: no data.	Occurs as xenomorphic inclusions in linnaeite and in Se-vaesite. Twin lamellae may be observed. Parag.: linnaeite, pyrite, clausthalite, Se-vaesite, molybdenite, ferroselite, cadmoselite. Ref.: 726, 1234, 1237, 1615.
H:<clausthalite,≪galena. VHN: 26-29 [1715]. VHN$_{10}$: 32-39 [810].	Occurs as allotriomorphic grains, sometimes enclosed by clausthalite. Parag.: clausthalite, umangite, berzelianite, penroseite, naumannite, other selenides, hematite. Ref.: 330, 726, 1237, 1392, 1397, 1552, 1615.

SELENIDES, isotropic or weakly anisotropic

Name Formula Crystal System	1: Colour 2: Bireflectance 3: Anisotropy 4: Internal reflections	Reflectance in % (Wavelength in nanometres)
AGUILARITE Ag_4SeS Ortho. $\xrightleftharpoons{133°C}$ Cubic	1. Light grey with a very faint greenish tinge. $\longrightarrow$ argentite, slightly darker. $\longrightarrow$ galena, olive-green. 2. Not observable or very weak. 3. Very weak: shades of grey. Sometimes isotropic. 4. Not present.	470 nm 35.5 550 35 590 33.5 650 32 Values estimated from the dispersion curve: [103a]
CROOKESITE $(Cu,Tl,Ag)_2Se$ Mono., pseudo-tetr.	1. Grey-white to pinkish cream and brown; colour is different for each locality, depending on the chem. comp. Very similar to many tetrahedrites. $\longrightarrow$ argentite, darker. $\longrightarrow$ tetrahedrite, lighter. $\longrightarrow$ eucairite, sometimes very similar. 2. Air: weak. Oil: distinct. 3. Weak, but discernible in oil: different shades of grey for each section. Extinction $/\!/$ cleavage. 4. Not present.	About 35 %.
PENROSEITE $(Ni,Cu,Co)Se_2$ Cubic	1. Air: white with a brownish tint; younger outer zones may be lighter. Oil: much darker, olive-green. $\longrightarrow$ clausthalite, distinctly brown tint, darker. $\longrightarrow$ naumannite, much darker; naumannite is tin-white. 2. + 3. Isotropic. 4. Not present.	About 35 %.
TROGTALITE $CoSe_2$ Cubic	1. Pinkish violet. $\longrightarrow$ breithauptite, very similar. 2. + 3. Isotropic. 4. Not present.	No data.

Hardness H: polishing hardness VHN: micro-indentation hardness in kg/mm^2	Miscellaneous
H: very low. VHN_{5-10}: 25-35 [103a].	Accompanied and replaced by argentite, electrum, stephanite, pearceite, Cu- and Ag-selenides. Replaces and veins galena; may enclose electrum. Ref.: 103a, 272, 330, 1343, 1397, 1603.
H: no data. VHN_{20-50}: 101-141 [810].	Occurs as aggregates of rounded grains. Cleavage has been observed. Forms myrmekitic intergrowths with berzelianite. Intergrown with umangite and klockmannite. Other ass.: clausthalite, Se-linnaeite. Ref.: 330, 1237, 1397, 1552.
H: >chalcopyrite. VHN_{20-50}: 407-550 [810].	Cleavage $/\!/$ (100) may be visible; sometimes as well as for galena. Zonal texture very common, the outer zones often being replaced; these are more pinkish and may show a slightly higher hardness than the inner parts of the grain. Replaced by clausthalite and selenium. Intergrown with naumannite. Parag.: clausthalite, naumannite, tiemannite, other selenides, Fe-hydroxides. Ref.: 69, 139, 330, 513, 611, 613, 1209, 1397.
H: very high. VHN: no data.	Occurs as idiomorphic, but strongly corroded grains in clausthalite. Often radially enveloped by hastite and bornhardtite. May occur as cocarde-textures in very fine intergrowths with clausthalite and gold. Other ass.: other Co-selenides, hematite. Ref.: 1237, 1246, 1552.

SELENIDES, isotropic or weakly anisotropic

Name Formula Crystal System	1: Colour 2: Bireflectance 3: Anisotropy 4: Internal reflections	Reflectance in % (Wavelength in nanometres)
TYRRELLITE* $(Co, Ni, Cu)_3Se_4$ Cubic	1. Light brassy bronze. 2. + 3. Isotropic. 4. Not present.	Depending on chem. comp.: about 35 to 45 %.
TRÜSTEDTITE Ni_3Se_4 Cubic	1. Yellow. 2. + 3. Isotropic. 4. Not present.	No data, but > sederholmite and penroseite.
BORNHARDTITE Co_3Se_4 Cubic	1. ⟶ clausthalite, distinctly pinkish. ⟶ hastite, trogtalite, freboldite, whiter. 2. + 3. Isotropic. 4. Not present.	No data.
CLAUSTHALITE PbSe (may contain some S) Cubic	1. Bright white. ⟶ galena, distinctly lighter; galena shows a violet-brown tint. ⟶ altaite, darker. ⟶ tiemannite, much lighter. ⟶ guanajuatite, bright white. 2. + 3. Isotropic. No complete extinction under crossed nicols. 4. Not present.	470 nm 55.4 546 49.5 589 48.1 650 47.1 [194].

* Has also been referred to as "selenidspinell".

221

Hardness H: polishing hardness VHN: micro-indentation hardness in kg/mm^2	Miscellaneous
H: no data. VHN$_?$: 336-469 [Harris, 1969, priv. comm.].	Mostly occurs as rounded grains, with sometimes subhedral development of the cube. Veined, embayed and replaced by umangite. Parag.: klockmannite, berzelianite, clausthalite, pyrite, hematite, chalcopyrite. Ref.: 782, 1276, 1397.
H: ~ penroseite. VHN: no data.	Occurs as euhedral crystals in clausthalite in association with penroseite and sederholmite. Forms a solid solution series with polydymite. Ref.: 1630, 1631. Note: Threadgold (1960) describes an unknown Ni-selenide and suggests it to be a member of the linnaeite-group. It is enclosed in veinlets of pitchblende. Probably this material is identical with trüstedtite because the a_o is about the same for both materials.
H: no data. VHN: no data.	Tendency to form idiomorphic crystals. Occurs intergrown with trogtalite, hastite, clausthalite. Replaces trogtalite; replaced by clausthalite. Other ass.: hematite, tiemannite, naumannite. Ref.: 1237, 1246, 1552.
H: slightly > tiemannite, ≪ galena. VHN: 46-72 [1715]. VHN$_{25}$: 48-74 [1715]. VHN$_{15}$: 43-54 [194]. VHN$_{10}$: 49-63 [810].	Forms a complete solid solution series with galena. Commonly occurs as allotriomorphic grains. Sometimes as colloform aggregates. Cubic cleavage less perfect than galena. Triangular pits usually occur. No indication of zoning or twinning. May be enclosed in bornite, chalcocite and digenite. Forms intergrowths with chalcocite. Replaces hematite, allopalladium, gold, Co-selenides, tiemannite, naumannite. Replaced by tiemannite. May occur in the core of zoned galena crystals. Other ass.: berzelianite, klockmannite, umangite, eucairite, paraguanajuatite, coloradoite, chalcopyrite, sphalerite, tetrahedrite, cobaltite, pitchblende, Bi-sulphosalts. Ref.: 3, 62, 277, 278, 599, 1205, 1237, 1333, 1377, 1392, 1397, 1552.

SELENIDES, anisotropy distinct to strong

Name Formula Crystal System	1: Colour 2: Bireflectance 3: Anisotropy 4: Internal reflections	Reflectance in % (Wavelength in nanometres)
UMANGITE Cu_3Se_2 Tetr. *	1. Brownish violet, similar to slightly tarnished bornite. 2. Very strong: O bright violet-red, E greenish blue-grey. 3. Very strong and characteristic. Colours in 45° position: in air orange-red to orange-yellow, in oil yellow to dark orange. Extinction parallel to the cleavage directions. 4. Not present.	$\quad R'_e \quad R_o$ 470 nm 18.6 15.8 546 16.0 12.0 589 16.0 12.1 650 14.6 20.1 [194]. The dispersion curves cross at about 620 nm.
KLOCKMANNITE CuSe Hex.	1. Greenish grey with a blue tint (strongly depending on orientation). 2. Very strong and characteristic: O greyish brown with a violet tint, E much lighter, bluish grey. Also described as varying from bluish green to white: [329]. 3. Very strong: creamy white to orange; (brownish blue to fiery orange: [329]. 4. Not present.	$\quad R_o \quad R'_e$ 470 nm 15.5 37.4 546 11.9 35.6 589 9.8 33.8 650 9.3 32.0 [194].
ESKEBORNITE $CuFeSe_2$ Yields a cubic powder pattern, resembling that of sulvanite, but otherwise behaves hexagonal.	1. Yellowish, sometimes with a brownish tint. 2. Distinct, yellowish white to brassy yellow. 3. Strong, but without pronounced polarization colours: greyish white with yellowish or greenish tints. 4. Not present.	$\quad R'_e \quad R_o$ 470 nm 20.4 30.6 546 24.4 34.6 589 26.7 36.2 650 29.0 37.9 [194].

* Earley (1950) determined the structure of umangite as orthorhombic; Earley's original crystal fragment, however, was later found to be twinned (Berry and Thompson, 1962).

Hardness H: polishing hardness VHN: micro-indentation hardness in kg/mm^2	Miscellaneous
H: ~klockmannite. VHN: 79-90 [1715]. VHN$_{25}$: 79-99 [1715]. VHN$_{10-20}$: 77-112 [194, 810].	Forms massive or fine-grained aggregates of needle-like and granular crystals. Two cleavage directions, poorly developed. Lamellar twinning not uncommon. May contain lamellae of klockmannite ∥ basal plane. Occurs as inclusions in coloradoite and weissite. Veins and replaces hematite, pitchblende, tyrrellite, eskebornite, berzelianite, clausthalite. Alters to klockmannite. Other ass.: selenides, altaite, chalcocite, antimony, pyrite, chalcopyrite. Diagn.: colour, bireflectance and anisotropy are very characteristic. Cannot be mistaken for any other mineral. Covellite shows blue colours in air and oil. Ref.: 62, 330, 1237, 1275, 1392, 1397, 1552, 1603, 1629.
H: ≦umangite, < eucairite. VHN$_{10-20}$: 57-86 [810].	A basal cleavage is commonly visible. Twinning not observed. Klockmannite lamellae may occur ∥ basal planes of umangite. Also as lamellae in chalcopyrite and as orientated intergrowths ∥ (111) of berzelianite. Replaces umangite and eucairite. Parag.: clausthalite and other selenides, pitchblende, hematite, pyrite. Ref.: 4, 117, 329, 1237, 1392, 1397, 1552, 1629.
H: slightly > clausthalite, naumannite, tiemannite, umangite, klockmannite, ≪ chalcopyrite. VHN$_{20-50}$: 141-165 [810]. VHN$_{15}$: 144-202 [194].	Good basal cleavage; sections ∥ (0001) may show triangular pits. Wedge-shaped twinning lamellae may be present due to deformations. Replaced by chalcopyrite, naumannite, clausthalite, tiemannite, umangite, eucairite. Rarely forms myrmekitic intergrowths with chalcopyrite and naumannite. May contain inclusions of ferroselite (?). Tischendorf (1960) distinguishes between more Cu-rich and more Fe-rich eskebornite. The Fe-rich eskebornite is somewhat lighter than the Cu-rich variety, and replaces the latter. Other ass.: selenides, hematite, pyrite. Ref.: 1237, 1248, 1397, 1547, 1552, 1553.

SELENIDES, anisotropy distinct to strong

Name Formula Crystal System	1: Colour 2: Bireflectance 3: Anisotropy 4: Internal reflections	Reflectance in % (Wavelength in nanometres)
EUCAIRITE α-$Cu_2Se \cdot Ag_2Se$ Ortho., pseudo-tetr.	1. Yellowish. 2. Air: very weak, yellow-white to white. Oil: distinct, creamy yellow with a pinkish brown tint to white with a greenish yellow tint. 3. Strong, olive-brown to steel-blue with a purplish tint. 4. Not present.	About 30 %.
WILKMANITE Ni_3Se_4 Mono.	1. Pale greyish yellow. 2. Distinct, pale yellow to greyish yellow. 3. Rather strong, pink to yellowish green. 4. Not present.	No data.

SELENIUM
Se
Hex. and Mono.
	1. Air: white. Oil: much darker, with a brownish grey tint, much like tenorite.
	2. Very strong.
	   Air: O creamy white, E darker, brownish.
	   Oil: O greyish white with blue tint, E dull brown.
	3. Very strong: green to greenish grey, rather vivid colour effects.
	4. Not present.
		About 25 to 35 %.

NAUMANNITE
β-$Ag_2Se \overset{133°C}{\rightleftharpoons} \alpha$-$Ag_2Se$
Ortho. (?) $\rightleftharpoons$ Cubic
	1. Air: greyish. Oil: much darker than in air.
	   $\longrightarrow$ clausthalite, greenish.
	   $\longrightarrow$ tiemannite, slightly lighter.
	   $\longrightarrow$ aguilarite, very similar.
	2. Air: not discernible. Oil: weak, but distinct: brownish grey to a darker greenish grey-brown.
	3. Distinct, light grey to dark grey.
	4. Not present.
		About 35 %.

Hardness H: polishing hardness VHN: micro-indentation hardness in kg/mm^2	Miscellaneous
H: slightly>klockmannite, ~clausthalite. VHN: 54-94 [1715]. VHN$_{25}$: 79-90 [1715]. VHN$_5$: 23-42 [810].	Usually granular. No evidence of cleavage, twinning or zoning. Forms myrmekitic intergrowths with klockmannite and umangite. Occurs as needles in umangite. With clausthalite as inclusions in chalcocite. Alters to klockmannite, similar to the alteration of stromeyerite to covellite. Other ass.: berzelianite, crookesite, eskebornite, tiemannite, weissite, coloradoite, hematite, occasionally Co-selenides. Ref.: 62, 278, 330, 1237, 1392, 1397, 1552.
H: no data. VHN: no data.	Alteration product of sederholmite; alters to selenium and ferroselite. Veined by Se-vaesite and Se-cattierite. Brecciates penroseite forming micro-veinlets and flame-like segregations in it. Ref.: 1630, 1631.
H:<clausthalite, galena. VHN: no data.	Forms acicular crystals or aggregates. Occurs as a decomposition product of selenides, forming bundles of crystals, single prismatic crystals, and gel-textures. A lamellar structure occurs. Often included by clausthalite. Replaces ferroselite, pitchblende, clausthalite. Other ass.: sulphur, tellurium, tiemannite, other selenides, occasionally silver, copper, argentite, pyrite, Fe-oxides, montroseite. Ref.: 3, 1209, 1237, 1248, 1377, 1397, 1487, 1521, 1603.
H:<clausthalite. VHN: 31-37 [1715]. VHN$_{10-15}$: 27-56 [Lebedeva, 1968, priv. comm.].	Usually shows traces of a cubic cleavage, probably due to parting associated with the α-β inversion. Mimetic twinning may be very distinct. May form myrmekitic intergrowths with clausthalite. Replaced by clausthalite. Parag.: clausthalite, tiemannite, umangite, other selenides, gold, electrum, pyrite, chalcopyrite, freibergite. Diagn.: very difficult to distinguish from aguilarite under the microscope, but the powder patterns are different. Ref.: 330, 1237, 1392, 1397, 1533, 1552.

Name Formula Crystal System	1: Colour 2: Bireflectance 3: Anisotropy 4: Internal reflections	Reflectance in % (Wavelength in nanometres)
MÄKINENITE γ-NiSe Hex.	1. Air: orange-yellow. Oil: pure yellow. 2. Strong, pure yellow to greenish yellow. 3. Oil: extremely strong, pale green to pale orange-yellow. Air: glowing cinder-red to blue-green or green. 4. Not present.	No data.
SEDERHOLMITE β-NiSe (two types: stoechiometric and Ni-deficient) Hex.	1. Stoech. type: orange-yellow. Ni-deficient: yellow. 2. Stoech. type: distinct, different hues of yellow. Ni-deficient type: weak, yellow to greyish yellow. 3. Stoech. type: strong, pale green to pink. Ni-deficient type: distinct, greenish to pink. 4. Not present.	No data, but lower than for trüstedtite.
KULLERUDITE $NiSe_2$ Ortho.	1. $\longrightarrow$ penroseite, similar, slightly paler. 2. Distinct, grey to pale grey. 3. Very strong, yellowish grey to grey or almost black. 4. Not present.	No data.
HASTITE $CoSe_2$ (may contain some Fe) Ortho.	1. Brownish red to reddish violet in air; in oil more vivid. Marginal zones of the crystals are more yellow and somewhat harder, possibly due to the Fe-content. 2. Air: very distinct. Oil: strong, brown-red to red-violet. 3. Oil: strong, with vivid colours, often in meat-red tints. 4. Not present.	No data.
FREBOLDITE CoSe Hex.	1. $\longrightarrow$ breithauptite, very similar. 2. Very weak, only visible on grain boundaries. 3. Strong, with vivid colours, but less than for breithauptite. 4. Not present.	No data.

Hardness H: polishing hardness VHN: micro-indentation hardness in kg/mm^2	Miscellaneous
H: ~clausthalite. VHN: no data.	Occurs as inclusions in clausthalite and Se-melonite. Ref.: 1630, 1631.
H: no data. VHN: no data.	Forms intimate intergrowths with wilkmanite. Occurs in clusters of penroseite. Brecciates penroseite forming micro-veinlets and flame-like segregations in it. Veined by Se-cattierite. Ref.: 1630, 1631.
H: rather soft. VHN: no data.	Occurs as an alteration product of wilkmanite and sederholmite. Contains relics of wilkmanite. Veined by Se-cattierite and Se-vaeseite. Ref.: 1630, 1631.
H: no data. VHN: no data.	Forms idiomorphic crystals, arranged in radial aggregates around a core of trogtalite or gold, the whole included in xenomorphic clausthalite. Lamellar twinning abundant. Other ass.: bornhardtite, freboldite, hematite. Ref.: 1237, 1246, 1552.
H: <trogtalite, bornhardtite, hastite. VHN: no data.	Parag.: other Co-selenides, clausthalite, hematite. Ref.: 1237, 1246, 1552.

SELENIDES, anisotropy distinct to strong

Name Formula Crystal System	1: Colour 2: Bireflectance 3: Anisotropy 4: Internal reflections	Reflectance in % (Wavelength in nanometres)
PLATYNITE $Pb_4Bi_7Se_7S_4$ Hex.	1. White. 2. Not indicated. 3. Strong, with a faint colour effect in greyish and brownish tints. A lamellar structure is visible. 4. Not present.	About 40 %.
PARAGUANAJUATITE $Bi_2(Se,S)_3$ with S<2.3 wt. % Hex.	1. ⟶ guanajuatite, air: darker and more greyish. oil: pinkish; guanajuatite is yellowish in contrast. 2. Air: distinct. Oil: strong: $\parallel$ (0001) somewhat lighter, $\perp$ (0001) darker and slightly bluish grey. 3. Very distinct, but practically no colour effects; not as strong as for guanajuatite. 4. Not present.	About 45 %.
FERROSELITE $FeSe_2$ (may contain up to 25 % $CoSe_2$) Ortho.	1. Pinkish cream, in between maucherite and niccolite. ⟶ pyrite, between grey-pink and yellow-pink. ⟶ rammelsbergite, no difference. ⟶ marcasite, pale pinkish. 2. Air: weak. Oil: more distinct, creamy to white with pinkish tints. 3. Air: distinct. Oil: strong, greenish grey to lilac. Reddish grey to brown colours have been reported [1377]. Straight extinction. 4. Not present.	546 nm 47-50 [1324].
LAITAKARITE Bi_4Se_2S Hex.	1. White. ⟶ bismuth, greyish. 2. Weak. 3. Moderate: dark grey to slightly brownish (nicols compl. crossed) or clearly brown (nicols not compl. crossed). 4. Not present.	About 50 %.

Hardness H: polishing hardness VHN: micro-indentation hardness in kg/mm^2	Miscellaneous
H: no data. VHN: no data.	Forms thin laminae or leaves. Cleavage $/\!/$ (0001) perfect. Along the cleavage and the periphery, platynite is replaced by a supergene mineral, probably molybdomenite. Parag.: chalcopyrite, arsenopyrite, cassiterite, semseyite, franckeite, guanajuatite. Ref.: 1397, 1475. Note: very doubtful mineral species.
H: $<$ guanajuatite. VHN_{15}: 30-100 [194]. VHN_{10}^{15}: 44-160 [810].	Paraguanajuatite is not a paramorph after guanajuatite. Occurs as lamellar bodies, granular or curved. Forms pseudomorphs after guanajuatite. Cleavage $/\!/$ (0001) perfect. Closely intergrown with guanajuatite, clausthalite, and selenium. Parag.: same as for guanajuatite. Ref.: 1237, 1377, 1615.
H: $\ll$ pyrite. VHN: 700-861 [200, 810]. VHN_{20-25}: 769-933 [733, 1324].	Occurs as idiomorphic prismatic to needle-like crystals or aggregates. Perfect cleavage $/\!/$ elongation. Twinning $/\!/$ (101). Occurs as inclusions in eskebornite. Enclosed by goethite. Other ass.: pyrite, selenium, chalcopyrite, clausthalite, cadmoselite, ilmenite, sphalerite, marcasite, galena. Diff.: may easily be mistaken for marcasite or loellingite. Can only be distinguished from rammelsbergite by chemical methods: the X-ray powder patterns are identical. Ref.: 200, 276, 278, 529, 733, 772, 1237, 1324, 1377, 1553.
H: $>$ bismuth, $<$ galena. VHN_{50}: 36-50 [194, 1623].	Forms plates and sheets with foliated structure, owing to the perfect cleavage $/\!/$ (0001). In places graphically intergrown with fine-grained chalcopyrite. Bismuth occurs as a graphic intergrowth in laitakarite. May contain thin lamellae of silver. Other ass.: sphalerite, molybdenite, pyrite, galena, arsenopyrite, cassiterite, stannite, wolframite, cosalite. Ref.: 118, 508, 1397, 1623.

SELENIDES, anisotropy distinct to strong

Name / Formula / Crystal System	1: Colour / 2: Bireflectance / 3: Anisotropy / 4: Internal reflections	Reflectance in % (Wavelength in nanometres)
IKUNOLITE $Bi_4(S,Se)_3$ Hex.	1. Faintly cream tint. 2. Not indicated. 3. Moderate to strong in sections normal to the basal cleavage: pale grey to grey. 4. Not present.	About 50 %.
KITKAITE NiTeSe Hex.	1. Pale yellow, sometimes with a reddish hue, approaching the colour of melonite. ⟶ Se-melonite, paler. 2. Not observable. 3. Distinct, pale grey to pink or purple (erroneously reported as pleochroism by Häkli et al., 1965). 4. Not present.	R_o R'_e 470 nm 54.7 52.0 546 58.3 55.8 589 61.0 58.1 650 62.8 60.0 [194].
GUANAJUATITE $Bi_2(Se,S)_3$ (max. Se-content is 26 wt. %) Ortho.	1. Creamy white. ⟶ paraguanajuatite, lighter, esp. in air. ⟶ bornite, the pinkish tint becomes indistinct: then very similar to tetradymite or melonite. 2. Air: distinct. Oil: strong. ∥ c white, somewhat more yellowish than clausthalite, ∥ b bluish grey-white, ∥ a pinkish white 3. Strong, more intense than for tetradymite but weaker than for bismuthinite; weak colour effects, sharp extinction. 4. Not present.	About 55 %.

Hardness H: polishing hardness VHN: micro-indentation hardness in kg/mm^2	Miscellaneous
H: no data. VHN: no data.	Resembles joseite-A in structure and optical properties. Cleavage $/\!/$ (0001) perfect. Parag.: ferberite, bismuth, bismuthinite, cassiterite. Ref.: 702. Note: Plumbian variety: $(Bi,Pb)_4(S,Se,Te)_3$. Shows a greyish white colour, intermediate between bismuthinite and joseite-A. Weak anisotropy. Contains along the cleavage inclusions of bismuthinite, joseite-A, gold and bismuth. Occurs with these minerals along the cleavage planes of molybdenite. Forms intergrowths with bismuthinite. Ref.: 865, 866.
H: no data. VHN_{15}: 109-119 [194]. VHN_2: 110 [564].	Forms a solid solution series with melonite. Cleavage $/\!/$ (001) often visible. Twinning or zoning not observed. Often enveloped by a narrow crust of clausthalite, sometimes together with Se-melonite. Other ass.: Ni-selenides, clausthalite, hematite, Se-polydymite, Se-linnaeite, Se-melonite, penroseite, Ni-Co-Se-bearing pyrite. Ref.: 564.
H: different with orientation; >paraguanajuatite, <melonite, galena. VHN_{15}: 102-210 [194]. VHN_{10}: 42-98 [810]. VHN : 75-150 [386].	Forms radiated or granular aggregates. Also occurs as idiomorphic crystals. Distinct cleavage in two directions. Some sections reveal a peculiar latticed internal structure. May be intergrown with clausthalite. Replaces tetradymite. Pseudomorphosed by paraguanajuatite. Other ass.: bismuthinite, bismuth, pyrite, cosalite, gold, other selenides, bornite, arsenopyrite, semseyite, tetrahedrite, cobaltite, tellurides. Ref.: 386, 1237, 1333, 1377, 1392, 1397.

TELLURIDES

TELLURIDES, isotropic or weakly anisotropic

Name Formula Crystal System	1: Colour 2: Bireflectance 3: Anisotropy 4: Internal reflections	Reflectance in % (Wavelength in nanometres)
WEISSITE $Cu_{2-x}Te$ Hex.	1. Light grey with a bluish tint. → chalcocite, about the same. → wittichenite, indistinct greyish green tint. 2. Weak or absent, 3. Weak to moderate: pink bluish grey to blue. 4. Not present.	About 30 %.
COLORADOITE HgTe (Hg,Cu)Te for Cu-coloradoite Cubic	1. Grey (with a pinkish or brownish tint when tarnished). → gold, dark violet-grey. → petzite, about the same. → calaverite and sylvanite, darker, greyish white. → galena, dull grey. → altaite, brownish grey. → nagyagite, pinkish brown. → gangue minerals, white. Cu-coloradoite is darker: light brownish grey, normal coloradoite being light pinkish grey in contrast. 2. + 3. Isotropic. 4. Not present.	480 nm 36.6 540 33.9 580 33.9 640 34.8 Dispersion curve: [1615].
PETZITE Ag_3AuTe_2 Cubic; two modifications possibly exist: a high-temp. cubic and a low-temp. ortho. one.	1. Greyish white with a violet or reddish tint. → hessite, whiter, slightly bluish. → altaite, dull grey with a violet tint. → galena, brownish. → coloradite and sylvanite, grey. → krennerite, more distinctly violet or bluish. → calaverite, distinctly greyish. 2. + 3. Isotropic, but a weak anisotr. may be shown, even in air. 4. Not present.	480 nm 42.1 540 37.1 580 34.9 640 33.3 Dispersion curve: [1615].

Hardness H: polishing hardness VHN: micro-indentation hardness in kg/mm^2	Miscellaneous
H: >Bi-tellurides, <rickardite. VHN: no data.	Tabular crystals show a straight and sharp extinction. Observed as lamellae $\!/\!/$ (111) in rickardite. Veins calaverite and krennerite along their cleavage directions. Encloses gold filaments. Parag.: vulcanite, rickardite, krennerite, calaverite, sylvanite, petzite, tellurium, coloradoite, Cu-selenides, Cu-sulphides, pyrite, gold, berthierite. Ref.: 62, 864, 1211, 1237, 1333, 1392, 1397, 1451, 1526, 1615. Note: The mineral described by Watanabe and reported by Ramdohr (1938a) as weissite with strong birefl. and anisotr. probably was vulcanite. Ramdohr (1960) points out that the mineral probably has the comp. CuTe. This view is supported by Cameron and Threadgold (1961).
H: differs strongly with orientation; <calaverite and sylvanite, usually >petzite, some sections <petzite. VHN: 23-28 [1715]. VHN$_{15}$: 26.5-27 [194]. VHN$_5$: 27-35 [810].	Coloradoite: Cleavage not observed. Appears included in, and moulded by granular hessite. Contains inclusions of chalcopyrite and lamellae of sylvanite. Occurs as inclusions in bornite. May replace pyrite. May be separated from krennerite by intergrowths of sylvanite with hessite. Other ass.: practically always gold; calaverite, petzite, altaite, nagyagite, tetrahedrite, galena. Diff.: much resemblance to petzite; the latter, however, is often weakly anisotropic. Ref.: 62, 146, 156, 218, 387, 463, 864, 871, 1237, 1392, 1441, 1451, 1527. Cu-coloradoite: Parag.: coloradoite, calaverite, petzite, sylvanite, gold, tetrahedrite. Ref.: 1194.
H: differs with orientation; <sylvanite, coloradoite, hessite, altaite, calaverite. VHN : 35-49 [Stumpfl, priv. comm.]. VHN$_{15}$: 43-74 [194]. VHN$_{10}$: 46-54 [810].	Cubic cleavage less perfect than for galena. Triangular pits may occur. Replaces krennerite, calaverite, pyrite, chalcopyrite. Replaced by hessite. May be intergrown with hessite, gold, electrum, wehrlite, altaite, calaverite. Forms exsolution-type intergrowths with sylvanite. Lamellae of petzite may occur in nagyagite $\!/\!/$ (010). Helke (1938) distinguishes between an isotropic, greyish violet variety (α-petzite) and an anisotropic modification (β-petzite) showing great resemblance to hessite. Parag.: pyrite, chalcopyrite, molybdenite, tellurobismuthite, frohbergite, sylvanite, krennerite, montbrayite, coloradoite, melonite, other tellurides, tetrahedrite, tennantite, gold, seligmannite, electrum, marcasite, stibioluzonite, sphalerite, galena, chalcocite, bornite, covellite, enargite. Ref.: 146, 207, 217, 218, 248, 454, 463, 476, 603, 803, 864, 1237, 1304, 1333, 1392, 1397, 1441, 1451, 1527, 1615, 1629.

TELLURIDES, isotropic or weakly anisotropic

Name Formula Crystal System	1: Colour 2: Bireflectance 3: Anisotropy 4: Internal reflections	Reflectance in % (Wavelength in nanometres)
NAGYAGITE $Au(Pb,Sb,Fe)_8(S,Te)_{11}$ (?)* Ortho. or Tetr.	1. Grey. ⟶ galena, about the same, appears pale lilac. ⟶ silver, much darker. ⟶ sylvanite, much darker. ⟶ hessite, brownish grey; hessite is darker. ⟶ petzite, about the same. ⟶ krennerite, grey with a creamy tint, darker. 2. Weak (as distinct from molybdenite): grey to brownish grey. 3. Distinct: light grey (bluish) to dark brownish or greenish grey in sections which traverse the cleavage; no definite extinction. In sections ∥ cleavage no appreciable anisotropy. 4. Not present.	480 nm 42.15 540 38.65 580 37.0 640 35.1 Dispersion curve: [1615].
JOSEITE-A and JOSEITE-B A: $Bi_{4+x}Te_{1-x}S_2$ B: $Bi_{4+x}Te_{2-x}S$ (x = 0-0.3, may contain some Se and Pb). Hex.	1. White with a yellow tint (esp. in oil). ⟶ galena, slightly whiter. ⟶ bismuthinite, lighter, creamy. Joseite-B is slightly lighter than joseite-A. 2. Not present (according to Gamyanin, 1968, the bireflectance is distinct). 3. Weak: grey to yellow-grey; distinct in sections ∥ c: air: light pinkish brown to dark brownish grey; oil: light grey to slate-grey with a pinkish tint. Basal sections appear isotropic. 4. Not present.	For joseite-A: 465 nm 47.7-51.7 555 48.8-53.1 593 49.9-53.9 665 50.0-54.4 [469]. For joseite-B: 465 nm 48.6-53.8 555 51.5-58.1 593 53.2-59.5 665 53.4-59.8 [469].

* Different authors at various times used some 20 different formulas for nagyagite. The formula given here is that of Berry (1946).

Hardness H: polishing hardness VHN: micro-indentation hardness in kg/mm^2	Miscellaneous
H: slightly < sylvanite. VHN: 39-110 [1715]. VHN$_{25}$: 48-116 [1715]. VHN$_{10-20}$: 56-129 [194, 810].	Usually occurs as thin-tabular laminated crystals, often bent. Perfect cleavage $/\!/$ (010). Polysynthetic twinning may occur in sections $/\!/$ (010). Contains orientated inclusions of petzite, sylvanite, altaite, bornite, pyrite. May contain inclusions of krennerite, calaverite, coloradoite, gold, galena, bournonite, tetrahedrite, hessite, tennantite. Replaced by krennerite, altaite, sylvanite, petzite, hessite, bournonite, tetrahedrite. Other ass.: seligmannite, chalcopyrite, covellite. Ref.: 114, 146, 387, 603, 864, 1237, 1333, 1340, 1392, 1397, 1451, 1475, 1527, 1615, 1629.
H: slightly < tetradymite, < bismuthinite. VHN$_{10}$(joseite-A): 40-87 [469, 922]. VHN$_{10}$(joseite-B): 29-67 [469]. According to Mintser et al. (1968) the data given by Lebedeva (1963) are erroneous.	Forms platy crystals. Perfect basal cleavage. A "spindle"-texture, due to folding, has been observed. Contains inclusions of ikunolite and bismuthinite. May be included in bismuth. Occurs as clusters of small crystals intergrown with bismuth. Observed as lamellae along the cleavage of ikunolite. Forms augen-like inclusions along the cleavage of molybdenite together with bismuth, ikunolite, bismuthinite and gold. Other ass.: hedleyite and other Bi-tellurides, arsenopyrite, pyrrhotite, galena, chalcopyrite, cosalite, hessite. Diagn.: difficult to distinguish from other Bi-tellurides. Ref.: 118, 318, 349, 440, 469, 470, 532, 537, 739, 806, 807, 836, 922, 1126, 1237, 1475, 1527, 1529, 1647. Note: "Gruenlingite" and "oruetite" are identical to joseite.

TELLURIDES, isotropic or weakly anisotropic

Name Formula Crystal System	1: Colour 2: Bireflectance 3: Anisotropy 4: Internal reflections	Reflectance in % (Wavelength in nanometres)
VOLYNSKITE $AgBi_{1.6}Te_2$ Ortho.	1. Pale purplish. ⟶ tellurobismuthite, darker. ⟶ galena, lighter. ⟶ petzite, lighter. 2. Very weak, only visible in oil (as distinct from tellurobismuthite). 3. Weak, no colour effects. 4. Not present.	486 nm 55.2 559 54.3 594 55.0 662 55.5 [23]. Dispersion curve of reflectance in oil: [130].
WEHRLITE BiTe Hex.	1. White. ⟶ tetradymite, more bluish grey and lighter; tetradymite appears pinkish in contrast. ⟶ tellurobismuthite, very similar. 2. Weak, more distinct in oil. 3. Weak, but distinct; 45° position: chocolate-brown. 4. Not present.	About 55 %.
MONTBRAYITE Au_2Te_3 (contains, however, essential Pb and Bi). Tricl.	1. Creamy white, like krennerite. ⟶ altaite, distinctly less white. ⟶ calaverite, very similar, perhaps faintly more creamy pink. ⟶ melonite, distinctly cream with a slight pinkish hue. 2. Weak. 3. Weak to moderate: light grey, light yellow-brown, blue-grey; varying somewhat with orientation. 4. Not present.	480 nm 55.8 540 63.5 580 66.13 640 67.3 Dispersion curve: [1615].

Hardness H: polishing hardness VHN: micro-indentation hardness in kg/mm^2	Miscellaneous
H:>tellurobismuthite. VHN$_?$: 98-103 [870]. VHN$_{5-10}$: 55-99 [130].	Triangular pits very common. Cleavage perfect in one direction, imperfect in 2 other directions. Forms complex intergrowths with tellurobismuthite in areas where hessite and altaite occur as inclusions in tellurobismuthite. Other ass.: Bi-tellurides, pyrite, tetrahedrite, galena, sphalerite. Ref.: 129, 130, 870.
H:≥tetradymite. VHN$_?$: 81 [869].	Occurs as aggregates of xenomorphic grains. Perfect cleavage in one direction. Under crossed nicols (with strong illumination) a mosaic structure may be shown. May occur as irregularly and finely desseminated grains in loellingite and arsenopyrite, or fills interstices between grains of these minerals. Contains inclusions of gold and delineates segregations of gold. Forms intergrowths with tetradymite. Very similar to hedleyite under the microscope. Parag.: other Bi-tellurides, altaite, petzite, hessite, bismuth, galena, gold, chalcopyrite, bornite, pyrite, sphalerite, bismuthinite, bournonite, arsenopyrite. Ref.: 869, 1237, 1397, 1495, 1527, 1529, 1532, 1647, 1651.
H: distinctly > tellurobismuthite,~melonite. VHN$_?$: 198-228 [1615].	Occurs as coarse, solid mosaics with optically continuous areas, often several mm in width. Cleavage occurs, no evidence of twinning. May contain inclusions of tellurobismuthite, petzite, altaite, melonite. Other ass.: gold, freibergite, pyrite, marcasite, chalcopyrite, chalcocite, covellite, calaverite. Very difficult to distinguish between calaverite, krennerite and montbrayite in polished section. Ref.: 1140, 1237, 1300, 1303, 1397, 1527, 1615.

TELLURIDES, isotropic or weakly anisotropic

Name Formula Crystal System	1: Colour 2: Bireflectance 3: Anisotropy 4: Internal reflections	Reflectance in % (Wavelength in nanometres)
CALAVERITE $AuTe_2$ (Ag may be present to a max. of 4 wt. %; may contain some Cu and Sb) Mono.	1. White with a creamy or yellowish tint. ⟶ pyrite, commonly lighter. ⟶ galena and sylvanite, distinctly lighter. ⟶ krennerite, in the lightest position slightly brighter with a golden yellow tint. ⟶ altaite, slightly pinkish. 2. Weak to distinct (depends on orientation, but more distinct than for krennerite): light yellow-white to light yellow-brown. 3. Weak to distinct: greyish red, green, greenish brown (or light grey to dark grey with brownish tints). Sharp oblique extinction. Sylvanite and krennerite show a stronger anisotropy. 4. Not present.	480 nm 61.0 540 64.4 580 65.8 640 67.3 Dispersion curve: [1615].
ALTAITE PbTe Cubic	1. Bright white. ⟶ galena, lighter and whiter, with a slight yellowish or greenish tint; galena shows a pinkish tint. ⟶ calaverite, light grey. ⟶ sylvanite, light grey, slightly greenish. ⟶ tellurobismuthite, bluish white. ⟶ petzite, white, much lighter. ⟶ gold, light bluish. 2. + 3. Isotropic. 4. Not present.	486 nm 69.7 559 70.5 594 68.8 662 64.4 [23].

Hardness H: polishing hardness VHN: micro-indentation hardness in kg/mm^2	Miscellaneous
H:>pyrargyrite, sylvanite, slightly< galena, < krennerite. VHN: 198-209 [1715]. VHN$_{20-50}$: 199-237 [194, 810, 1715].	Usually occurs as idiomorphic needle-shaped crystals sometimes radially developed, penetrating the other tellurides, or as irregular lath-shaped crystals. Cleavage generally absent. Multiple twinning may occur. Replaces petzite, altaite, gold, chalcopyrite, pyrite. Replaced by gold: this may result in pseudomorphs of gold after calaverite. May contain inclusions of pyrite, coloradoite, altaite, melonite, gold. Occurs as inclusions in chalcopyrite and pyrrhotite. Other ass.: other tellurides, freibergite, sulphides of Fe, Cu, Pb, Zn, tetrahedrite, stibnite, molybdenite, arsenopyrite, bismuthinite, jamesonite. Diff.: nagyagite shows a perfect cleavage and a different colour. Calaverite is virtually indistinguishable from montbrayite under the microscope. Ref.: 146, 210, 213, 218, 387, 463, 464, 602, 803, 864, 871, 1303, 1323, 1327, 1333, 1391, 1397, 1451, 1527, 1570, 1572, 1615.
H: slightly > tellurobismuthite and petzite, <galena. VHN: 34-57 [159, 1715]. VHN$_{10-15}$: 39-60 [194, 810].	Usually occurs as granular aggregates of xenomorphic grains. Cubic cleavage commonly visible, less marked than in galena. Triangular pits not common, but may occur. May be found as tiny idiomorphic disseminations in galena; also as rounded inclusions //(100) of galena. Intergrown with galena, tetradymite, tellurobismuthite, tellurium. Forms orientated intergrowths with galena and nagyagite. Occurs as inclusions in bornite, hessite and coloradoite, or may be found along the grain boundaries of these minerals. May contain inclusions of calaverite, krennerite or melonite. Forms rims together with hessite and galena around pyrite crystals. Other ass.: petzite, montbrayite, aguilarite, tetrahedrite, tennantite, chalcopyrite, sphalerite, chalcocite, covellite, enargite, arsenopyrite, jamesonite, boulangerite, bournonite, molybdenite, gold, silver, bismuth, antimony, pyrrhotite. Ref.: 62, 146, 156, 217, 218, 248, 387, 463, 464, 476, 602, 603, 850, 864, 1140, 1237, 1392, 1397, 1451, 1456, 1527, 1629, 1647, 1710.

TELLURIDES, anisotropy distinct to strong

Name Formula Crystal System	1: Colour 2: Bireflectance 3: Anisotropy 4: Internal reflections	Reflectance in % (Wavelength in nanometres)
RICKARDITE Cu_7Te_5* Tetr.	1. Bright purplish red to bright violet; violet hues predominate. The colours differ from those of all other minerals. ⟶ bornite, similar. ⟶ umangite, brighter. 2. Very strong and characteristic: see at colour. 3. Extraordinarely high: canary-yellow to deep brown-red [1209, 409], fiery orange [1392], white blue-grey, dark blue, fiery orange. Basal sections appear isotropic. 4. Not present.	About 25 %.
STUETZITE $Ag_{5-x}Te_3$ Hex.	1. Light grey. 2. Weak to distinct. 3. Strong: brownish to strong bluish tint. 4. Not present.	546 nm 37.2-38.9 589 36.7-38.3 [1486].
HESSITE Ag_2Te 155°C Mono. ⇌ Cubic**	1. Air: grey-white. Oil: grey with a brown tint. ⟶ galena, light brown. ⟶ sylvanite, darker, shows more scratches. ⟶ argentite, distinctly brown. ⟶ altaite, brown-grey to reddish grey. ⟶ petzite, pinkish. ⟶ krennerite, much darker and olive-green tint. ⟶ tellurium, greyer. 2. Air: barely visible. Oil: distinct. 3. Strong; at normal temp. mono.** with cubic relics; hence the extinction often has a spotted appearance, esp. well visible in larger grains with strong illumination. Colours: dark gold-brown or dark orange to dark blue-grey or dark blue. 4. Not present.	480 nm 39.9-40.7 540 38.7-40.85 580 37.8-41.3 640 37.3-43.0 Dispersion curves: [1615].

* Vlasov (1966) gives the results of recent studies in the Cu-Te-system; the homogeneous phase closest to rickardite, Cu_7Te_5, is stoechiometric and cannot have structural defects.

** According to Rowland and Berry (1951) and to Markham (1960) the low-temp. polymorph of synthetic hessite is ortho., and according to the latter the inversion range is between 105°-145°C depending on the excess of Ag or Te in the system.

Hardness H: polishing hardness VHN: micro-indentation hardness in kg/mm^2	Miscellaneous
H: low, but>weissite, Bi-tellurides. VHN$_5$: 133-167 [Lebedeva, 1968, priv. comm.].	May show a complex pattern of polysynthetical twinning along two sets of planes probably at right angles to each other. Occurs as exsolution lamellae in two sets in vulcanite. Forms coatings on tellurium. Replaces vulcanite, tellurium, Au-tellurides, chalcocite, melonite. Replaced by tellurium and weissite. Forms intergrowths with weissite, other tellurides, pyrite, berthierite, tellurium. Rickardite can hardly be mistaken for any other mineral. Other ass.: magnetite, chalcopyrite, petzite, stibnite. Ref.: 222, 410, 411, 1209, 1392, 1397, 1527, 1534, 1615.
H: <empressite. VHN$_2$: 75-90 [1486].	Occurs as massive compact granular aggregates. Parag.: empressite, hessite, petzite, sylvanite, tellurium, coloradoite, frohbergite, altaite, galena, sphalerite, tetrahedrite, tennantite. Diff.: very similar to hessite, but without inversion twinning. Ref.: 639, 1303, 1486, 1499. Note: Studies on synthetic material revealed that between $Ag_{5-x}Te_3$ (stuetzite) and Ag_2Te (hessite) there exists a "gamma-phase" which contains 61.3 to 61.7 wt. % Ag. It is similar to petzite in optical properties. Ref.: 210, 211.
H: lowest of all tellurides; >argentite, slightly <altaite, ≪galena, gold. VHN: 24-41 [159, 1715]. VHN$_{10-15}$: 28-44 [194, 810].	Occurs as rounded aggregates or as allotriomorphic crystals; also as skeleton-shaped crystals. Cleavage not observed. When deposited above 155°C, inversion twin lamellae in 1, 2, or 3 sets are present. Practically always visible under crossed nicols. Replaces gold, chalcopyrite, pyrite, sphalerite, galena, tennantite. May be replaced by sylvanite and gold. Forms intergrowths with sylvanite and petzite. Occurs as inclusions in galena and petzite, sometimes together with tetradymite; also in pyrite and bornite. May contain inclusions of gold, silver, pyrite, altaite. Forms rims around parkerite. Occurs as interstitial veins in massive maucherite. May be transsected by very fine veins of gold. Other ass.: argentite, electrum, antimony, melonite, coloradoite, joseite, tellurides of Au and Ag, wehrlite, magnetite, tetrahedrite, scheelite, cassiterite, bismuthinite, hematite, ilmenite, copper, chalcocite, tellurium, arsenopyrite, boulangerite. Ref.: 62, 146, 156, 217, 218, 339, 455, 463, 476, 588, 824, 836, 864, 1237, 1292, 1333, 1392, 1397, 1441, 1451, 1456, 1527, 1529, 1603, 1647, 1679.

TELLURIDES, anisotropy distinct to strong

Name Formula Crystal System	1: Colour 2: Bireflectance 3: Anisotropy 4: Internal reflections	Reflectance in % (Wavelength in nanometres)
EMPRESSITE AgTe Ortho.	1. + 2. Strong bireflectance: white to brownish grey. 3. No polarization colours, only distinct change from light to dark. 4. Not present.	546 nm 34.1-49.9 589 34.4-50.1 [1486].
CSIKLOVAITE $2Bi_2S_3 \cdot Bi_2Te_3$ Hex.	1. Slightly bluish grey. ⟶ tetradymite, bluish. 2. Air: not perceptible. Oil: weak, bluish grey to grey. 3. Air: grey to dark grey with a brownish tint. Oil: light to dark bluish grey. 4. Not present.	About 45 %.
FROHBERGITE $FeTe_2$ Ortho.	1. Pinkish lilac. ⟶ melonite, purplish pink; melonite shows yellowish pink tints. 2. Air: not perceptible. Oil: extremely weak; only visible on grain boundaries. 3. Strong: orange-red to inky-blue. 4. Not present.	480 nm 50.9 540 45.3 580 44.6 640 45.75 Dispersion curve: [1615].
VULCANITE CuTe Ortho.	1. + 2. Very strong bireflectance: bright yellow or yellow-white to medium blue. 3. Very strong: brilliant yellow-white, greyish yellow-white, yellow-orange, grey. Extinction // cleavage and // elongation of the laths. 4. Not present.	About 35 to 60 % (due to bireflectance).

Hardness H: polishing hardness VHN: micro-indentation hardness in kg/mm^2	Miscellaneous
H: >stuetzite. VHN$_2$: 108-133 [1486].	Occurs as coarse granular aggregates. No cleavage. Forms irregular intergrowths with stuetzite. Parag.: tellurium, sylvanite, petzite, hessite, rickardite, altaite, stuetzite, galena. Ref.: 639, 1486, 1499, 1667; erroneous information is given in 311, 1535.
H: = tetradymite. VHN: no data.	Occurs as a reaction rim between tetradymite and replaced minerals. Also in patches rimmed by fine-grained chalcopyrite. Parag.: bismuthinite, Bi-Te-minerals, Cu-minerals. Ref.: 532, 739, 1495.
H: >chalcopyrite, <tetrahedrite. VHN$_2$: 250-297 [1615].	Occurs as inclusions in gold, petzite, chalcopyrite, melonite. Cleavage and twinning not observed. Forms intergrowths with gold, petzite, montbrayite. Forms a reaction rim between chalcopyrite and the younger tellurides, but practically always separated from the chalcopyrite by a narrow rim of melonite. Other ass.: tellurobismuthite, altaite, coloradoite, sylvanite, stuetzite, other tellurides, pyrite, marcasite, sphalerite, chalcocite, covellite. Ref.: 1237, 1300, 1303, 1397, 1525, 1527.
H: no data. VHN: no data.	Occurs as elongated irregular laths or as equidimensional grains. Two cleavages respectively ∥ elongation (prominent) and at right angles to the first one. Extremely thin twinning occurs at 45° to the prominent cleavage. Shows two types of intergrowths with rickardite: a) due to exsolution: thin blades of rickardite occur in two sets, orientated at a high angle to one another, and roughly at 45° to the length of the vulcanite laths; b) due to replacement by rickardite ∥ crystallographic directions of vulcanite. Ref.: 222, see also note at weissite.

TELLURIDES, anisotropy distinct to strong

Name Formula Crystal System	1: Colour 2: Bireflectance 3: Anisotropy 4: Internal reflections	Reflectance in % (Wavelength in nanometres)
TELLURIUM Te (may contain some Se, Au, Ag and Fe). Hex.	1. White. ⟶ platinum, silver and gold, slightly grey, darker. ⟶ galena, lighter, creamy tint; galena is greyer. ⟶ hessite, distinctly lighter; hessite is slightly greyer. ⟶ sylvanite, slightly creamier. ⟶ krennerite, somewhat lighter. ⟶ calaverite, somewhat lighter, less yellow. ⟶ arsenic, brighter. 2. Distinct: O white, E darker, in air with a slight brownish tint. 3. Strong: colours in bluish grey and brownish grey tints. Similar to stibnite, but not so bright. 4. Not present.	546 nm 43.4-53.5 [194].
HEDLEYITE $Bi_{14}Te_6$* Hex.	1. White. ⟶ silver and galena, slightly darker. ⟶ joseite-B, lighter. 2. Very weak, yellowish grey to bluish. 3. Distinct, yellowish grey with brown tinges to brown and dark grey. 4. Not present.	546 nm 48.0-51.2 [194].

* According to studies on synthetic material by Godovikov et al. (1966) the exact formula should be Bi_5Te_3. The formula obtained on natural material, $Bi_{14}Te_6$, may be due to very fine-grained contaminations of bismuth.

Hardness H: polishing hardness VHN: micro-indentation hardness in kg/mm^2	Miscellaneous
H: low; <empressite, montbrayite, krennerite, calaverite, >sylvanite, tellurobismuthite, altaite. VHN_{10-20}: 25–87 [194, 810].	Occurs as prismatic or acicular crystals and as dense columnar, sometimes fine-grained masses. Cleavage occasionally observed. Twinning not observed. May form microscopic segregations in galena. Forms graphic intergrowths with galena. Parag.: gold, silver, selenium, tellurides of Au and Ag, rickardite, tetradymite, altaite, pyrite, galena, bismuth, bismuthinite, molybdenite, chalcopyrite, tetrahedrite, stannite, cosalite, scheelite, marcasite. Ref.: 64, 227, 461, 464, 850, 864, 1237, 1397, 1603, 1615, 1656. Note: Selentellurium is Te containing 29.32 % Se. Ref.: 1100, 1397.
H: ~tetradymite. VHN_{15}: 30–48 [194]. $VHN_?$ (on synthetic material): 89 [509].	Can hardly be distinguished from other Bi-tellurides under the microscope. Parag.: bismuth, joseite, gold, hessite, altaite, sulphides as arsenopyrite, pyrrhotite, molybdenite. Ref.: 339, 509, 1237, 1527, 1531, 1651.

TELLURIDES, anisotropy distinct to strong

Name Formula Crystal System	1: Colour 2: Bireflectance 3: Anisotropy 4: Internal reflections	Reflectance in % (Wavelength in nanometres)
TETRADYMITE Bi_2Te_2S Hex.	1. White with a creamy or light yellow tint. ⟶ chalcopyrite, much lighter. ⟶ galena, yellow-white. ⟶ pyrite, less yellow. ⟶ wehrlite, slightly darker. ⟶ tellurobismuthite, slightly darker, faintly greyish green. ⟶ krennerite, greyish white, the creamy yellow tint is less distinct. 2. Weak in greenish grey tints. 3. Distinct: bluish grey or yellow-grey (pinkish light grey to brownish dark grey: [739]. 4. Not present.	546 nm 49.8-51.7 [194].
KITKAITE NiTeSe Hex.	1. Pale yellow, with sometimes a reddish hue, approaching the colour of normal melonite. ⟶ Se-melonite, paler. 2. Not observable. 3. Distinct, pale grey to pink or purple; erroneously reported by Häkli et al. (1965) as pleochroism. 4. Not present.	R_o R_e 470 nm 54.7 52.0 546 58.3 55.8 589 61.0 58.1 650 62.3 60.0 [194].
SYLVANITE $AuAgTe_4$ Mono.	1. Creamy white. ⟶ tellurium, slightly creamier. ⟶ hessite, light creamy. ⟶ galena and nagyagite, lighter, creamy white. ⟶ calaverite, darker. ⟶ altaite, slightly pinkish. 2. Distinct, esp. in larger twinned spec.: creamy white to creamy brown. 3. Strong: light bluish grey to dark brown. No definite extinction. 4. Not present.	480 nm 49.1-55.7 540 49.7-58.2 580 49.3-59.4 640 48.5-60.7 Dispersion curves: [1615].

Hardness H: polishing hardness VHN: micro-indentation hardness in kg/mm^2	Miscellaneous
H: <tellurobismuthite, wehrlite. VHN: 28-52 [1715]. VHN$_{15}$: 37-47 [194]. VHN$_{10}$: 25-76 [810].	May show anhedral granular texture. A "spindle"-like texture occurs due to folding. Forms idiomorphic needle-shaped crystals with straight extinction and hexagonal cross-sections. A perfect basal cleavage is always present. A fine lamellar twinning is rarely visible. Contains exsolution lamellae of tellurobismuthite and bismuthinite. Forms intergrowths, some of them orientated, with gold, bismuth, tellurium, altaite, melonite, chalcopyrite, goldfieldite, digenite, bornite. Occurs as inclusions in pyrrhotite, hessite, galena, chalcopyrite. Elongated lamellae of tetradymite may occur along the cleavage in galena. May contain inclusions of pyrrhotite, sphalerite and coloradoite. Other ass.: other tellurides, esp. Bi-tellurides, cosalite, gersdorffite, scheelite, rutile, magnetite, parkerite, galena, pyrite, tetrahedrite, arsenopyrite, jamesonite, cobaltite, stibnite, cassiterite. Ref.: 412, 440, 461, 588, 654, 850, 864, 1047, 1091, 1162, 1237, 1324a, 1333, 1397, 1495, 1529, 1538, 1629, 1641, 1646, 1647, 1649, 1663.
H: no data. VHN$_2$: 110 [564]. VHN$_{15}$: 109-119 [194].	Forms a solid solution series with melonite. Twinning or zoning not observed. Cleavage $/\!/$ (001) often visible. Often enveloped by a narrow crust of clausthalite, sometimes together with Se-melonite. Other ass.: hematite, Se-polydymite, Se-linnaeite, Ni-Co-Se-bearing pyrite, penroseite, sederholmite, trüstedtite, Se-molybdenite. Ref.: 564.
H: >argentite, hessite, >altaite, Bi-tellurides, slightly > nagyagite, <pyrargyrite. VHN: 102-221 [159, 1715]. VHN$_{50}$: 193-250 [1715]. VHN$_{10-20}$: 60-203 [194, 810].	Often forms skeleton-shaped crystals. Cleavage $/\!/$ (010) perfect, $/\!/$ (100) less distinct. A characteristic polysynthetic twinning is always present (as distinct from krennerite and calaverite). Forms intergrowths with hessite and nagyagite. May contain inclusions of gold. Replaces galena and hessite; may be enclosed by argentite, sphalerite, bornite, pyrite, chalcopyrite, galena. Other ass.: petzite, altaite, calaverite, krennerite, coloradoite, weissite, rickardite, melonite, tellurium, stuetzite, seligmannite, tetrahedrite, stibioluzonite, stibnite, orpiment, tellurobismuthite, bismuthinite, arsenopyrite, tennantite, boulangerite. Ref.: 62, 146, 156, 210, 213, 217, 218, 387, 463, 603, 864, 871, 1237, 1391, 1397, 1451, 1456, 1527, 1573, 1603, 1615, 1629, 1679.

TELLURIDES, anisotropy distinct to strong

Name Formula Crystal System	1: Colour 2: Bireflectance 3: Anisotropy 4: Internal reflections	Reflectance in % (Wavelength in nanometres)
KOSTOVITE $AuCuTe_4$ Probably mono.	1. Creamy white. ⟶ tellurium, creamier. ⟶ sylvanite, very similar. 2. Air: distinct, accentuated by twinning. Oil: stronger, light creamy white to light creamy brownish. 3. Strong, clear colour effects in pinkish greyish brown, yellowish grey, brownish grey, and reddish hues. 4. Not present.	482 nm 52.1-55.2 559 54.9-60.1 589 53.0-57.9 668 49.3-55.2 Dispersion curves: [1517].
TELLUROBISMUTHITE Bi_2Te_3 (may contain some Sb or Se). Hex.	1. White with a creamy or pinkish tint. ⟶ altaite, faintly reddish white. ⟶ tetradymite, slightly lighter. 2. Very weak (creamy white to greyish white for the Sb-bearing variety). 3. Distinct: grey or greyish blue to yellowish grey. 4. Not present.	486 nm 61.5 559 63.6 594 64.5 662 64.6 [23].

Hardness H: polishing hardness VHN: micro-indentation hardness in kg/mm^2	Miscellaneous
H: <tellurium. VHN$_2$: 35-43 [1517].	Forms polygonal grains, seldom rounded, and veins in tennantite and chalcopyrite. A characteristic fine lamellar twinning is always present. Parag.: tennantite, chalcopyrite, pyrite, tellurium, gold, tellurobismuthite, nagyagite, sylvanite, altaite, schirmerite, sphalerite, këstertite, colusite. Ref.: 1517.
H: >tetradymite. VHN$_{10-20}$: 32-93 [194, 810].	Forms exsolution intergrowths with tetradymite. May occur as idiomorphic inclusions in chalcopyrite. Perfect cleavage $/\!/$ (0001). Lamellar twinning has been observed. Gold may occur as a very fine film $/\!/$ cleavage of tellurobismuthite. Parag.: other Bi-tellurides, altaite, calaverite, montbrayite, petzite, bismuth, bismuthinite, gold; occasionally sulphides, arsenopyrite, tetrahedrite. Ref.: 440, 541, 1047, 1133, 1140, 1317, 1332, 1381, 1397, 1507, 1527, 1529, 1647, 1649. Note: Aksenov et al. (1968) described an unnamed new bismuth telluride, Bi_2Te_5, crystal system unknown. It occurs as desseminations in pyrite and pyrrhotite. The colour is creamy white, with weak birefl. (only visible in oil) and strong anisotr. (bluish to yellowish). The compound has a mosaic structure. VHN$_{5-10}$: 35-50; Refl. (559 nm): 60.9 %.

Name Formula Crystal System	1: Colour 2: Bireflectance 3: Anisotropy 4: Internal reflections	Reflectance in % (Wavelength in nanometres)
MELONITE $NiTe_2$ (member of the $NiTe$-$NiTe_2$ isomorphous series; forms a solid solution series with merenskyite; may contain some Se). Hex.	1. Light pink to creamy pink. The pinkish tint is very distinct (but not as typical as for niccolite) and becomes more pronounced with increasing Se-content. ⟶ tellurobismuthite, distinctly pinkish. ⟶ frohbergite, yellowish pink. 2. Very weak: O more creamy, E more pinkish. 3. Distinct, greyish mauve to yellowish brown. Straight extinction. Basal sections may show a weak anomalous anisotropy. 4. Not present.	R_o R'_e 470 nm 56.3 54.5 546 60.6 57.0 589 64.0 60.4 650 66.9 62.9 [194].
KRENNERITE $AuAgTe_4$ (may contain some Cu) Ortho.	1. Creamy white. ⟶ calaverite, somewhat less yellow. 2. Weak: yellowish creamy, about same as for calaverite, to yellowish creamy with a violet-grey tint, about same as for petzite, but much lighter. 3. Strong, stronger than for calaverite: light grey, yellow, brown. No definite extinction, as opposed to calaverite. 4. Not present.	480 nm 64.9 540 71.9 580 74.6 640 76.0 Dispersion curve: [1615].

Hardness H: polishing hardness VHN: micro-indentation hardness in kg/mm^2	Miscellaneous
H: differs with locality and composition; usually $\geq$chalcopyrite,$>$gold and krennerite, sometimes $\ll$gold and$<$montbrayite, $\sim$tellurobismuthite, $<$frohbergite. VHN$_{20-50}$: 63-156 [1615]. VHN$_{15}$: 66-119 [194].	Occurs as isolated euhedral or subhedral crystals enclosed in krennerite, altaite or montbrayite, with or without partial rims of petzite; as rounded or distorted hexagonal sections or irregular patches in intergrowths of tellurobismuthite and altaite. Cleavage $/\!/$ elongation. Forms rims between frohbergite and chalcopyrite. As inclusions with idiomorphic outlines in chalcopyrite, bornite and pyrite. Forms orientated intergrowths with digenite, bornite, tellurobismuthite. Other ass.: marcasite, sphalerite, tetrahedrite, coloradoite, rickardite, calaverite, sylvanite, tellurium, gold, chalcocite. Ref.: 62, 156, 564, 864, 1139, 1140, 1209, 1237, 1392, 1397, 1452, 1615, 1638. Note: According to Yushko-Zakharova (1964) the NiTe end-member of the NiTe-NiTe$_2$ series is called "imgreite". This compound occurs as oval grains in hessite associated with sylvanite, calaverite and other tellurides. VHN$_{20}$: 210-220; Refl. (589 nm) 52.4 %. The optical properties are very close to those of melonite. The name has been disapproved by the IMA-CNMMN.
H: slightly $>$petzite and pyrargyrite,$>$sylvanite and calaverite. VHN$_{10}$: 36-88 [810]. VHN$_{25}$: 117-130 [1715]. (Probably these data apply to different minerals)	Idiomorphic thin-tabular crystals are rare. Cleavage, in two directions, less perfect than for sylvanite. Multiple twinning may occur. May contain inclusions of pyrite and magnetite. May be intergrown with tellurium, other tellurides, esp. sylvanite. Forms myrmekitic intergrowths with freibergite and pyrrhotite. Replaced by hessite, petzite, sylvanite, gold, chalcopyrite. Weissite penetrates along the cleavage directions of krennerite. Replaces calaverite, nagyagite, pyrite. Occurs as an alteration product of sylvanite together with hessite. Forms pseudomorphs after nagyagite. Other ass.: bornite, chalcocite, tetrahedrite, coloradoite. Ref.: 62, 146, 210, 213, 387, 463, 464, 603, 864, 871, 1209, 1237, 1333, 1391, 1397, 1451, 1527, 1571, 1574, 1615.

Ag - SULPHOSALTS
AND
Ag–Fe - SULPHIDES

Ag-SULPHOSALTS

Name Formula Crystal System	1: Colour 2: Bireflectance 3: Anisotropy 4: Internal reflections	Reflectance in % (Wavelength in nanometres)
ARGYRODITE- CANFIELDITE $4Ag_2S.GeS_2 \rightleftharpoons 4Ag_2S.SnS_2$ High temp. form: cubic Low temp. form: ortho.	1. Air: greyish white with a pinkish brown tint, which is strongest for argyrodite. Oil: much darker, pinkish brown, about as enargite in darkest position. $\longrightarrow$ pyrite, distinctly violet tint. $\longrightarrow$ sphalerite, pinkish brown. $\longrightarrow$ galena, more distinctly pinkish brown. $\longrightarrow$ pyrargyrite, much darker, different tint. $\longrightarrow$ stephanite, pinkish. $\longrightarrow$ tetrahedrite, darker and greyer. 2. Commonly observable, more distinct in oil than in air: greyish white to yellowish grey (esp. besides galena or pyrargyrite). 3. Some spec. are isotropic, others are weakly anisotropic, with bluish-grey polarization colours. 4. Not present.	470 nm 24.1-24.4 546 21.8-22.4 589 21.5-22.0 650 21.5-22.0 [1398]. Dispersion curve for canfieldite: [231].
STEPHANITE $5Ag_2S.Sb_2S_3$ Ortho.	1. Grey with a distinctly pinkish violet tint. $\longrightarrow$ galena, darker, pinkish grey. $\longrightarrow$ polybasite, lighter pinkish. $\longrightarrow$ pyrargyrite, lighter grey. $\longrightarrow$ argentite, slightly lighter with a violet-pinkish tint. 2. Weak, but distinct: grey, grey with a brownish pink tint, pink; more distinct than for polybasite. 3. In transverse sections (in oil) strong: dark violet to dull greyish green (nicols not completely crossed, strong illumination). Sharp extinction. Basal sections appear isotropic. 4. Not present: as distinct from proustite, polybasite, pyrargyrite.	About 25-30 %.

Hardness H: polishing hardness VHN: micro-indentation hardness in kg/mm^2	Miscellaneous
H:>pyrargyrite,<galena*, tetrahedrite. VHN$_{50}$: 154-172 [1715]. VHN$_{25}$: 133-171 [1398].	Usually occurs as granular aggregates, sometimes idiomorphically. Individual grains often show lamellar twinning. May show coarse-lamellar structure. Zonation occurs. May contain inclusions of pyrite, galena or pyrargyrite. Occurs along the cleavages of galena. Replaces argentite, pyrargyrite, pyrite; may be replaced by argentite, silver, marcasite. Parag.: arsenopyrite, pyrite, chalcopyrite, sphalerite, galena, stannite, argentite, Ag-Sb-minerals, hocartite, silver, bismuth, bismuthinite, tetrahedrite, cassiterite. Diff.: enargite, stibioluzonite, stephanite and hocartite show a much stronger anisotropy; the first two minerals are harder and lighter. Ref.: 7, 14, 16, 91, 231, 339, 395, 648, 805, 1146, 1237, 1248, 1461, 1532, 1615.
H:≫argentite,>polybas- ite and pyrargyrite, <tetrahedrite. VHN: 45-107 [1715]. VHN$_{25}$: 50-124 [1715]. VHN$_{15}$: 31-37 [194]. VHN$_{10}$: 26-47 [810].	Occurs as columnar crystals and as xenomorphic aggregates. A very fine compound twinning, sometimes developed in two systems, is not uncommon. As reaction rims between silver and pyrargyrite. Forms intergrowths with pyrargyrite and polybasite. Replaces pyrite, chalcopyrite, sphalerite, silver, pyrargyrite, galena, Ni-Co-Fe-arsenides. May be replaced by polybasite, silver, dyscrasite, andorite, owyheeite. May contain inclusions of silver, argentite, sternbergite or other replaced minerals. Other ass.: tetrahedrite, freibergite, arsenopyrite, loellingite, stibioluzonite, electrum, miargyrite. Ref.: 14, 223, 557, 866, 997, 1145, 1147, 1237, 1379, 1392, 1531, 1629, 1679, 1687, 1699, 1732.

* Stillwell and Edwards (1943) report H > galena.

Ag-SULPHOSALTS

Name Formula Crystal System	1: Colour 2: Bireflectance 3: Anisotropy 4: Internal reflections	Reflectance in % (Wavelength in nanometres)
XANTHOCONITE- PYROSTILPNITE $3Ag_2S \cdot As_2S_3 \rightleftharpoons 3Ag_2S \cdot Sb_2S_3$ Mono.	1. Bluish grey, but less blue than proustite and pyrargyrite. 2. + 3. Similar to proustite and pyrargyrite. 4. Yellowish or yellowish brown instead of the red ones of proustite and pyrargyrite.	About as for pyrargyrite and proustite.
PROUSTITE $3Ag_2S \cdot As_2S_3$ Hex.	1. Bluish grey. ⟶ pyrargyrite, almost the same, slightly darker, pyrargyrite looks white with a creamy tint. 2. Air: O white with a yellowish tint, E bluish grey, darker. Oil: O greyish blue with a brownish tint, E greyish blue, darker. 3. Strong, in oil masked by int. refl. 4. Always visible, scarlet-red, brighter and more intense than in pyrargyrite and miargyrite.	About 25-28 %.
PYRARGYRITE $3Ag_2S \cdot Sb_2S_3$ Hex.	1. Bluish grey. ⟶ proustite, very similar, slightly lighter. ⟶ galena, greyish blue. ⟶ miargyrite, distinctly bluish. ⟶ bournonite and boulangerite, grey with a blue to violet tint. 2. Distinct to strong, no great colour differences. 3. Strong, masked by int. refl. (in oil); in air pale grey to dark grey. 4. Intense, carmine-red, less pronounced than in proustite.	About 28-31 %.

Hardness H: polishing hardness VHN: micro-indentation hardness in kg/mm^2	Miscellaneous
H: ~proustite and pyrargyrite. VHN: no data.	Very similar to each other and to proustite and pyrargyrite. Only differing from proustite and pyrargyrite in int. refl. The paragenesis may give an uncertain indication whether xanthoconite or pyrostilpnite is present: if arsenic and proustite are present, the mineral is most probably xanthoconite; if pyrargyrite is present, the mineral may be pyrostilpnite. Pyrostilpnite may be distinguished from pyrargyrite by its pronounced cleavage $/\!/$ (010). Pyrostilpnite has been observed in argentopyrite, as aggregates of slender prismatic crystals, elongated along the c-axis. It was veined by argentite. Has also been observed as single, flattened crystals and as radiating clusters of blade- or needle-like crystals, associated with miargyrite and pyrargyrite. Parag.: for xanthoconite the same as for proustite; for pyrostilpnite see pyrargyrite. Ref.: 779, 851, 866, 966, 1024, 1027, 1131, 1237, 1372, 1556, 1687.
H:~pyrargyrite. VHN: 50-156 [159, 1715]*. VHN$_{15}$: 106-127 [194]*.	Occurs as irregular grains, as allotriomorphic aggregates or as idiomorphic needle-shaped crystals. Twinning and zonal texture may occur. Replaces arsenic, bismuth, freibergite, rammelsbergite, sphalerite, argentite, pyrargyrite, silver. May be surrounded by uraninite. Intergrown with arsenic, galena, pyrargyrite, rathite, marcasite. As inclusions in galena. May enclose silver. Other ass.: Ni-Co-Fe-arsenides, realgar, Ag-sulphosalts, chalcopyrite, arsenopyrite, tetrahedrite. Ref.: 85, 91, 360, 475, 632, 701, 960, 997, 1237, 1260, 1310, 1379, 1392, 1441, 1556, 1561, 1603, 1629, 1705.
H: $>$argentite,$>$polybasite,$<$stephanite, galena, and most of the accompanying minerals. VHN: see at proustite.	Occurs as idiomorphic crystals and as aggregates of irregular grains. Often twinned and with zonal texture. May contain small inclusions of argentite, bismuth, arsenopyrite, pyrite, tetrahedrite, chalcopyrite, sternbergite, canfieldite. Forms intergrowths, some of them myrmekitic or orientated, with galena, polybasite, tetrahedrite, chalcopyrite, bournonite, boulangerite, stibnite, argentite, proustite. Occurs as inclusions in galena, pyrite, sphalerite, freibergite, meneghinite. Replaces most of the accompanying minerals, e.g. tetrahedrite, boulangerite, bournonite, arsenopyrite, pyrite, sphalerite, silver, pyrrhotite. May be replaced by polybasite, owyheeite, stephanite, dyscrasite, silver, argentite. Other ass.: Ni-Co-Fe-arsenides, pearceite, miargyrite, argyrodite. Ref.: 14, 28, 31, 36, 322, 342, 346, 348, 360, 762, 851, 997, 1047, 1050, 1146, 1153, 1164, 1212, 1227, 1237, 1328, 1379, 1392, 1435, 1441, 1561, 1629, 1699, 1732.

* For the proustite-pyrargyrite series.

Name Formula Crystal System	1: Colour 2: Bireflectance 3: Anisotropy 4: Internal reflections	Reflectance in % (Wavelength in nanometres)
SAMSONITE $2Ag_2S \cdot MnS \cdot Sb_2S_3$ Mono.	1. Bluish grey, similar to pyrargyrite. 2. Distinct: olive-green-grey to blue-grey (similar to the darkest position of polybasite). 3. Weak, greenish grey to purple. 4. Deep-red, very common; not as abundant as in proustite or pyrargyrite.	About 28 %.
ANDORITE General formula: $Pb(Ag,Cu)Sb_3S_6$ Ortho.	1. White (in air); in oil much darker, yellowish grey. $\longrightarrow$ galena, much darker, with a bluish green tint. $\longrightarrow$ bismuthinite, much darker, more bluish. $\longrightarrow$ jamesonite in lightest position, distinctly darker. 2. Weak (in air); in oil distinct on grain boundaries and twin lamellae. 3. Moderately strong: grey-green to grey-blue, violet-grey or pinkish grey. 4. Red, rarely visible (with strong illumination).	About 30 %.
BILLINGSLEYITE $Ag_7(As,Sb)S_6$ Ortho.	No optical data were published!	About 30 % (?).

Hardness H: polishing hardness VHN: micro-indentation hardness in kg/mm^2	Miscellaneous
H: no data. VHN: no data.	Occurs as single crystals and as radial aggregates. Triangular pits may occur. Twinning not observed. Very similar to pyrargyrite. Parag.: pyrolusite, pyrargyrite, argyropyrite, sternbergite, galena, chalcopyrite, dyscrasite. Ref.: 442, 1237, 1392, 1687.
H: >pyrargyrite, stibnite, slightly <semseyite, ≦jamesonite, distinctly <bournonite, ≪sphalerite, stannite. VHN: 140-193 [1113, 1715]. VHN$_{50}$: 192-206 [1715].	Occurs as idiomorphic prismatic crystals and as aggregates of rounded grains. Lamellar twinning very common, often in different directions; a parquet-like texture may also occur. Replaces stephanite, freieslebenite, diaphorite, freibergite. May contain replacement relics of stibnite, plagionite and zinkenite. A reaction rim of antimony may occur between andorite and inclusions of these minerals. Replaced by zinkenite, pyrargyrite or other Pb-Sb-minerals. Parag.: arsenopyrite, sphalerite, galena, cassiterite, stannite, franckeite, freibergite, Pb-Sb-minerals, stibnite, Pb-Ag-minerals, pyrargyrite, realgar, antimony, lead, silver, enargite, bournonite, berthierite. Ref.: 6, 14, 22, 219, 224, 239, 629, 655, 841, 1032, 1210, 1237, 1392, 1441, 1647, 1698, 1699. Note: The minerals "fizelyite", "nakaseite" and "ramdohrite" are structural variants of andorite. Most specimens of andorite were shown to consist of syntaxic intergrowths of two andorite species: andorite-IV and andorite-VI. "Ramdohrite" = andorite-VI; "nakaseite" = andorite-XXIV. Ranges in specific gravity determinations of the various species suggest different chem. comp. A general formula may be: $Pb(Ag,Cu)Sb_3S_6$. The names "fizelyite", "nakaseite" and "ramdohrite" should be dropped. Ref.: 22, 312, 655, 1032.
H: no data. VHN: no data.	Occurs as fine-grained aggregates. Contains inclusions of argentite. Parag.: argentite, tennantite, bismuthinite, galena, pyrite. Ref.: 450.

Name Formula Crystal System	1: Colour 2: Bireflectance 3: Anisotropy 4: Internal reflections	Reflectance in % (Wavelength in nanometres)

PEARCEITE-
ANTIMONPEARCEITE
ARSENPOLYBASITE-
POLYBASITE
$(Ag,Cu)_{16}(As \lessgtr Sb)_2 S_{11}$
(Cu is an essential compound; may contain some Se)
Mono.-pseudohex.
	1. Grey, often with a greenish tint.
	⟶ galena, much darker.
	⟶ argentite, basal sections about the same, transverse sections darker.
	⟶ pyrargyrite and stephanite, commonly darker, brownish.
	⟶ tetrahedrite, about the same.
	2. Air: very weak, occasionally visible on grain boundaries. Oil: distinct in sections⊥ (001), weak in basal sections:
	    O greenish tint,
	    E darker, dark grey with violet-blue-grey tint.
	3. Air: moderate. Oil: strong, esp. in sections⊥ (001): blue, grey, yellowish green and brown polarization colours. Basal sections may appear isotropic. The As-rich members are apparently more strongly anisotropic than the Sb-rich ones.
	4. A deep-red int. refl. is nearly always visible (on close examination). With higher Cu-contents, the int. refl. decrease in number and intensity. Antimonpearceite has no int. refl.
		Pearceite: about 30 %.
		Polybasite: about 35 %.

x MIARGYRITE $Ag_2S \cdot Sb_2S_3$ Mono.	1. In air white, almost as galena. In oil much darker, greyish, often with a bluish tint. ⟶ galena, darker, with a greenish grey tint. ⟶ freibergite, lighter, bluish. ⟶ pyrargyrite, whiter. 2. Stronger than for similar silver minerals: white, light bluish grey, grey. 3. Stronger than for similar silver minerals. In air: light grey, blue-grey, brownish. In oil often masked by the int. refl. 4. Deep-red, less intense and less common than in proustite and pyrargyrite.	About 30-35 %.

Hardness H: polishing hardness VHN: micro-indentation hardness in kg/mm^2	Miscellaneous
H:>argentite, ~pyrargyrite, <stephanite, ≪ tennantite Pearceite: VHN: 142-164 [159, 1715]. VHN$_{50}$: 148-155 [1715]. Polybasite: VHN: 131-139 [1715]. VHN$_{50}$: 116-141 [194, 1715]. VHN$_{20}$: 108-114 [810].	NOMENCLATURE: Two solid solution series exist: pearceite-antimonpearceite, and arsenpolybasite-polybasite. Polybasite and antimonpearceite have Sb>As in at. %. Arsenpolybasite and pearceite have As>Sb in at. %. The polybasite-arsenpolybasite has cell dimensions: a~26 Å, b~15, c~24, β=90°. The pearceite-antimonpearceite has cell dimensions: a~13 Å, b~7.4, c~12, β=90°. The two solid solution series are not isodimorphous, but are separated from each other by a narrow two-phase field. One specimen has been observed with an intermediate cell: a~26 Å, b~15, c~12, β=90°. It may be related to order-disorder effect or due to a mixture of members of two series. The minerals occur as idiomorphic pseudohexagonal plates and tablets, sometimes arranged in subparallel and rosette-like groups. Also as irregular xenomorphic inclusions in galena, tetrahedrite, sphalerite, pyrite. Cleavage ∥ (001). Lamellar twins were observed only once. May be veined by chalcopyrite and tetrahedrite along the cleavage. May form myrmekitic intergrowths with stephanite, pyrargyrite, galena, stromeyerite, and orientated intergrowths with argentite, galena, chalcopyrite, stephanite. Replaces arsenopyrite, pyrite, chalcopyrite, sphalerite, freibergite, pyrargyrite, dyscrasite, galena, stromeyerite, sternbergite, uraninite (pitchblende), bornite, chalcocite. May be replaced by silver, gold, electrum and argentite. By dissociation silver may be produced. Different members of the series may be present at the same time. It is virtually impossible to distinguish between the different members of the two series under the microscope. Other ass.: safflorite, rammelsbergite, enargite, sphalerite, stibioluzonite, tennantite, magnetite. Ref.: 6, 14, 85, 342, 447, 458, 566, 581, 672, 762, 799, 805, 1153, 1237, 1343, 1379, 1392, 1435, 1441, 1529, 1531, 1629, 1732.
H: slightly> pyrargyrite, <stephanite, galena, ≪freibergite. VHN: 88-123 [159, 1715]. VHN$_{50}$: 102-130 [194, 1715].	Forms granular aggregates. Twinning has been observed, occasionally polysynthetic. May contain inclusions of pyrostilpnite and sphalerite. Occurs as inclusions in tetrahedrite and galena. Commonly intergrown with pyrargyrite or other Ag-Sb-minerals. Replaces arsenopyrite, sphalerite, freibergite, tetrahedrite, pyrargyrite. May be replaced by silver, polybasite, pyrargyrite. Other ass.: chalcopyrite, bournonite, argentite, stephanite, loellingite, rammelsbergite. Diff.: stephanite and stromeyerite do not show int. refl. and have other colours. Ref.: 14, 85, 91, 339, 377, 526, 737, 1237, 1345, 1379, 1392, 1676, 1687, 1732.

Name Formula Crystal System	1: Colour 2: Bireflectance 3: Anisotropy 4: Internal reflections	Reflectance in % (Wavelength in nanometres)
ARAMAYOITE $Ag_2S.(Sb,Bi)_2S_3$ Tricl.	1. Grey, in oil much darker than in air. ⟶ galena, darker. 2. Distinct. 3. Strong: light pink to steel-blue. 4. Rare, dark red.	About 30-35 %.
FREIESLEBENITE $4PbS.2Ag_2S.2Sb_2S_3$ Mono.	1. White-grey. ⟶ galena, in air about the same, in oil darker with yellowish green tint. Galena shows a reddish tint in contrast to freieslebenite. ⟶ freibergite, somewhat lighter. 2. Weak. 3. Distinct, without colour effects. 4. Not present.	About 35 %.
DIAPHORITE $4PbS.3Ag_2S.3Sb_2S_3$ Mono.	1. White. ⟶ galena, slightly olive or lemon-yellow. ⟶ tetrahedrite, similar, but paler. 2. Distinct on grain boundaries and at contacts of twin lamellae: white to lemon-grey. 3. Weak to moderate, brownish grey to dull purplish grey. Oblique extinction. 4. Not present.	480 nm 43.8 600 38.0 640 37.5 [651].

Hardness H: polishing hardness VHN: micro-indentation hardness in kg/mm^2	Miscellaneous
H: < pyrargyrite. VHN: no data.	Occurs as xenomorphic grains. Sections ∥ (010) may show quadrangular pits. Cleavage ∥ (100) and ∥ (001) typical. Twin lamellae ∥ (110) may be visible. Parag.: galena, tetrahedrite, Ag-Sb-minerals, Pb-Sb-minerals, bismuthinite, stannite, cassiterite, arsenopyrite, chalcopyrite, marcasite, sphalerite, pyrite. Ref.: 7, 103, 526, 1237, 1392.
H: ≪ galena. VHN_{20}: 85-140 [810].	Occurs as irregular aggregates of crystals with idiomorphic boundaries. Cleavage in two directions, ∥ (110) and ∥ (001), may be visible. Polysynthetic twinning ∥ (100) occurs. Observed as small exsolution bodies in galena. Replaces sphalerite, replaced by andorite. Parag.: galena, sphalerite, pyrite, freibergite, argentite, Pb- and Ag-Sb-sulphosalts. Difficult to distinguish freieslebenite from other sulphosalts. Ref.: 91, 219, 1104, 1146, 1237, 1379, 1392, 1676, 1699, 1732.
H: < galena. VHN: 197-242 [1715]. VHN_{20}: 200 [651].	Occurs as aggregates of rounded or slightly elongated grains. A well-developed twinning is generally present and diagnostic. Veined and surrounded by pyrargyrite and galena; as segregated lenses in galena; fills interstices between sphalerite grains. Replaces owyheeite, andorite; replaced by andorite. Forms intergrowths with pyrargyrite, galena and andorite. Parag.: galena, pyrargyrite, tetrahedrite, arsenopyrite, sphalerite, boulangerite, stannite, owyheeite, silver, other Ag-Pb-sulphosalts, freibergite. Diff.: bournonite has a lower refl. and shows quite different polarization colours. Ref.: 605, 651, 799, 805, 1104, 1379, 1676, 1699.

Name Formula Crystal System	1: Colour 2: Bireflectance 3: Anisotropy 4: Internal reflections	Reflectance in % (Wavelength in nanometres)
SMITHITE $AgAsS_2$ Mono.	1. Air: white. Oil: white with a bluish tint. → trechmannite, very similar. → proustite, distinctly less bluish. 2. Moderate, in oil bluish white to bluish grey. 3. Moderate, but obscured by the int. refl. 4. Strong and abundant: bright orange (in contrast to the orange-red of trechmannite).	About 40 % (?).
TRECHMANNITE $AgAsS_2$ Hex.	1. Air: white. Oil: white with a bluish tint. → smithite, very similar. → proustite, distinctly less bluish. 2. Moderate, in oil bluish white to bluish grey. 3. Moderate, but obscured by int. refl. 4. Strong and abundant: orange-red (in contrast to the bright orange of smithite).	About 40 % (?).
OWYHEEITE $5PbS.Ag_2S.3Sb_2S_3$ Ortho.	1. Greyish white with a greyish green or olive tint. → tetrahedrite, lighter. → arsenopyrite, distinct bluish shade. → boulangerite, less grey. 2. Distinct, greenish grey-white to olive-grey-white; not as strong as for jamesonite. 3. Very strong: straw or brownish white, greyish white and pale blue-grey to dark blue just before extinction; straight extinction. 4. Not present.	About 40 %.

Hardness H: polishing hardness VHN: micro-indentation hardness in kg/mm^2	Miscellaneous
H:<proustite. VHN: no data.	Occurs as fine-bladed, tabular crystals, sometimes of pseudohexagonal form. Nearly indistinguishable from trechmannite under the microscope, and also very similar to schapbachite and to several Pb-As-sulphosalts. Ref.: 522, 607, 1021, 1284, 1373. Note: All microscopic data were obtained on synthetic material (Roland, 1968b).
H:<proustite. VHN: no data.	See smithite. Ref.: 522, 1021, 1024, 1026, 1284. Note: All microscopic data were obtained on synthetic material (Roland, 1968b).
H:≧pyrargyrite, ~boulangerite, diaphorite, slightly <galena, ≪ tetrahedrite. VHN: 98-129 [798]. VHN$_{20}$: 128-214 [652, 1490]. VHN$_{10}$: 29-75 [810]. VHN$_?$: 72-140 [1550].	Occurs as acicular and fusiform crystals. Also as irregular fine-grained aggregates. Triangular pits may occur in not well-polished surfaces. Cleavage // (001) in prismatic sections. Occasionally twinned // elongation. Replaces galena, tetrahedrite, chalcopyrite, sphalerite, arsenopyrite, freibergite, diaphorite, pyrargyrite. Replaced by diaphorite, jamesonite, galena. Fills interstices between euhedral arsenopyrite crystals. May be replaced by pyrargyrite along the prismatic cleavage. Forms intergrowths with boulangerite and pyrargyrite (also pseudo-eutectic). Often contains inclusions of galena. Other ass.: pyrite, stephanite, miargyrite, boulangerite, jamesonite, silver, stannite, bournonite, geocronite, proustite. Difficult to identify in polished section; very similar to jamesonite. Ref.: 28, 36, 652, 798, 1209, 1274, 1392, 1490, 1529, 1532, 1647, 1699. Note: Timofeyevskiy (1967a) describes several varieties of owyheeite: normal owyheeite, Ag-poor owyheeite, and an As-analogue of owyheeite. The Ag-poor owyheeite is described as a separate species ("teremkovite"), but it might be merely a variety. It is grey to white with grey-green or olive tints, shows a weak birefl., a distinct anisotr. and dark brown int. refl. VHN = 83-155, and refl. is about 44 %. The formula is $Ag_2Pb_7Sb_8S_{20}$.

Ag-Fe-SULPHIDES

Name Formula Crystal System	1: Colour 2: Bireflectance 3: Anisotropy 4: Internal reflections	Reflectance in % (Wavelength in nanometres)

ARGENTOPYRITE
$AgFe_2S_3$
Ortho.
	1. Greyish white to yellowish.
	   → argyropyrite, more greyish.
	   → pyrite, more yellow.
	   → chalcopyrite, less yellow.
	2. Distinct, but not as strong as for argyropyrite: pale yellow to brownish or greyish yellow.
	3. Strong to intense: very deep-blue, paler shades of blue-grey, bright greyish white, reddish.
	4. Not present.
		About 45 %.

STERNBERGITE
$AgFe_2S_3$
Ortho.
	1. Brown, no yellow tint as in argyropyrite.
	   → pyrrhotite, somewhat darker.
	   → argyropyrite, slightly darker, brown.
	2. Strong. Air: $\parallel$ c dark brown,
	                 $\perp$ c light brown.
	        Oil: $\parallel$ c dark chocolate-brown,
	                 $\perp$ c cocoa-brown.
	3. Strong, vivid bluish and reddish to lilac polarization colours.
	4. Not present.
		About 45 %.

ARGYROPYRITE
$Ag_3Fe_7S_{11}$
Ortho.
	1. Yellowish brown or brownish yellow.
	   → pyrrhotite, about the same.
	   → sternbergite, yellowish brown tint, slightly lighter.
	2. Strong, stronger than for pyrrhotite.
	   In oil: $\parallel$ c brown,
	            $\perp$ c much lighter, brownish yellow.
	3. Strong, with vivid polarization colours.
	4. Not present.
		About 45 %.

Hardness H: polishing hardness VHN: micro-indentation hardness in kg/mm^2	Miscellaneous
H: >pyrrhotite, <pyrite. VHN?: 250-252 [866].	Occurs as prismatic crystals, but mostly as pseudo-hexagonal crystals formed by interpenetrating twinning. Lamellar twinning is also present. Forms intricate intergrowths of numerous individuals. Forms intergrowths with pyrostilpnite, the pyrostilpnite being present in the argentopyrite as clusters of slender prismatic crystals. Breaks down to pyrite, argentite and pyrrhotite. Occurs as inclusions in pyrite. Parag.: pyrostilpnite, argyropyrite, arsenic, proustite, pyrite, skutterudite, stephanite, pyrargyrite, argentite, chalcopyrite, arsenopyrite, pyrrhotite. Ref.: 289, 289a, 824, 866, 970, 1237.
H: >bismuth, ~pyrargyrite, proustite, slightly < argyropyrite, < galena, silver. VHN: 5-72 [1715]. VHN$_{25}$: 40-74 [1715]. VHN$_?^{25}$: 32-54 [866].	Occurs as thin-tabular crystals, mostly of pseudo-hexagonal outlines due to interpenetrating twinning. Sometimes very fine-grained. Perfect basal cleavage // (001). Mimetic twinning very common. Usually occurs as inclusions in polybasite, pearceite, pyrargyrite, proustite, galena, schwazite; also observed as thin lath-shaped sections in antimony. May show orientated intergrowths with argyropyrite and pyrite. Often contains inclusions of pyrite and marcasite; often replaced; during this process pyrite or marcasite and pyrargyrite or proustite may be formed. Replaces many of the accompanying minerals. May alter to linear bands of small pyrite grains. Breaks down to pyrite, argentite and pyrrhotite. Other ass.: arsenopyrite, sphalerite, chalcopyrite, silver, tetrahedrite, tennantite, jamesonite, bournonite, arsenic, bismuth, pitchblende, niccolite, Ni-Co-arsenides. Ref.: 6, 289, 289a, 632, 686, 866, 1047, 1237, 1261, 1379, 1392, 1629, 1687.
H: >pyrargyrite, proustite, slightly >sternbergite. VHN: no data.	Occurs as short-prismatic or thick-tabular crystals, mostly of pseudo-hexagonal outlines, due to mimetic twinning. Basal cleavage may be visible, but less distinct than in sternbergite. May show intergrowths with sternbergite, chalcopyrite, bournonite. Difficult to distinguish between argyropyrite and sternbergite under the microscope. Parag. and occurrence as for sternbergite. Ref.: 686, 1237, 1261, 1392, 1603, 1687. Note: There is no clear definition of the mineral frieseite. Chem. comp. ~ $Ag_2Fe_5S_8$; cryst.: ortho-pseudohex. Unstable, breaks down to sternbergite and pyrite. Relation with argyropyrite is possible, but not assured. Ref.: 1128, 1237, 1475.

Pb–Sb SULPHOSALTS

Name Formula Crystal System	1: Colour 2: Bireflectance 3: Anisotropy 4: Internal reflections	Reflectance in % (Wavelength in nanometres)
BOULANGERITE $5PbS.2Sb_2S_3$ Mono. (According to Born and Hellner (1960): $Pb_{18}Sb_{18}S_{44}$, ortho.)	1. Oil: white with a bluish green tint; much darker than in air. → galena, darker, greenish grey. → stibnite, slightly lighter. → jamesonite, darker. 2. Distinct, weaker than for jamesonite and stronger than for bournonite. Oil: $\parallel$ c and b: grey-white or white with a bluish green tint. $\parallel$ a: darker, green-grey. 3. Air: distinct. Oil: strong, much stronger than for bournonite: light tan, brown, bluish grey. Polarization colours in air more variegated than in oil (as opposed to jamesonite). 4. Rare, red.	$R_p \parallel c$ $R_g \perp c$ 472 nm 36.0 39.8 550 37.5 41.5 579 36.5 40.4 640 34.3 38.0 Dispersion curves: [959].
DADSONITE $11PbS.6Sb_2S_3$ Mono.	1. White with a greenish tint. 2. Not observable. 3. Distinct to strong: greenish grey. 4. Blood-red.	470 nm 34.8-39.7 546 34.9-40.0 589 34.2-39.6 650 32.7-37.4 [663a].
FÜLÖPPITE $3PbS.4Sb_2S_3$ Mono.	1. White. 2. Not indicated. 3. Moderate, bluish green to reddish brown. 4. Not present.	About 30-35 %.

Hardness H: polishing hardness VHN: micro-indentation hardness in kg/mm^2	Miscellaneous
H: slightly < galena, < bournonite. VHN: 90-183 [159, 194, 522, 1113, 1591, 1715]. VHN$_{50}$: 116-182 [194, 810, 1715]. VHN$_{20}$: 113-179 [959].	Usually occurs as granular or fibrous aggregates or as needle-shaped or tabular crystals enclosed in accompanying minerals, such as galena, sphalerite, bournonite, chalcopyrite, tetrahedrite, stannite. The needle-shaped crystals are often arranged in sub-parallel groups. Cleavage not discernable. Twinning or zonal texture not observed. May be replaced by galena, tetrahedrite, bournonite, chalcopyrite, pyrargyrite, and many other Sb-minerals. If replaced by chalcopyrite, a reaction rim of bournonite may be formed. Shows very fine graphic-like intergrowths with galena, due to replacement by galena. Forms orientated intergrowths with stibnite and jamesonite. Replaces pyrite, stannite, cassiterite, sphalerite, stibnite. Other ass.: arsenopyrite, pyrrhotite, marcasite, gold. Diff.: meneghinite and geocronite are less anisotropic and have different colour; jamesonite commonly shows a basal cleavage // elongation. Ref.: 111, 239, 295, 314, 316, 322, 346, 522, 587, 663, 841, 891, 1047, 1099, 1133, 1174, 1237, 1272, 1379, 1392, 1416, 1433, 1542, 1545, 1629, 1653.
H: no data. VHN: no data.	Occurs as needle-shaped crystals. Ass.: jamesonite, robinsonite. Ref.: 663a.
H: no data VHN: no data.	No cleavage (as opposed to plagionite and semseyite). Occurs as clusters or as irregularly intergrown crystals associated with needles of zinkenite, with sphalerite, dolomite and sulphur. Ref.: 663b, 1034, 1497.

Name Formula Crystal System	1: Colour 2: Bireflectance 3: Anisotropy 4: Internal reflections	Reflectance in % (Wavelength in nanometres)
GEOCRONITE $27PbS.7(Sb,As)_2S_3$ (forms an isomorphous series with jordanite). Mono.	1. White with a bluish green tint. → galena, darker, greenish. → lightest position of jamesonite, about the same. → darkest position of jamesonite, distinctly lighter and whiter. 2. Air: very weak, only visible on grain boundaries and on twin lamellae. Oil: more distinct, but still weak: light yellow-white to greenish white. 3. Air: weak. Oil: distinct, no strong or vivid colour effects: light grey to dark grey, bluish grey or steel blue; creamy tan to brownish grey. Oblique extinction. 4. Not present.	472 nm 36.0-39.5 550 38.0-41.6 579 37.7-41.2 640 35.0-38.3 Dispersion curves: [959].
GUETTARDITE $9PbS.8(Sb,As)_2S_3$ Mono.	1. White. 2. Relatively strong. 3. Strong. 4. Not indicated.	470 nm 36.3-44.2 546 34.8-42.0 589 34.0-40.8 650 32.2-39.0 [662].
HETEROMORPHITE $7PbS.4Sb_2S_3$ Mono.	1. White with a greenish tint. 2. Distinct to strong. 3. Strong. 4. Not indicated.	472 nm 35.9-39.8 550 37.0-41.0 579 36.5-40.3 640 34.4-38.0 Dispersion curves: [959].
LAUNAYITE $22PbS.13(Sb,As)_2S_3$ Mono.	1. See at birefl. 2. Fairly strong: white to grey. 3. Strong. 4. Not indicated.	470 nm 38.6-46.2 546 36.9-43.8 589 36.2-42.7 650 35.5-40.9 [662].
MADOCITE $17PbS.8(Sb,As)_2S_3$ Ortho.	1. See at birefl. 2. Strong, but slightly less than for boulangerite: white to grey. 3. Strong, slightly less than for boulangerite. 4. Not present.	Only maximum values: 470 nm 44.5 546 42.3 589 40.0 650 37.9 [661].

Hardness H: polishing hardness VHN: micro-indentation hardness in kg/mm^2	Miscellaneous
H: slightly < galena, > boulangerite, ~franckeite. VHN: 119-206 [522, 1113, 1591, 1715]. VHN$_{50}$: 134-161 [1715]. VHN$_{20}$: 95-186 [810, 959].	Occurs as tabular crystals and as granular aggregates. More or less rounded grains are not uncommon. Cleavage may be visible. Twinning lamellae very common and characteristic; usually developed in one direction (as distinct from bournonite), exactly parallel, but not always of uniform thickness. A coarser twinning after another law may be present. May be replaced by galena in a direction transverse to the twinning lamellae. Also replaced by bournonite. Replaces galena. Occurs as exsolution lamellae in meneghinite. Inclusions of antimony, bournonite, jamesonite, cassiterite, pyrite may occur. Rhythmic intergrowths with wurtzite have been observed. Commonly associated with Pb- and Cu-sulphosalts, sulphides of Fe, Cu, Pb and Zn, arsenic, antimony, franckeite, cylindrite, cassiterite. Ref.: 295, 316, 339, 369, 473, 474, 522, 628, 663, 780, 781, 891, 1168, 1212, 1237, 1244, 1382, 1591, 1597, 1626.
H: no data. VHN$_{50}$: 180-197 [662].	Typically occurs as isolated anhedral grains. Perfect cleavage. Polysynthetically twinned. Ref.: 662, 663.
H: no data. VHN$_{20}$: 137-187 [959].	Forms intergrowths with plagionite and semseyite. Very similar to jamesonite under the microscope. Ref.: 663b, 959, 1237.
H: no data. VHN$_{50}$: 171-197 [662].	Two perfect cleavages // (100) and // (001). Ass.: veenite, boulangerite. Ref.: 662, 663.
H: no data. VHN$_{50}$: 141-171 [661].	Perfect cleavage // (010). Ass.: jamesonite, boulangerite. Ref.: 661, 663.

Name Formula Crystal System	1: Colour 2: Bireflectance 3: Anisotropy 4: Internal reflections	Reflectance in % (Wavelength in nanometres)
PLAGIONITE $5PbS.4Sb_2S_3$ Mono.	1. White to greyish white. $\longrightarrow$ galena, darker. 2. Distinct in oil: white or light grey to brownish pink. 3. Distinct: pinkish, brownish, blue. Oblique extinction. 4. Dark red, occasionally visible.	472 nm 33.8-37.6 550 34.5-38.0 579 33.6-37.5 640 31.8-34.5 Dispersion curves: [959].
PLAYFAIRITE $16PbS.9(Sb,As)_2S_3$ Mono.	1. See at birefl. 2. Strong: white to brownish grey. 3. Strong. 4. Not indicated.	470 nm 38.3-42.3 546 36.4-40.3 589 35.4-39.2 650 34.0-37.7 [662].
ROBINSONITE $7PbS.6Sb_2S_3$* (may contain about 20 wt. % Bi). Tricl.	1. Very similar to boulangerite. 2. Not indicated. 3. Very strong: blue-grey, creamy white, brown and grey (in air). 4. Not indicated.	No data, but about 40 %.
SEMSEYITE $9PbS.4Sb_2S_3$ Mono.	1. White, with a greenish tint. $\longrightarrow$ galena, darker, distinct yellowish tint; galena shows a pinkish tint. $\longrightarrow$ bournonite, lighter. 2. Distinct: white with a yellow-green tint to greenish grey, about as tetrahedrite, but lighter. 3. Strong: light grey, bluish grey, brown, dark grey. Straight, but no complete extinction; sometimes undulatory extinction. 4. Not present.	About 40 %.
SORBYITE $17PbS.11(Sb,As)_2S_3$ Mono.	1. White. 2. Relatively strong. 3. Strong. 4. Not indicated.	470 nm 39-45 546 37-43 589 36-41 650 34-40 [662].

* Electron-microprobe analysis gives $3PbS.2Sb_2S_3$.

Hardness H: polishing hardness VHN: micro-indentation hardness in kg/mm^2	Miscellaneous
H: no data. VHN: 147-165 [1113]. VHN_{20}: 120-165 [959].	May occur as tabular crystals or as coatings; forms aggregates of needle-shaped or prismatic developed grains. Cleavage observable only in coarse-grained specimens. Parag.: other Pb-Sb-minerals, franckeite. Ref.: 239, 339, 663b, 967, 1043, 1237, 1392, 1497.
H: no data. VHN_{50}: 150-171 [662].	May occur at the periphery of other sulphosalt grains, particularly of boulangerite, and commonly extends into these minerals in the form of irregular veinlets. Perfect cleavage $\#$ (100). Twin lamellae were observed. Ref.: 662, 663.
H: no data. VHN: 118-123 [194].	Occurs as inclusions in boulangerite. Parag.: pyrite, sphalerite, stibnite, jamesonite, kobellite. Ref.: 120, 662, 663, 664.
H: distinctly < galena. VHN: 109-173 [1715]. VHN_{50}: 116-153 [1715].	Occurs as lath-like, may be prismatic, coarsely crystalline aggregates, or as fine granular aggregates or veins, cutting coarser crystalline, older semseyite. The coarser crystalline material may be slightly whiter than the fine-grained younger substance. Cleavage oblique to the elongation of the grains. Twinning or zonal texture not observed. Forms crusts around galena, encloses sphalerite. Replaced by bournonite. May contain granules, wedges, droplets and lamellae of galena. Other ass.: jamesonite, pyrargyrite, jordanite, tetrahedrite, other Pb-Sb-minerals. Very difficult to distinguish between semseyite and plagionite. Ref.: 31, 33, 191, 663, 663b, 967, 1043, 1176, 1237, 1281, 1392, 1497.
H: no data. VHN_{50}: 172-186 [662].	Occurs as thin-tabular, always elongate, crystals. Irregular twin lamellae are present. Ref.: 662, 663.

Name Formula Crystal System	1: Colour 2: Bireflectance 3: Anisotropy 4: Internal reflections	Reflectance in % (Wavelength in nanometres)
STERRYITE $12PbS.5(Sb,As)_2S_3$ Ortho.	1. See at birefl. 2. Strong: white to grey. 3. Strong. 4. Not indicated.	470 nm 37.6-40.4 546 36.0-38.7 589 35.1-37.7 650 33.9-36.3 [662].
TWINNITE $PbS.(Sb,As)_2S_3$ Mono, pseudo-ortho.	1. White. 2. Strong. 3. Strong. 4. Not indicated.	470 nm 38.7-45.6 546 36.9-43.0 589 35.9-41.6 650 34.6-39.6 [662].
VEENITE $2PbS.(Sb,As)_2S_3$ Ortho.	1. White. 2. Weak, white to pale pinkish grey. 3. Moderate, dark grey. 4. Not present.	470 nm 39.5-45.5 546 37.6-43.2 589 36.3-42.0 650 34.3-39.9 [661].
ZINKENITE $PbS.Sb_2S_3$ or $6PbS.7Sb_2S_3$ (some spec. give a comp. in between these two formulae). Hex.	1. White to greyish white. → galena, grey-white. → tetrahedrite, lighter grey-white; no brownish tint. → jamesonite, about the same, slightly lighter. 2. Weak, but perceptible. 3. Moderate: light grey to dark grey. Straight extinction. Sections ⊥ elongation may appear isotropic. 4. Rare, dark red.	air R_o R_e 470 nm 38.90 44.63 546 37.67 42.66 589 36.76 41.38 650 35.10 39.33 oil R_o R_e 470 nm 23.36 28.80 546 22.20 27.09 589 21.29 25.96 650 19.75 24.23 Dispersion curves: [849], also: [959].

Hardness H: polishing hardness VHN: micro-indentation hardness in kg/mm^2	Miscellaneous
H: no data. VHN: no data.	Occurs as needle-shaped crystals in veenite parallel to the polysynthetic twinning. Also as single anhedral crystals associated with veenite. Very fine lamellar twinning occasionally occurs. Ref.: 662, 663.
H: no data. VHN$_{50}$: 131-152 [662].	Perfect cleavage $\parallel$ (100). Polysynthetically twinned. Ref.: 662, 663.
H: no data. VHN$_{50}$: 156-172 [661].	Occurs massive or as disseminated grains. Polysynthetically twinned. Parag.: boulangerite, jamesonite. Ref.: 661, 663. Note: Veenite is the Sb-analogue of dufrenoysite. This mineral also occurs in the amorphous Pb-As-S glasses of Cerro de Pasco, Peru, as microcrystalline aggregates. It was named "stibiodufrenoysite" by Burkart-Baumann et al. (1966), but the name veenite has priority. Ref.: 90, 191, 915.
H: no data. VHN: 123-207 [159, 268, 1113, 1715]. VHN$_{20}$: 137-187 [959].	Occurs as felted masses of tiny acicular or hair-like crystals; as coatings; as vein- and fracture-fillings, or as radial fibrous aggregates. As small drop-like inclusions, resembling exsolution bodies, or as acicular aggregates in sphalerite. Replaces cassiterite, pyrite, stannite, freibergite, andorite. Other ass.: arsenopyrite, pyrrhotite, chalcopyrite, galena, tetrahedrite, bournonite, jamesonite, stibnite, other Pb-Sb-sulphosalts, gold. Ref.: 239, 268, 269, 518, 561, 579, 663, 721, 841, 957, 1034, 1237, 1281, 1392, 1599, 1653.

Bi - SULPHOSALTS

282 Bi-SULPHOSALTS

Name Formula Crystal System	1: Colour 2: Bireflectance 3: Anisotropy 4: Internal reflections	Reflectance in % (Wavelength in nanometres)
SCHAPBACHITE* $AgBiS_2$ (solid solution with galena is complete above 215°C). High-temp.: cubic. Low-temp.: hex.	1. White. $\rightarrow$ galena, in air very similar, in oil yellowish. 2. Air: very weak; oil: weak: $\parallel$ elongation: yellowish white, $\perp$ elongation: darker, white with greenish tint. 3. Distinct to strong, esp. in oil: light to dark grey. 4. Not present.	470 nm 44.2-45.3 546 43.8-44.9 589 43.4-44.4 650 42.6-44.0 Dispersion curves: [582].
PAVONITE $Ag_2S.3Bi_2S_3$ Mono.	1. + 2. Bireflectance in air distinct, in oil strong: sections $\perp$ elongation: $\parallel$ cleavage: white, $\perp$ cleavage: white with a greyish pink tinge; sections $\parallel$ elongation: white to bluish grey-white or white with a greyish pink tinge to bluish grey-white. 3. Strong and rather vivid: pale to intense blue, and light tan to brown. 4. Not present.	About 42 %.
SCHIRMERITE $PbS.2Ag_2S.2Bi_2S_3$ Cubic	1. White with a creamy tint. 2. Weak, even in oil; distinct at grain boundaries. 3. Weak, at grain boundaries more distinct, esp. with strong illumination and not completely crossed nicols. 4. Not observed.	About 35 %.

* The $AgBiS_2$ phase is generally referred to as "matildite" in American literature, as "schapbachite" in European literature. Consistent with European usage the name "schapbachite", which moreover has priority over "matildite", is used here.

Hardness H: polishing hardness VHN: micro-indentation hardness in kg/mm^2	Miscellaneous
H: slightly > galena, < chalcopyrite. VHN: 68-91 [159]. VHN$_{25}$: 59-76 [582].	Below 215°C characteristic Widmanstätten structure-like textures of schapbachite lamellae in a groundmass of galena and vice-versa are formed by exsolution. Also occurs in irregular intergrowths with, and as irregular blebs in, galena. Lamellar twinning produces a lattice texture. Fills interstices between pyrite grains. As inclusions in pyrite, sphalerite. May contain inclusions of bismuth. Parag.: always galena; arsenopyrite, pyrite, pyrrhotite, chalcopyrite, sphalerite, tetrahedrite, cassiterite, stannite, bismuthinite, valleriite, electrum, silver, bismuth, argentite, marcasite, tetradymite. Ref.: 286, 344, 490, 526, 547, 582, 588, 1208, 1213, 1237, 1334, 1392, 1532. Note: Banás and Ottemann (1967) described a mineral which is very similar to schapbachite in reflectance, hardness and X-ray data. It occurs intergrown with clausthalite. The composition is probably $AgBiSe_2$. The name "bohdanowiczyte" must be considered as highly preliminar, since no sufficient data were presented.
H: ≧ chalcopyrite, < sphalerite. VHN: no data.	Occurs as tiny bladed crystals elongated ∥ b. Cleavage is present. Parag.: bismuthinite, chalcopyrite, aikinite. Ref.: 1041, 1237. Note: Originally, the Cerro Bonete (Bolivia) material was thought to be "alaskaite". All specimens labelled "alaskaite" including the type material proved to be mixtures of different minerals (Thompson, 1950b). Since the type-"alaskaite" had been discredited, and this name as well as the supposed composition were not applicable to the Cerro Bonete mineral, Nuffield (1954) has given this species the new name pavonite.
H: no data. VHN: no data.	Forms granular aggregates of elongated grains. Cleavage may be visible. Replaces bismuth. Diff.: schapbachite shows a stronger birefl. and anisotr. Cosalite may show a more pronounced cleavage. Ref.: 168, 390, 1210.

Bi-SULPHOSALTS

Name Formula Crystal System	1: Colour 2: Bireflectance 3: Anisotropy 4: Internal reflections	Reflectance in % (Wavelength in nanometres)
BENJAMINITE $Pb_2(Ag,Cu)_2Bi_4S_9$ Mono.	1. White with a yellowish tint. 2. Weak in air, distinct in oil. 3. Strong, blue to pale brown. 4. Not present.	About 41-43 %.
BERRYITE $Pb_2(Cu,Ag)_3Bi_5S_{11}$ [1042] $Pb_3(Cu,Ag)_5Bi_7S_{16}$ [699] Mono. [1042] Ortho. [699]	1. Air: white to grey-white. ⟶ galena, no difference. Oil: white to grey-white with a weak creamy tint. ⟶ galena, distinctly creamy. In comparison to berryite, aikinite is distinctly cream to lemon-yellow. 2. Air: weak to distinct. Oil: stronger. 3. Air: distinct to strong. Oil: stronger, green or red-brown to greyish white. Aikinite is more strongly anisotropic (more vivid colours). 4. Not present.	546 nm 41.8-43.0 [699].
NEYITE $Pb_7(Cu,Ag)_2Bi_6S_{17}$ Mono.	1. White. ⟶ galena, light tan. 2. Not perceptible. 3. Moderate: light grey, yellow-green, light reddish brown to grey-black. 4. Not present.	No data.

Hardness H: polishing hardness VHN: micro-indentation hardness in kg/mm^2	Miscellaneous
H: no data. VHN_{20-50}: 183-194 [810]. $VHN_?$: 161-179 [921].	Occurs as irregular and spindle-shaped aggregates. No cleavage. Forms intergrowths with emplectite. Ass.: chalcopyrite. Ref.: 921. Note: The material from Andrasman (Central Asia) described by Mintser (1967) yielded the same data as given by Shannon for material from Round Mountain mine, Nye Co., Nevada. This type material is associated with chalcopyrite, molybdenite, pyrite, chalcocite and covellite. Nuffield (1953) discredited the mineral because he obtained the X-ray powder pattern of aikinite and an unknown mineral, which he called benjaminite in turn. Apparently, he did not examine type material, as Mintser (1967) obtained the same chem. comp. and X-ray powder pattern as Shannon.
H: >galena, <aikinite. VHN: 131-171 [699].	Occurs as regularly developed tablets, some of them slightly bent. When present in large amounts, the tablets may be grouped in bunches. Poor cleavage. A repeated twinning // tabular plane of the crystals is present only in the largest grains. May contain lamellae of galena // elongation of the tablets, due to replacement. Parag.: cuprobismutite, chalcopyrite, wolframite, galena, tetrahedrite, emplectite, aikinite, benjaminite, cosalite, pyrrhotite, marcasite, pyrite, bismuth. Ref.: 699, 1042.
H: no data. VHN: no data.	Occurs as needles and as acicular aggregates. Under crossed nicols the mineral consists of solid coarse mosaics with optically continuous areas up to 1 mm^2. Twinning was not observed. Forms intergrowths with pyrite, sphalerite and chalcopyrite. Other ass.: galena, aikinite, cosalite, tetrahedrite, molybdenite, nuffieldite. Ref.: 317a.

Name Formula Crystal System	1: Colour 2: Bireflectance 3: Anisotropy 4. Internal reflections	Reflectance in % (Wavelength in nanometres)
GALENOBISMUTITE $PbS \cdot Bi_2S_3$ Ortho.	1. Creamy white. → galena, very similar. → cosalite, darker. → bismuthinite, darker; bismuthinite is slightly yellowish. 2. Strong, yellowish white to pinkish grey or bluish grey. 3. Strong, yellow to dark brown. 4. Not present.	About 42 %.
COSALITE $2PbS \cdot Bi_2S_3$ (may contain some Cu and Ag, and up to 6.5 % Se) Ortho.	1. White with pinkish or grey tints. → galena, yellowish to green tint. 2. Air: weak. Oil: more distinct: $\parallel$ elongation: lighter, creamy, $\perp$ elongation: darker, greenish. 3. Weak, but visible (Warren, 1939, reports strong anisotr.). Pinkish yellow, bluish or violet-grey. Se-cosalite is more strongly anisotr. 4. Not present.	About 43 %.
LILLIANITE $3PbS \cdot Bi_2S_3$ Ortho.	1. White with a creamy tint. → galena, creamy. 2. Air: weak. Oil: distinct: $\parallel$ cleavage: creamy white, $\perp$ cleavage: darker and less creamy. 3. Distinct, with colours more distinct in air than in oil. Sections $\parallel$ cleavage are almost isotropic. 4. Not present.	About 45 %.

Hardness H: polishing hardness VHN: micro-indentation hardness in kg/mm^2	Miscellaneous
H: ≪chalcopyrite. VHN: 88-113 [1113]. VHN$_?$: 142-150 [988].	Usually occurs as radial aggregates of fibrous and feathery crystals, intergrown with bismuthinite. Cleavage $/\!/$c and $\perp$c may be visible. May show simple and polysynthetic twinning. Forms skeletal intergrowths with bismuthinite. Narrow lamellae of galenobismutite may be included $/\!/$ cleavage of molybdenite. Occurs as exsolved lamellae $/\!/$ (100) of galena. May contain inclusions of galena orientated $/\!/$ (001), cosalite and bismuth. Alters to bismuth and other Bi-sulphosalts. Included in sphalerite, pyrite. Replaces pyrite and pyrrhotite. Forms rims around gold. Rimmed by bismuthinite. Other ass.: tetradymite, aikinite, emplectite, pyrargyrite, chalcopyrite, freibergite, wolframite, cassiterite, scheelite, arsenopyrite. Ref.: 21, 112, 188, 202, 381, 650, 805, 806, 844, 988, 1107, 1232, 1237, 1392, 1529, 1629, 1644, 1649, 1684.
H: slightly>galena. VHN: 74-152 [1113, 1715]. VHN$_{50}$: 83-161 [1715]. VHN$_?$: 109-118 [988].	Occurs as minute needle-shaped crystals and fibers or as short-prismatic crystals, often forming bundle- or sheave-like aggregates. Also occurs as granular aggregates. Cleavage very rare, similar to that of bismuthinite. Twinning not observed. Replaces pyrite, glaucodot; may be replaced by Bi-tellurides. Shows inclusions of chalcopyrite, pyrrhotite, gold, hessite, galenobismutite, bismuth, pyrite. Found as inclusions in galena, sphalerite, pyrite, pyrrhotite, galenobismutite, bismuthinite. Rimmed by bismuthinite. May be intergrown with or intimately associated with gold, bismuthinite, galenobismutite. Other ass.: tetradymite, aikinite, joseite, emplectite, molybdenite, arsenopyrite, gersdorffite, niccolite, skutterudite, tetrahedrite, huebnerite, wolframite, uraninite, Mn-minerals, scheelite, marcasite. Ref.: 28, 71, 106, 109, 202, 366, 596, 643, 691, 692, 721, 739, 806, 844, 919, 988, 1047, 1107, 1133, 1237, 1322, 1557, 1586, 1644, 1645, 1647.
H: slightly <silver. VHN$_?$: 120-195 [734, 859].	Usually occurs as aggregates of tabular, needle-like or platy crystals. The mineral is unstable and decomposes with preservation of the over-all composition to galena, bismuthinite and argentite or to galenobismutite and galena, or to cosalite and galena. A prominent cleavage is present. Parag.: silver, bismuth, sphalerite, cosalite, arsenopyrite, pyrite, galena. Ref.: 734, 776, 859, 1064, 1237, 1493, 1586. Note: The redefinition based on data provided by Ramdohr (1960), and by Syritso and Senderova (1964) was rejected by the IMA-CNMMN (1964); afterwards Otto and Strunz (1968) obtained a synthetic compound with formula 3PbS.Bi$_2$S$_3$ and the same optical properties as given by Ontoev (1959), and by Syritso and Senderova (1964). This is confirmed by data of Malakhov et al. (1968), and of Klyakhin and Dmitriyeva (1968). Lillianite should be regarded as a valid mineral species. The material described by Kupčik et al. (1961) as homogeneous lillianite has been proved to be a mixture, one of the components being a Bi-rich variety of jamesonite (Kupčik et al. 1969).

Name Formula Crystal System	1: Colour 2: Bireflectance 3: Anisotropy 4: Internal reflections	Reflectance in % (Wavelength in nanometres)
BONCHEVITE $PbS \cdot 2Bi_2S_3$ Ortho.	1. Creamy white. $\longrightarrow$ aikinite, very similar. 2. Weak in air, distinct in oil. 3. Distinct to strong, no colour effects. 4. Not present.	527 nm 46.6 589 46.2 686 45.8 [989].

BURSAITE
$5PbS \cdot 2Bi_2S_3$
Mono.
	1. White.
	2. Air: weak, only visible on grain boundaries.
	Oil: distinct:
	    $/\!\!/$ elongation: white with a brownish tint,
	    $\perp$ elongation: light grey.
	3. Air: distinct.
	Oil: very distinct: $/\!\!/$ elongation: dark schist-grey with a distinct brown tint to brownish olive-green. In sections $/\!\!/$ (010): oblique extinction.
	4. Not present.
		About 45 %.

CANNIZZARITE $Pb_3Bi_5S_{11}$ (?) Mono.	No optical data available.	No data.

Hardness H: polishing hardness VHN: micro-indentation hardness in kg/mm^2	Miscellaneous
H: no data. VHN$_{15-20}$: 129-205 [989].	Occurs as acicular and prismatic crystals, and as granular aggregates. Perfect cleavage $\parallel$ (100), i.e. $\parallel$ elongation. May contain small inclusions of bismuth, tetradymite and cosalite. Parag.: scheelite, wolframite, pyrite, sphalerite, molybdenite, cassiterite, bismuthinite, cosalite. Ref.: 752, 989. Note: The type locality bonchevite (Kostov, 1958) was shown to be a mixture of galenobismutite and a new sulphosalt with formula $Pb_3Bi_2S_6$ (Kupčik et al. 1969).
H: $\gg$ bismuth, $\ll$ chalcopyrite VHN: no data.	Occurs as short- or long-columnar (elongation $\parallel$ c-axis) individuals, often gathered together in radiating groups. Cleavage very distinct $\parallel$ (100) and (010), distinct $\parallel$ (h0l), probably $\parallel$ (001). Twinning very common, mostly in broad lamellae $\parallel$ (110). Contains exsolution grains of bismuth, more rarely of bismuthinite. Intimately associated with sphalerite and chalcopyrite. Slightly replaced by sphalerite. Other ass.: scheelite, pyrrhotite, pyrite, molybdenite. Diff.: cosalite has straight extinction, weak anisotr. and no twinning; lillianite has straight extinction and is isotropic in sections $\parallel$ (010). Ref.: 687, 1315, 1559, 1709. Note: The validity of this mineral is doubtful. Type locality material is exhausted (priv. comm. A. Kraëff, 1970).
No data.	Occurs as leafy crystals associated with bismuthinite and galenobismutite in the sublimation products of fumaroles on the island Vulcano. Forms twins and trillings. Also found associated with galena. Ref.: 528, 1029.

Bi-SULPHOSALTS

Name / Formula / Crystal System	1: Colour / 2: Bireflectance / 3: Anisotropy / 4: Internal reflections	Reflectance in % (Wavelength in nanometres)
KOBELLITE-TINTINAITE $5PbS.4(Bi \gtrless Sb)_2S_3$ (Some Se may be present) Ortho.	1. White. → galena, slightly darker, yellowish green tint in lightest position; in darkest position brownish grey tint. 2. Distinct: $R_g \parallel$ elongation: greenish white, R_p: violet-grey. 3. Distinct: steel-grey to grey-brown. Apparently straight extinction. 4. Not present.	For kobellite: $\quad R_p \quad R_m$ 546 nm 36.6 39.0-41.1 $\quad R_g$ 546 nm 45.4 [580]. Dispersion curves: [990]. For tintinaite: $\quad\quad R_p \quad R_m \quad R_g$ 470 nm 37.4 38.3 43.9 546 36.3 37.4 43.3 589 35.0 36.2 41.9 650 35.1 35.8 42.0 [194].
USTARASITE $PbS.3(Bi,Sb)_2S_3$?	1. Pure white. 2. Not indicated. 3. Strong. 4. Not present.	About 42 %.
GIESSENITE $Pb_9CuBi_6Sb_{1.5}S_{30}$ Ortho.	1. White. → galena, reddish brown tint. 2. Fully absent. 3. Air: very weak; oil: more distinct. Straight extinction. 4. Not present.	About 35 %.

Hardness H: polishing hardness VHN: micro-indentation hardness in kg/mm^2	Miscellaneous
H: $\gg$ bismuth, slightly $<$ galena*. For kobellite: VHN: 69-173 [159, 194, 1715]. VHN$_{50}$: 124-168 [1715]. VHN$_{20-50}$: 128-172 [558b]. VHN$_{25}$: 142-151 [194]. VHN$_{10}$: 84-169 [990]. For tintinaite: VHN$_{50}$: 149-157 [194].	Occurs as granular aggregates and as radial aggregates of columnar and needle-like crystals. Good cleavage $\parallel$ (010), i.e. $\parallel$ elongation. Twinning not uncommon. Chalcopyrite and galena fill interstices between kobellite grains. Surrounds and replaces arsenopyrite, pyrite, pyrrhotite, marcasite. Replaced by chalcopyrite and tetrahedrite. May contain exsolution bodies of bismuth and inclusions of tellurides. Forms myrmekitic intergrowths with tetrahedrite. Other ass.: electrum, sphalerite, magnetite, hematite, jamesonite, bismuthinite, boulangerite, molybdenite, scheelite. Ref.: 558b, 562, 580, 842, ·990, 1037, 1047, 1237, 1342a, 1534, 1564, 1565, 1566, 1727.
H: no data. VHN: no data.	Parag.: pyrite, chalcopyrite, bismuth. Ref.: 1316.
H: close to galena. VHN$_{10}$: 65 [520].	Occurs as fine needles on galena in dolomite (Binnental). Parag.: pyrite, rutile, tennantite, molybdenite. Ref.: 520, 522.

* Háber and Streško (1969) report H $>$ galena.

Name Formula Crystal System	1: Colour 2: Bireflectance 3: Anisotropy 4: Internal reflections	Reflectance in % (Wavelength in nanometres)
AIKINITE $2PbS.Cu_2S.Bi_2S_3$ Ortho.	1. Air: creamy white. $\longrightarrow$ galena, distinctly cream. Oil: more pinkish. $\longrightarrow$ galena, distinctly pink. $\longrightarrow$ berryite, distinctly cream to lemon-yellow. 2. Air: distinct. Oil: very distinct: $\parallel$ c creamy white, $\perp$ c white or light brown. 3. Air: distinct. Oil: strong, blue-grey-white, pinkish grey-white and a characteristic white. Sharp and complete extinction, except in some orientations where slight dispersion of extinction is indicated by a deep indigo-blue colour. 4. Not present.	470 nm 38.0-44.0 546 39.2-45.7 589 39.5-46.0 [723].
NUFFIELDITE $10PbS.2Cu_2S.5Bi_2S_3$ Ortho.	1. Pale creamy white. 2. Not perceptible. 3. Very weak: bluish grey to greyish red. 4. Not present.	470 nm 39.8-45.6 546 39.0-44.9 589 38.6-44.5 [723].
WITTICHENITE $3Cu_2S.Bi_2S_3$ Ortho.	1. Creamy grey with an olive tint. $\longrightarrow$ aikinite, brownish grey. $\longrightarrow$ emplectite, slightly darker. $\longrightarrow$ tennantite, lighter, pinkish. $\longrightarrow$ tetrahedrite, lighter. $\longrightarrow$ chalcocite, slight greenish or brownish tint. $\longrightarrow$ annivite, brownish grey. 2. Weak, only visible in oil at grain boundaries. 3. Air: weak; oil: more pronounced, dull brown colours*. 4. Not present.	About 35-40 %.

*Stillwell and Edwards (1942b) report strong anisotr. for wittichenite.

293

Hardness H: polishing hardness VHN: micro-indentation hardness in kg/mm^2	Miscellaneous
H: $>$galena, $\geqq$bismuthinite, galenobismutite, $<$bournonite VHN: 165-227 [1113, 1715]. VHN$_{50}$: 201-246 [1715]. VHN$_{15}$: 201-218 [723].	Forms idiomorphic columnar or acicular crystals. May occur in bundles of prismatic crystals. Perfect cleavage $/\!/$ (010). Occasionally intimately intergrown with galena, cosalite, galenobismutite, tetradymite, chalcopyrite, gold, tetrahedrite, bournonite. Replaces tetrahedrite; may be replaced by argentite, silver, wittichenite. Occurs as inclusions in bornite, galena, chalcopyrite. Rimmed by covellite, bismuth, Other ass.: jamesonite, pyrrhotite, sphalerite, other sulphides of Fe, Cu, Zn, Pb, bismuthinite. Ref.: 29, 507, 709, 721, 776a, 801, 884, 1127, 1162, 1237, 1270, 1392, 1423, 1455, 1662, 1685. Note: The structure of aikinite is very similar to that of bismuthinite. "Gladite","hammarite", "lindströmite", and "rezbanyite" have an uncertain composition intermediate between aikinite and bismuthinite. Re-investigation of these compounds on type-locality material proved them to be intermediate members between aikinite and bismuthinite. They should be considered as such, and all four names should be dropped. A new nomenclature has been proposed by Moore, 1967b. Ref.: 776a, 1093, 1095, 1162, 1673.
H: no data. VHN$_{15}$: 149-178 [723].	Occurs as prismatic to acicular crystals and as compact aggregates. Cleavage $\perp$(001) indistinct, $/\!/$ (001) excellent. No twinning observed. Ass.: aikinite. Ref.: 723.
H: $\gg$bismuth, $>$emplectite, $\geqq$chalcocite, $<$bornite, chalcopyrite, tetrahedrite, tennantite, annivite. VHN: 161-206 [1311, 1715]. VHN$_{50}$: 167-216 [194, 1715].	Texture commonly coarse granular; also as euhedral crystals of acicular and elongated rectangular outlines; as long slender blades or as irregular grains. No cleavage or twinning observed. As inclusions in bornite and chalcopyrite, or at contacts between bornite and chalcocite. Inclusions in chalcocite are corroded and have been inherited from replaced bornite. As reaction rims between chalcopyrite and bismuth or bismuthinite. Rarely observed in magnetite and hematite. May show inclusions of tennantite, annivite and bismuth. Replaced by bismuth, bismuthinite, chalcopyrite, emplectite. Replaces bismuth, aikinite, annivite, "klaprothite"; the replacement of aikinite may result in a parquet-like texture of aikinite criss-crossed by a multutide of veinlets of wittichenite. Forms intergrowths with digenite and bornite. Chalcocite forms pseudomorphs after wittichenite. Other ass.: enargite, stibioluzonite, silver and Ag-minerals, gold, pyrite, sphalerite, covellite. Ref.: 6, 147, 584, 764, 801, 840, 900, 978, 1036, 1057, 1214, 1237, 1461, 1529. Note: The mineral described by Krieger (1940) as "klaprothite" has been shown to be wittichenite (Nuffield, 1947a); thus, the properties given by Krieger are cited here for wittichenite.

Name Formula Crystal System	1: Colour 2: Bireflectance 3: Anisotropy 4: Internal reflections	Reflectance in % (Wavelength in nanometres)
CUPROBISMUTITE $CuBiS_2$, dimorphous with emplectite. Mono.	No optical data available.	No data.
EMPLECTITE $CuBiS_2$ Ortho.	1. Creamy or yellowish white, sometimes with a light brown tint. ⟶ chalcopyrite, darker. ⟶ galena, brownish cream. ⟶ bismuthinite, slightly darker and more yellowish. ⟶ other Cu-Bi-sulphides, reddish brown. 2. Visible in oil: ∥ elongation: darkest, bluish green tint, ⊥ elongation: light greenish or creamy. 3. Strong: colours darker than for bismuthinite: brown to bluish, dark violet. 4. Not present.	Values estimated from the dispersion curves: 470 nm 35.0-39.6 550 36.2-42.2 590 36.0-42.1 650 36.0-40.8 [1430].
"KLAPROTHITE" $CuBiS_2$?	1. Creamy grey with olive-green tint. ⟶ wittichenite, darkest position similar to the lightest of wittichenite. ⟶ emplectite, olive. 2. Strong (in oil): ∥ elongation, lighter, ⊥ elongation, darker. 3. Very strong: blue to yellowish white; straight extinction. 4. Not present.	Values estimated from the dispersion curves: 470 nm 30.6-36.3 550 32.7-40.2 590 33.0-40.2 650 34.3-40.0 [1430].

Hardness H: polishing hardness VHN: micro-indentation hardness in kg/mm^2	Miscellaneous
No data.	Occurs as tiny prismatic crystals elongated $/\!/$ b, and as massive grains, associated with chalcopyrite, wolframite, cassiterite, galena, tetrahedrite, emplectite, aikinite, benjaminite, berryite, tetradymite and bismuthinite. May contain inclusions of gold. Ref.: 1039, 1042, 1197. Note: cuprobismutite may be identical with "klaprothite"; see "klaprothite".
H: $\gg$ bismuth, slightly $<$ bismuthinite and chalcocite*. VHN: 168-238 [159, 1715]. VHN$_{20-50}$: 158-249 [810, 1715].	Usually developed in fibrous or needle-shaped crystals or in crystal aggregates. Cleavage $/\!/$ (001) ($\perp$ elongation) may be visible. Bismuthinite shows a cleavage $/\!/$ elongation. Twinning occurs. Replaces annivite, wittichenite, chalcopyrite, "klaprothite". May be replaced by bismuthinite and Bi-tellurides. May show inclusions of bismuth. Forms intergrowths with bismuthinite. Other ass.: Cu-sulphides, gold. Ref.: 488, 688, 732, 739, 801, 844, 900, 1204, 1232, 1237, 1392, 1430. Diff.: wittichenite commonly occurs as rounded grains; annivite is distinctly harder and isotropic and shows a much lower refl.
H: $>$ chalcopyrite, emplectite, $\sim$ bornite, $<$ annivite. VHN: no data.	Occurs as long prismatic crystals. Cleavage $/\!/$ (100) distinct. Twinning rarely observed. Replaced by bismuth, emplectite, wittichenite. Replaces wittichenite. Other ass.: annivite, aikinite. Ref.: 900, 1237, 1430. Note: "Klaprothite" has been a doubtful mineral since Nuffield (1947b) found all "klaprothite" material available to him to consist mostly of wittichenite and emplectite. Only Ramdohr stated to have observed a mineral the properties of which corresponded exactly with those given for "original klaprothite"; wittichenite, emplectite, annivite and bismuth were also present in the same polished section. Springer and Demirsoy (1969) examined material from the same locality. "Klaprothite" has the same chem. comp. as emplectite, but the dispersion curves of the reflectance of these minerals are different. Springer and Demirsoy (1969) suggest that "klaprothite" is identical to cuprobismutite, which is a dimorph of emplectite. If this is confirmed the name "klaprothite" has priority over cuprobismutite.

---

* Mehnert (1949) described a spec. of emplectite with H $>$ chalcopyrite.

Pb–As - SULPHOSALTS

Name Formula Crystal System	1: Colour 2: Bireflectance 3: Anisotropy 4: Internal reflections	Reflectance in % (Wavelength in nanometres)
HUTCHINSONITE $(Tl,Pb)_2As_5S_9$ Ortho.	1. Greyish white. → galena, jordanite, bluish with a green-grey tint. → other Pb-As-sulphosalts, distinctly grey. In oil: the colour is more bluish. → gel-pyrite, violet-grey. → schalenblende, white with a grey-blue tint. → smithite, very similar. → orpiment, weakly grey-blue. 2. Weak, but distinct: bluish violet-white to bluish white with a green tint. 3. Distinct to strong, violet to deep-blue. In oil: not visible in sections ∥ elongation: obscured by int. refl.; sections ⊥ elongation show less int. refl. Straight extinction. 4. Abundant, carmine-red.	530 nm 30.0-31.0 [522].
MARRITE $PbAgAsS_3$ Mono.	1. White. → galena, distinct yellow-brown. 2. Not observable. 3. Air: distinct. Oil: strong, with a remarkable blue tint in lightest position. 4. Abundant, red.	530 nm 31.5-34.0 [522].
GRATONITE $9PbS.2As_2S_3$* Hex.	1. White. → jordanite, slightly pinkish. → galena, pinkish. 2. Very weak, only visible on grain boundaries, even in oil. 3. Distinct, but weaker than in any other Pb-As-sulphosalt**. Straight extinction. 4. Red, rarely visible.	530 nm 33.4-34.4 [194].

* Burkart-Baumann et al. (1968) reported gratonite from Rio Tinto with a composition similar to that of jordanite.
** Anderson (1948) reported a strong anisotr. for gratonite, with yellow and blue polarization colours.

299

Hardness H: polishing hardness VHN: micro-indentation hardness in kg/mm^2	Miscellaneous

H: no data.
VHN: 170-171 [522].
	Forms radiating aggregates of needle-like crystals. No twinning or zoning observed. Good cleavage $/\!/$ (010).
	Replaces most of the other Pb-As-minerals. Fills cracks in baumhauerite, even in the core of the crystals. May form lamellae in jordanite. Forms orientated intergrowths with sartorite and rathite. Occurs in aggregates of well developed needle-like crystals with rectangular to square sections in schalenblende, together with orpiment.
	Ref.: 505, 522, 1023, 1035, 1237, 1372, 1373, 1505, 1615, 1629.
	
	Notes:
	1. Hatchite, PbTl(Ag, Cu)(As, Sb)$_2$S$_5$, tricl. Forms tabular crystals.
	   Ref.: 119, 522, 876, 1018, 1019, 1021, 1023, 1026.
	2. Wallisite, PbTlCuAs$_2$S$_5$, isomorphous with hatchite. VHN$_{10}$: 113-165; Refl. (white light): 38%.
	   Ref.: 1016, 1018, 1020, 1024.

H: no data. VHN: 161-171 [522].	Twin lamellae may occur: two systems cutting each other at small angles, the lamellae often being remarkably bent. Parag.: proustite, xanthoconite, pyrite. Diagn.: the polarization colours in oil and the unusual twin lamellae (if present) are characteristic. Ref.: 522, 1027, 1706.
H: $\sim$galena, slightly <jordanite. VHN: 123-156 [194, 1715].	Always occurs in sheaf-like aggregates of idiomorphic crystals. The cross-sections have hexagonal or trigonal outlines. Zonal texture may occur and be developed by etching with NaClO during 1 min. (saturated sol. in water, boil 10 min.). May be replaced by jordanite and galena. Replaces galena and chalcopyrite. Forms pseudomorphs after galena; forms intergrowths with galena. Commonly occurs in "schalenblende". Other ass.: pyrite, arsenopyrite, sphalerite, freibergite, chalcocite, covellite, bornite, bismuthinite, realgar. Ref.: 34, 193, 606, 1024, 1102, 1103, 1218, 1237, 1285, 1309.

Name Formula Crystal System	1: Colour 2: Bireflectance 3: Anisotropy 4: Internal reflections	Reflectance in % (Wavelength in nanometres)
LIVEINGITE* $Pb_{19}As_{13}S_{58}$ Mono.	1. White. $\longrightarrow$ galena, slight brownish green tint. 2. Not observed. 3. Distinct, weaker than for rathite-I and dufrenoysite; different with orientation; in oil somewhat stronger than in air. 4. Rather rare, clear red.	530 nm 34.0-36.0 [522].
LENGENBACHITE $Pb_{37}Ag_7Cu_6As_{23}S_{78}$ Tricl.	1. White. $\longrightarrow$ galena, weakly reddish brown. 2. Not observable. 3. Air: weak, but distinct on grain boundaries. Oil: stronger, olive-green to blue-green. 4. Not present.	530 nm 34.0-37.0 [522].
RATHITE-I** $(Pb,Tl)_3As_4(As,Ag)S_{10}$ Mono. (true symmetry possibly tricl.).	1. Pure white. 2. Air: distinct on twin lamellae. Oil: strong. 3. In oil extremely strong, strongest of all Binnatal minerals: olive-green, yellow, violet-blue, blue. 4. Very common, brown-red to clear red.	530 nm 34.0-38.5 [522].
BAUMHAUERITE acentric modification: $Pb_{12}As_{16}S_{36}$ or $Pb_{10}As_{17}(Ag,Tl)S_{35}$ centric modification: $Pb_{11}As_{17}S_{36}$ Both modifications are tricl.	1. Air: pure white. $\longrightarrow$ galena, pinkish brown tint, sometimes distinctly blue. Oil: blue-green tint. $\longrightarrow$ sartorite and liveingite, very similar. 2. Distinct (Graeser, 1965, did not observe any birefl.). 3. Distinct to strong: remarkable blue-grey colour in lightest position. 4. Not uncommon, dark red.	530 nm 34.0-39.0 [522].

* All specimens labelled "liveingite" proved to be identical to rathite-II. The name rathite-II being already quite familiar, "liveingite" should be dropped (Nowacki, 1967a). However, Graeser (1965) proposed to give rathite-II another name in view of the marked differences in properties between rathite-I and rathite-II. The authors agree with this proposal, especially on account of the confusion around the nomenclature of rathite-I, rathite-II, and rathite-III. For reasons of priority, the name rathite-II should be dropped in favour of liveingite.
** According to the nomenclature of Berry (1953) and Graeser (1965).

301

Hardness H: polishing hardness VHN: micro-indentation hardness in kg/mm^2	Miscellaneous
H: no data. VHN: 173-183 [522].	Shows no twinning. Forms prismatic needles. Very difficult to distinguish from other sulphosalts, but it is possible with precise measurements of VHN and reflectance. Ref.: 57, 505, 779. See also rathite-I.
H: no data. VHN$_3$: 29-40 [522].	Forms aggregates of foliated grains. Twinning lamellae may occur. Forms intergrowths with, and replaces jordanite. Parag.: other Pb-As-sulphosalts. Ref.: 57, 505, 522, 1018, 1019, 1021, 1022, 1025, 1031, 1392.
H: no data. VHN: 159-163 [522].	Shows polysynthetic twinning, similar to plagioclase twinning. Cleavage ∥ (010) often distinct. May be replaced along the cleavage planes by liveingite, baumhauerite, sartorite, realgar, hutchinsonite. Ref.: 57, 116, 505, 522, 813, 814, 874, 875, 1018, 1022, 1028, 1133, 1392.
H: no data. VHN: 128-182 [522, 1113].	Often occurs as granular aggregates. Twin lamellae ∥ (100) very common, generally of regular appearance and of equal thickness. Replaces rathite-I and liveingite along cleavage and parting planes; may be replaced by sartorite and hutchinsonite. Ref.: 57, 116, 505, 522, 662, 779, 812, 814, 1018, 1021, 1028, 1104, 1133, 1392.

Name Formula Crystal System	1: Colour 2: Bireflectance 3: Anisotropy 4: Internal reflections	Reflectance in % (Wavelength in nanometres)
SARTORITE $PbS.As_2S_3$ (may contain some Tl). Mono.	1. Pure white. ⟶ galena, yellow-brown tinge. ⟶ baumhauerite, liveingite, very similar. 2. Not observable. 3. Air: distinct. Oil: slightly stronger: grey-blue to yellowish grey. Somewhat weaker than for dufrenoysite. 4. Very common, deep-red.	530 nm 35.0-39.0 [522].
DUFRENOYSITE $2PbS.As_2S_3$ Mono.	1. Pure white. ⟶ galena, weak grey-blue tint. ⟶ seligmannite and baumhauerite, lighter. 2. Distinct, but only on grain boundaries and twin lamellae. 3. Strong, brownish violet to dark green, darker than for baumhauerite. 4. May occur, dark red.	530 nm 36.5-40.5 [522].
JORDANITE $27PbS.7As_2S_3$ Mono.	1. White with a faint greenish tint. ⟶ baumhauerite, pinkish; baumhauerite is greenish. ⟶ galena, about the same, darker, slightly greenish. ⟶ gratonite, whiter; gratonite is yellowish. 2. Distinct (in oil): white, yellowish grey-white, grey-white with a faint greenish tint. 3. Strong, with vivid colours: dark grey, yellowish grey, greenish grey, dark brownish grey. 4. Not present.	530 nm 38.0-39.5 [522].

Hardness H: polishing hardness VHN: micro-indentation hardness in kg/mm^2	Miscellaneous
H: no data. VHN: 194-197 [522].	Often irregularly developed. Abundantly twinned $/\!/$ (100), commonly broad lamellae of equal width. Replaces rathite, liveingite, baumhauerite. May be replaced by hutchinsonite. Forms orientated intergrowths with baumhauerite and hutchinsonite. Ref.: 505, 522, 779, 967, 1025, 1133, 1629.
H: no data. VHN: 145-156 [522].	Occurs as idiomorphic and xenomorphic grains. Abundant polysynthetic twin lamellae of varying width. Parag.: seligmannite, lengenbachite, other Pb-As-sulphosalts, realgar. Ref.: 57, 116, 505, 522, 876, 1392. Note: For "stibiodufrenoysite", see veenite.
H: slightly>semseyite, gratonite, ~galena. VHN: 149-198 [522, 1113, 1715]. VHN_{50}: 172-204 [194, 1715].	Rarely occurs as idiomorphic acicular or reticulate crystals. Usually as concentric or botryoidal masses: these show undulatory extinction and weak anisotropy. Layers of jordanite may alternate with layers of galena. Cleavage may be present. A regular twinning is very common. Forms pseudomorphs after gratonite due to replacement. May be replaced by dufrenoysite, baumhauerite, enargite. Replaces pyrite. Other ass.: Pb-As-sulphosalts, proustite, pyrargyrite, semseyite, bournonite, tennantite, marcasite, wurtzite, sphalerite, chalcopyrite, bornite, digenite, chalcocite, covellite, geocronite, realgar, orpiment, hutchinsonite, meneghinite. Ref.: 31, 33, 57, 173, 192, 299, 316, 482, 505, 522, 573, 668, 779, 863, 1027, 1030, 1104, 1133, 1228, 1237, 1283.

Name Formula Crystal System	1: Colour 2: Bireflectance 3: Anisotropy 4: Internal reflections	Reflectance in % (Wavelength in nanometres)
SELIGMANNITE $2PbS \cdot Cu_2S \cdot (As,Sb)_2S_3$ Ortho.	1. Oil: greyish white, often with a pinkish tint. ⟶ galena, reddish brown. ⟶ petzite, slightly more greyish. ⟶ tetrahedrite, much less grey. ⟶ tennantite, lighter, distinctly pink. ⟶ enargite, greyish green. ⟶ jordanite, pinkish. ⟶ dufrenoysite, darker, pinkish. No bluish green colour as for bournonite. 2. Air: weak, distinct on twin lamellae. Oil: somewhat stronger. 3. Rather strong, in oil stronger than in air, and stronger than for bournonite: brown, green, greenish blue. 4. Not present.	530 nm pure seligmannite: 36.0-40.0, 530 nm Sb-seligmannite: 36.0-42.0 [522].

Hardness H: polishing hardness VHN: micro-indentation hardness in kg/mm^2	Miscellaneous
H: slightly > galena, ≪ tennantite. VHN: pure seligmannite: 160-167; Sb-seligmannite: 149-161 [522].	Usually forms xenomorphic grains. Needle-like idiomorphic grains have been observed in stromeyerite. Commonly polysynthetically twinned, but not as regular as for jordanite. Parquet-like twinning occurs, very similar to that of bournonite. Replaces jordanite, dufrenoysite; may be replaced by galena, naumannite. Forms intimate intergrowths with petzite. Occurs as orientated blebs in altaite. Where tennantite replaces galena, seligmannite may occur along cleavage planes and cracks; also as isolated blebs at the border between both minerals. At these contacts seligmannite may contain minute inclusions of galena. Forms reaction rims between tennantite and galena, chalcopyrite, stromeyerite, bornite. Forms orientated rims around circular tennantite-aggregates. Diff.: jordanite and geocronite show another birefl., and more regular twinning. Ref.: 57, 442, 505, 522, 864, 1260, 1372, 1452, 1494, 1496.

Sn - SULPHOSALTS
AND
Sn - SULPHIDES

Sn-SULPHOSALTS and Sn-SULPHIDES

Name Formula Crystal System	1: Colour 2: Bireflectance 3: Anisotropy 4: Internal reflections	Reflectance in % (Wavelength in nanometres)
SAKURAIITE $(Cu,Ag)_2(In,Sn)(Zn,Fe)S_4$ In-analogue of kësterite Tetr.	1. Purplish olive-grey. 2. Not present. 3. Weak. 4. Not present.	About 20 %.
HOCARTITE Ag_2SnFeS_4 Ag-analogue of stannite Tetr.	1. Brownish grey. $\longrightarrow$ stannite, shows a violet tint. 2. Weak, greyish brown to violet-grey. 3. Distinct to strong: orange to green (very similar to enargite, but colours are not as intense as for enargite). 4. Not observed.	$\quad\quad\quad R_o \quad R'_e$ 480 nm 24.8 22.6 540 $\quad$ 24.3 22.6 580 $\quad$ 24.0 22.2 640 $\quad$ 23.8 22.4 Dispersion curves: [231].
BERNDTITE SnS_2 Hex.	1. See at bireflection. 2. $\parallel$ basal plane: grey to slightly brown. $\perp$ basal plane: multicoloured. Depending on the quality of polishing, the characteristic yellow colour is more or less predominant. 3. Masked by abundant int. refl. 4. Very intense: yellowish brown to orange-yellow.	About 25 %.

Hardness H: polishing hardness VHN: micro-indentation hardness in kg/mm^2	Miscellaneous
H: slightly > stannite. VHN: no data.	Occurs in exsolution textures with stannite. Parag.: stannite, sphalerite, chalcopyrite, cassiterite, schapbachite, "danaite". Ref.: 703.
H: = stannite. VHN: no data.	Occurs as inclusions in stannite, wurtzite, sphalerite and argyrodite-canfieldite. Shows polysynthetic twinning. Forms orientated intergrowths, with stannite. Previously described as "stannite argentifère" by Périchaud et al. (1966). Diff.: canfieldite is isotropic. Ref.: 231.
H: <ottemannite. VHN: no data.	Forms tabular crystals. A pronounced basal cleavage is present. Occurs as small inclusions in pyrite. Replaced by cassiterite; replaces ottemannite. Other ass.: sphalerite, chalcopyrite, stannite, varlamoffite. Ref.: 270a, 934, 935, 936.

Name / Formula / Crystal System	1: Colour / 2: Bireflectance / 3: Anisotropy / 4: Internal reflections	Reflectance in % (Wavelength in nanometres)
STANNITE JAUNE $Cu_{2+x}Sn_{1-x}FeS_4$ $x = 0-0.44$ Intermediate member in the stannite-idaite series. Tetr.	1. + 2. Strong bireflection: yellowish brown to orange brown. 3. Strong: reddish, bluish and greenish tints. 4. Not present.	R_o 480 nm 20.0-21.2 540 22.7-25.0 580 25.2-27.3 640 28.8-29.6 R'_e 480 nm 22.2-24.4 540 25.2-27.5 580 27.4-29.4 640 30.0-31.9 Dispersion curves: [832].
STANNOIDITE $Cu_5Sn(Fe,Zn)_2S_8$ Ortho.	1. Brownish. ⟶ bornite (fresh surface), very similar. 2. Distinct, light salmon-brown to brown. 3. Strong: brownish, yellowish and greyish tints. 4. Not present.	470 nm 20.6-22.6 546 24.4-26.4 589 25.7-27.8 650 27.6-29.6 [194].
KËSTERITE* Cu_2SnZnS_4 Zn-analogue of stannite Tetr.	1. Greyish. ⟶ stannite, somewhat darker, no brownish tint. ⟶ mawsonite, grey. 2. Not present. 3. Weak. 4. Not present.	Range due to chem. comp. 470 nm 22.7-24.6 546 24.8-26.2 589 25.4-26.6 650 26.4-27.1 [1398]. Dispersion curve: [832].

* The name is also spelled: kösterite, köstérite, kusterite, custerite and kesterite; the mineral was named after the type locality, Këster, Yakutia, Siberia. In view of the pronunciation of the Russian letter ë, it should have been written as "kjosterite". Here the orthography of Fleischer (1966) is followed.

Hardness H: polishing hardness VHN: micro-indentation hardness in kg/mm^2	Miscellaneous
H: >bornite. VHN: no data.	The mineral is identical to the stannite-like minerals described by Orcel (1943) and Lévy (1956). According to Lévy (1967) the "stannite (?) I" of Ramdohr (1944), which afterwards has been named "hexastannite" (Ramdohr, 1960), may also be stannite jaune. The "red stannite" of Schermerhorn (1956) also seems to belong to this group. See also stannoidite. Data for stannite jaune and "hexastannite": Occurs as exsolution lamellae in stannite. Also as reaction rims between between chalcopyrite and stannite, and between cassiterite and bornite. Replaced by, and decomposes to an intergrowth of chalcopyrite and stannite. May contain inclusions of chalcopyrite, stannite and digenite. Other ass.: mawsonite, chalcocite, tetrahedrite, tennantite, pyrite, galena, enargite, arsenopyrite. Ref.: 143, 831, 832, 867, 932, 937, 1220, 1248, 1427.
H: >chalcopyrite, somewhat < stannite. VHN_{50}: 232-271 [194].	The composition is close to that of some stannite jaune spec. Ass.: chalcopyrite, stannite, galena. Ref.: 703a.
H: ~stannite. VHN_{50}: 320-322 [810].	According to Lévy (1967) këserite may be identical to the "stannite (?) II" of Ramdohr (1944), which also has been named isostannite. This "stannite (?) II" shows rare int. refl. and occurs as exsolution and replacement intergrowths with stannite and "hexastannite". It is replaced by cassiterite. Other ass.: Fe-hydroxides, chalcopyrite, covellite, digenite. Ref.: 657, 832, 932, 937, 1078, 1220, 1398, 1427.

Name Formula Crystal System	1: Colour 2: Bireflectance 3: Anisotropy 4: Internal reflections	Reflectance in % (Wavelength in nanometres)
STANNITE Cu_2SnFeS_4 Tetr.	1. Brownish olive-grey. ⟶ tetrahedrite, slightly darker; brownish grey. ⟶ sphalerite, lighter, yellow-brown to olive-green. ⟶ chalcopyrite, much darker, greenish brown. ⟶ cassiterite, brownish olive-green. ⟶ cylindrite, brown with a pinkish tint. ⟶ franckeite, yellowish brown. 2. Distinct, light brown to brownish olive-green. 3. Rather strong: yellowish brown, greyish olive-green, bluish or violet-grey. 4. Not present.	R_o 480 nm 24.4-25.2 540 27.4-28.2 580 27.9-28.8 640 27.9-29.5 R'_e 480 nm 25.7-26.7 540 27.4-29.3 580 29.0-29.4 640 28.0-29.8 Dispersion curves: [832]. See also: [1398].

Hardness H: polishing hardness VHN: micro-indentation hardness in kg/mm^2	Miscellaneous
H:>chalcopyrite,~tetrahedrite,<sphalerite. VHN: 140-326 [159, 1398, 1427, 1715].	Cleavage ∥ (110), rarely also ∥ (001) may appear by triangular pits in badly polished sections. Very fine compound twinning, sometimes with microcline-pattern, not uncommon. May also show fine patchy or coarse lamellar twinning. Zonal texture may be shown, often due to orientated exsolution products (chalcopyrite, sphalerite). Forms orientated intergrowths with sphalerite, tetrahedrite, chalcopyrite, "hexastannite", and isostannite; occurs as reaction rims between cassiterite and pyrrhotite, chalcopyrite, tetrahedrite, sphalerite; canfieldite and franckeite may form reaction rims on stannite. Shows inclusions of chalcopyrite (often very minute particles) and sphalerite, both commonly due to exsolution; of arsenopyrite, cassiterite, pyrite, tetrahedrite, bismuth, bismuthinite, boulangerite, rutile, gold, galena. May be enclosed in sphalerite, pyrrhotite, galena. Replaces galena, sphalerite, chalcopyrite, cassiterite, pyrrhotite, pyrite, arsenopyrite, bismuthinite, Co-Ni-Fe-arsenides. May be replaced by sphalerite, tetrahedrite, chalcopyrite, marcasite, bismuth, bismuthinite, zinkenite, franckeite, andorite, galena, covellite, Ag-minerals. Other ass.: stannite-like minerals, wolframite, scheelite, silver, jamesonite, bournonite, geocronite, cylindrite, teallite, herzenbergite, argyrodite, argentite, Co-Ni-Fe-arsenides, cubanite, parkerite, bornite, tetradymite, pentlandite, hematite. Ref.: 7, 9, 11, 12, 172, 224, 239, 336, 339, 341, 369, 474, 588, 648, 797, 832, 841, 951, 1014, 1046, 1069, 1070, 1073, 1192, 1206, 1207, 1220, 1237, 1320, 1328, 1358, 1365, 1379, 1392, 1398, 1451, 1462, 1603, 1629, 1732.

Note:
Several unnamed varieties of stannite have been reported:
- by Ramdohr (1944): "stannite (?) I" (see stannite jaune), "stannite (?) II" (see kësterite, "stannite (?) III" and "stannite (?) IV" (see below);
- by Orcel (1943) and Lévy (1956): see stannite jaune;
- by Ramdohr (1960) and Moh and Ottemann (1962): "brown stannites": they are isotropic or weakly anisotropic, and show no olive colour; no further data;
- by Picot, Troly and Vincienne (1963): see mawsonite;
- by Périchaud et al. (1966): "argentiferous stannite": see hocartite.

Name Formula Crystal System	1: Colour 2: Bireflectance 3: Anisotropy 4: Internal reflections	Reflectance in % (Wavelength in nanometres)
"STANNITE (?) III" (Ramdohr, 1944) Mineral of the $CuSnS_2$-$AgSnS_2$ series, with usually some Zn and Fe present. Hex.	1. Brownish to pinkish grey. ⟶ stannite, somewhat lighter and more brownish. 2. Weak (in oil). 3. Weak, but distinct. 4. Not present.	No data.
"STANNITE (?) IV" (Ramdohr, 1944) Probably a Sn-Ag-Zn-tetrahedrite. Cubic	1. ⟶ stannite, much lighter, more bluish. ⟶ kësterite, distinctly lighter. ⟶ "stannite (?) III", somewhat lighter. 2. Not present. 3. Very weak (in oil). 4. Not present.	No data.
RHODOSTANNITE $Cu_2Sn_3FeS_8$ Hex.	1. Reddish in comparison to stannite. 2. Not indicated. 3. Distinct, not as strong as for stannite: bluish grey to dark brown. 4. Not observed.	520 nm 27.8 (mean refl.) [1427].

Hardness H: polishing hardness VHN: micro-indentation hardness in kg/mm^2	Miscellaneous
H: no data. VHN: no data.	Takes a bad polish; always shows a characteristic "porosity". May show parquet-like twinning. Replaces "stannite (?) IV". May contain inclusions of cassiterite and sphalerite. Parag.: pyrite, marcasite, sphalerite, wurtzite, andorite and other Pb-Ag-sulphosalts, galena, chalcocite, Zn-stannite, isostannite, herzenbergite. Ref.: 932, 937, 1220.
H: no data. VHN: no data.	Occurs as cubic crystals and as allotriomorphic grains. Shows twin lamellae. Replaced by "stannite (?) III" along the twin lamellae. A rim of sphalerite is always present at the contact between "stannite (?) III" and "stannite (?) IV". Parag.: galena, pyrite, pyrargyrite, andorite and other Pb-Ag-sulphosalts. Ref.: 932, 937, 1220.
H: no data. VHN: 243-266 [1427].	Occurs as fine-grained alteration product of stannite. The replacement is accompanied by a porous texture due to shrinkage in volume. Ref.: 1427.

Name Formula Crystal System	1: Colour 2: Bireflectance 3: Anisotropy 4: Internal reflections	Reflectance in % (Wavelength in nanometres)
CYLINDRITE $6PbS \cdot 6SnS_2 \cdot Sb_2S_3$ (?) Ortho.	1. Greyish white. ⟶ galena, slightly darker. ⟶ stannite, very light grey. ⟶ sphalerite and wurtzite, yellowish white. ⟶ pyrrhotite, greenish grey. ⟶ cassiterite, grey-white. ⟶ franckeite, greyer. 2. Distinct: ∥ elongation: creamy white, ⊥ elongation: bluish grey-white. 3. Weak to distinct, but not as distinct as for franckeite: dark blue-black to yellow-white. 4. Not present.	470 nm 30.4-32.9 546 28.2-30.9 589 28.1-30.9 650 27.9-30.6 [1398].
OTTEMANNITE Sn_2S_3 Ortho.	1. Mouse-grey. ⟶ herzenbergite, significantly darker. 2. Weak to distinct, slight variations in luster, not in colour. 3. Strong: reddish brown to bluish grey. 4. Orange-brown.	About 30 %.
FRANCKEITE $5PbS \cdot 3SnS_2 \cdot Sb_2S_3$ Mono.	1. Greyish white. ⟶ galena, slightly greyer. ⟶ teallite, darker and less yellow. ⟶ stannite, greyish white. ⟶ sphalerite and wurtzite, yellowish white. ⟶ pyrite and marcasite, whitish grey. ⟶ boulangerite, greyer. ⟶ cylindrite, less grey. 2. Weak, grey-white to grey-white with a brown tint. 3. Distinct: light grey with a brownish tint to dark grey. 4. Not present.	470 nm 31.6-35.1 546 26.6-34.3 589 29.9-33.8 650 29.1-33.8 [1398].

317

Hardness H: polishing hardness VHN: micro-indentation hardness in kg/mm^2	Miscellaneous
H: slightly > franckeite, < chalcopyrite, sphalerite, stannite. VHN$_{25}$: 31-131 [1398, 1715].	Transverse sections show a concentric texture with layers of different thickness. In longitudinal sections these layers appear as twinning lamellae, approaching each other towards the end of the crystals. Between the cylindrite-layers, layers of franckeite may appear (with a higher refl. and a more distinct anisotr.), mostly occurring as circle-segments. The cylindrite-layers may show radial twinning formed by pressure and interrupted by the franckeite-layers. The cores of the cylinders commonly consist of a fine-grained mixture of cylindrite, sphalerite, quartz and occasionally cassiterite. Cylindrite may show inclusions of stannite, sphalerite, cassiterite, pyrite. May occur as drops or subhedral grains in chalcopyrite and in the cores of wurtzite-globules. Alters to an aggregate of a graphic-like intergrowth of galena, boulangerite and cassiterite. Ref.: 7, 35, 259, 951, 967, 984, 1192, 1237, 1392, 1398.
H: > berndtite, < herzenbergite. VHN: no data.	Commonly occurs as small laths. Many crystals are twinned. Replaces stannite, herzenbergite. Replaced by cassiterite, berndtite. Other ass.: isostannite, pyrite, chalcopyrite, covellite, Fe-oxides. Ref.: 934, 935, 936.
H: slightly > galena, distinctly < teallite, ≪ sphalerite. VHN: 13-108 [1715]. VHN$_{25}$: 32-85 [1398]. VHN$_{10}$: 23-52 [810].	May form tabular crystals, prisms, fibers (with feathery appearance) or spheroidal aggregates with radial texture. Cleavage ∥ (010) often visible. The possibility of displacement ∥ (001) which results in a flexibility causing twins, may be the most typical difference between franckeite and teallite. Wedge-shaped twins also occur. Replaces stannite, galena, sphalerite; replaced by cassiterite. May contain orientated exsolutions of pyrrhotite. Encloses stannite, pyrrhotite, canfieldite. Alters to a graphic-like intergrowth of cassiterite-jamesonite or galena-boulangerite-cassiterite aggregates. Other ass.: cylindrite, plagionite, zinkenite, pyrite, marcasite, sphalerite, wurtzite, andorite, tetrahedrite, geocronite. Ref.: 7, 224, 239, 284, 369, 656, 951, 1038, 1169, 1192, 1237, 1392, 1398, 1462, 1576.

Name Formula Crystal System	1: Colour 2: Bireflectance 3: Anisotropy 4: Internal reflections	Reflectance in % (Wavelength in nanometres)
HERZENBERGITE SnS Ortho.	1. Greyish white; well polished surfaces very similar to teallite and franckeite. 2. In oil: weak, but distinct on grain boundaries. $\parallel$ tabular planes: bluish white, $\perp$ tabular planes: somewhat darker, yellowish white. 3. Strong, straight extinction; 45° (nicols exactly crossed): bright red to yellowish red colours; blue and violet in other positions. 4. A deep red-brown int. refl. may be visible in oil.	546 nm 42.1-44.3 [194].
TEALLITE PbS.SnS (may contain a large amount of Zn and Ag). Ortho.	1. White with a creamy pink tint. $\longrightarrow$ galena, yellowish. $\longrightarrow$ franckeite, slightly lighter, yellowish. $\longrightarrow$ cassiterite, white. $\longrightarrow$ sphalerite, yellowish white. $\longrightarrow$ pyrite and marcasite, grey-white. 2. Weak, but distinguishable: white ($\perp$001) to yellowish ($\parallel$ 001). 3. Very distinct: light grey, brownish grey, dark grey, steel-blue, violet. 4. Not present.	470 nm 40.1-48.3 546 40.2-46.6 589 40.1-45.8 650 39.9-44.4 [1398].

Hardness H: polishing hardness VHN: micro-indentation hardness in kg/mm^2	Miscellaneous
H: no data. VHN$_2$: 48-114 [653].	Occurs as thin-tabular crystals in subparallel aggregates; the crystals may be deformed. Zonal crystals are common (well visible with crossed nicols). Parag.: pyrite, pyrrhotite, cassiterite, stannite, sphalerite, chalcopyrite, cubanite, bismuth, bismuthinite, jamesonite, galena. Diff.: teallite shows a slightly higher refl.; franckeite has a weaker anisotr.; both are commonly coarser crystalline. Ref.: 633, 653, 927, 935, 1206, 1207. Note: "Montesite", $PbSn_4S_5$, described by Herzenberg (1949) as a part of the isotypic series between teallite and herzenbergite, has never been proved to be a valid mineral, neither by X-ray nor by optical methods.
H: $>$franckeite, $\ll$sphalerite. VHN: 31-83 [1715]. VHN$_{25}$: 66-125 [1398, 1715]. VHN$_{15}$: 52-65 [194].	Occurs as coarse laths; also platy, elongated $\parallel$ (001); the crystals may be deformed. Forms radiating aggregates of bladed crystals. Basal cleavage often distinct. May be polysynthetically twinned. Parallel displacement not uncommon. Wave-like textures and twinning occur. May be replaced by pyrite, marcasite, cassiterite, sphalerite, wurtzite, galena; may occur as inclusions in the replacing minerals. The replacements by galena and cassiterite may result in a graphic-like intergrowth. Diff.: franckeite has a slightly lower refl. and a weaker anisotr. Other ass.: cylindrite, franckeite, geocronite, other sulphosalts, arsenopyrite. Ref.: 7, 633, 951, 1192, 1237, 1392, 1398, 1576, 1578.

PLATINOID MINERALS

PLATINOID MINERALS, isotropic or weakly anisotropic

Name Formula Crystal System	1: Colour 2: Bireflectance 3: Anisotropy 4: Internal reflections	Reflectance in % (Wavelength in nanometres)
COOPERITE PtS (may contain some Pd). Tetr.	1. Brownish. → pyrrhotite, very similar, in oil more coffee-brown to olive-leather-brown. 2. Very weak, only visible in oil on grain boundaries. 3. Air: very weak (as distinct from pyrrhotite), grey-pink to grey-green. Oil: relatively strong. 4. Not present.	460 nm 40.5 550 39.0 600 37.0 650 36.5 [1382a].
LAURITE RuS_2 (may contain some Os, Ir). Cubic	1. White, in oil with a bluish grey tint. 2. + 3. Isotropic. 4. Not present.	Range due to differences in chem. comp. 470 nm 47.0-48.0 546 41.8-42.5 589 40.3 650 37.2-38.2 [820a].
HOLLINGWORTHITE (Rh,Pt,Pd,Ru,Ir)AsS Cubic	1. Greyish white. → Rh-sperrylite, slightly more bluish. → irarsite, somewhat lighter, no bluish tint. 2. + 3. Isotropic. 4. Not present.	Hollingworthite: about 40-45 %. Ru-hollingworthite: 520 nm 49.4 [496]. Ir-hollingworthite: 540 nm 52.5 [1382a].
IRARSITE (Ir,Ru,Rh,Pt)AsS Cubic	1. Greyish white with a characteristic blue tint. → hollingworthite, somewhat darker. 2. + 3. Isotropic. 4. Not present.	490 nm 47.6 520 47.8 589 46.1 650 45.2 Dispersion curve: [496].
ALLOPALLADIUM Pd Hex.	1. Yellowish white. → clausthalite, white-yellow. → chalcopyrite, bright white. 2. Very weak or not perceptible. 3. Weak to distinct. 4. Not present.	About 50 %.

Hardness H: polishing hardness VHN: micro-indentation hardness in kg/mm^2	Miscellaneous
H:<platinum, ≪sperrylite. VHN$_?$: 505-588 [1299a].	Occurs as xenomorphic grains and as prismatic crystals. Shows simple and polysynthetic twinning. Forms intergrowths with platinum, sperrylite and braggite. Forms myrmekitic intergrowths with platinum. Occurs as minute tabular inclusions in platinum and braggite. Sperrylite is replaced by intergrowths of platinum and cooperite. Alters to platinum. Parag.: other Pt-minerals. Ref.: 2, 68, 1237, 1339, 1637.
H: highest of all sulphides; >pyrite, sperrylite, chromite. VHN: 1393-2167 [820a, 1299a, 1715].	Occurs as diamond-shaped inclusions in chromite. Replaced by platinum. Other ass.: gold, sperrylite, pyrrhotite, Ni-Fe-sulphides. Ref.: 68, 820a, 1096, 1237.
H:>sperrylite. VHN$_?$: (Ru-hollingworthite): 848 [496]. VHN$_?$: (Ir-hollingworthite): 657 [820a].	Occurs as small grains closely intergrown with sperrylite, Rh-sperrylite, geversite. Forms rims around irarsite. Ref.: 496, 1301, 1302, 1484.
H: no data. VHN$_?$: 976 [496].	Always rimmed by hollingworthite. Also intergrown with hollingworthite and platinum. Other ass.: chromite, ilmenite, sperrylite, laurite, chalcopyrite, chalcocite. Ref.: 496, 1301, 1302. Note: The Rh-sperrylite described by Stumpfl and Clark (1965a) probably is a Pt-rich variety of irarsite (Rucklidge, 1969a).
H:≫ clausthalite, >gold, chalcopyrite, <pyrrhotite, ≪iridosmium VHN: no data.	Occurs as thick-tabular crystals with six-sided cross-sections. Cleavage // (0001) often visible. Replaced by clausthalite, gold, naumannite, tiemannite. Forms intergrowths with gold and naumannite. Parag.: Pt-minerals, gold, Se-minerals. Ref.: 260, 1237, 1552.

PLATINOID MINERALS, isotropic or weakly anisotropic

Name Formula Crystal System	1: Colour 2: Bireflectance 3: Anisotropy 4: Internal reflections	Reflectance in % (Wavelength in nanometres)
STIBIOPALLADINITE Pd_3Sb Probably ortho.	1. White; in air with a yellowish tint; in oil with a pinkish tint. ⟶ pyrrhotite, much lighter. ⟶ chalcopyrite, distinctly pink. ⟶ platinum, brownish. 2. Commonly not perceptible. 3. Weak, but distinct on grain boundaries. 4. Not present.	460 nm 49.0 540 54.6 580 54.6 660 57.2 [1382a].
SPERRYLITE $PtAs_2$ Cubic	1. White, in oil with a faint creamy or bluish tint. ⟶ pyrite, whiter. ⟶ platinum, distinctly darker. ⟶ stibiopalladinite, slightly darker. 2. + 3. Isotropic. 4. Not present.	460 nm 55.0 540 55.5 580 55.5 660 52.0 [1382a].
MICHENERITE (Pd,Pt)BiTe Cubic	1. Greyish white. 2. + 3. Isotropic. 4. Not present.	About 56 %.
POTARITE PdHg Tetr. [1516]. Cubic [260, 1129].	1. Pure white. 2. + 3. Isotropic. 4. Not present.	About 60 %.

Hardness H: polishing hardness VHN: micro-indentation hardness in kg/mm^2	Miscellaneous
H: ~platinum, ≪ sperrylite. VHN: no data.	Commonly occurs as granular aggregates and as irregular rounded grains. May show the same crystal-forms and twins as dyscrasite. A basal cleavage may be visible. May be replaced by sperrylite. Alters readily to an intergrowth of PdO, Sb-oxides and palladium, which is characteristic. Diff.: niccolite is much more anisotropic; platinum and palladium are isotropic; palladium and allopalladium do not show pinkish tints; sperrylite is isotropic. Parag.: platinum, sperrylite, cooperite, braggite, pyrrhotite, pentlandite, chalcopyrite. Ref.: 1, 1237, 1248, 1339. Notes: 1. Arsenopalladinite, Pd_3As, hexagonal. Ref.: 263. 2. Stannopalladinite, Pd_3Sn_2, hexagonal. Occurs as rounded and elongated grains closely associated with ferroplatinum. Ref.: 877.
H: ≫ platinum, >braggite, geversite ≧ pyrite, <laurite, hollingworthite. VHN: 960-1277 [194, 1715].	Usually occurs as idiomorphic cubic crystals. Cleavage ∥ (100) occasionally visible. No twinning or zoning observed. Forms intergrowths, sometimes myrmekitic, with platinum and magnetite. May occur as orientated rods in platinum. Replaces stibiopalladinite. May be replaced by chalcopyrite, pentlandite, cubanite, sphalerite, galena, stannite, platinum, cooperite, braggite. Occurs as inclusions in chalcopyrite, pyrrhotite, pentlandite, chromite. May contain inclusions of chalcopyrite, pyrrhotite, sphalerite, gold, Pd-minerals, hessite. Other ass.: parkerite, Ni-arsenides, iridosmium, Rh-sperrylite, geversite, hollingworthite, copper. Ref.: 51, 348, 588, 592, 1237, 1323, 1342, 1392, 1484, 1629. Note: Rh-sperrylite contains up to 11.6 % Rh. Shows a medium grey colour with a bluish tint and is much darker than sperrylite (about 40-45 %). H: > sperrylite, geversite, <hollingworthite. Ass.: sperrylite, geversite, hollingworthite. Ref.: 1484. According to Rucklidge (1969a) this Rh-sperrylite probably is a Pt-rich variety of irarsite.
H: >chalcopyrite, ~kotulskite, <moncheite. VHN: no data.	Parag.: hessite, maucherite, moncheite, kotulskite. Ref.: 495, 588, 589.
H: low. VHN: no data.	Shows a columnar or fibrous texture. Contains inclusions of a distinctly anisotropic compound of Pd and Hg which is light grey in contrast. Ref.: 260, 583, 1129, 1237, 1425, 1516.

PLATINOID MINERALS, isotropic or weakly anisotropic

Name Formula Crystal System	1: Colour 2: Bireflectance 3: Anisotropy 4: Internal reflections	Reflectance in % (Wavelength in nanometres)
ZVYAGINTSEVITE $(Pd,Pt)_3(Pb,Sn)$ Cubic	1. White with a creamy tint. 2. + 3. Isotropic. 4. Not present.	486 nm 61.4 550 63.6 589 65.4 650 66.7 Dispersion curve: [214].
GEVERSITE $PtSb_2$ Cubic	1. Light grey. 2. + 3. Isotropic. 4. Not present.	About 65 %.
PALLADIUM Pd Cubic	1. White with a creamy tint. ⟶ platinum, slightly more yellowish. 2. + 3. Isotropic. 4. Not present.	About 70 %.
PLATINUM Pt Cubic	1. White. ⟶ antimony, slightly creamy. ⟶ silver, whiter; silver is slightly creamier, but lighter. ⟶ palladium, slightly more bluish; palladium is yellower. ⟶ iridosmium, lighter and more yellowish. ⟶ sperrylite, much lighter. ⟶ iridium, more yellowish. Iridosmium-rich Pt is slightly creamy. 2. + 3. Isotropic, but no complete extinction. 4. Not present.	About 70 %.

Hardness H: polishing hardness VHN: micro-indentation hardness in kg/mm^2	Miscellaneous
H: > cubanite, < platinum, sperrylite. VHN$_{15}$: 241-318 [214]. VHN$_?$: 316 [494].	Occurs in veins of pentlandite-chalcopyrite and pentlandite-cubanite-chalcopyrite as small grains and as skeletal forms rimmed by ferro-platinum. Other ass.: talnakhite, valleriite, electrum, magnetite. Ref.: 153, 214, 491, 494.
H: ≧ platinum, ≪ sperrylite VHN: no data.	Contains, or forms myrmekitic intergrowths with other platinoid minerals. Parag.: sperrylite, Rh-sperrylite, hollingworthite. Ref.: 1481, 1484.
H: distinctly < platinum. VHN: no data.	Occurs as small octahedral single crystals and as xenomorphic grains. Forms rims around Pt-sulphides embedded in platinum. May occur as an isomorphous constituent in some sulphide minerals. Occasionally enclosed by platinum. May be formed by the alteration of stibiopalladinite. Difficult to distinguish from platinum. Ref.: 348, 1237, 1367, 1680.
H: varying with the amount of admixtures: Fe-rich spec. show lowest H; > sphalerite, < pyrrhotite. VHN: 125-127 [159] VHN$_{50}$: 114-274 [810, 1715]. VHN$_{50}$: (for polyxen): 329-397 [810].	Commonly forms irregular xenomorphic grains, seldom idiomorphic crystals. Zonal texture not uncommon. May contain exsolution bodies of iridium, or irregular laths or grains of iridosmium, osmiridium and osmium, also due to exsolution. Replaces sperrylite and chromite. Contains inclusions of sperrylite, cooperite, braggite, laurite, stibiopalladinite, chromite, hematite. Occurs as inclusions in chromite. May occur interstitial to chromite grains. Other ass.: Cu-sulphides, bornite, chalcopyrite, magnetite. Ref.: 2, 125, 348, 833, 1237, 1482, 1485, 1603, 1637. Note: Ferroplatinum contains 16-21 % Fe; polyxen contains 6-11 % Fe and in addition some Ir, Os, Rh, Pt, Ru; platiniridium is Ir-rich platinum.

Name Formula Crystal System	1: Colour 2: Bireflectance 3: Anisotropy 4: Internal reflections	Reflectance in % (Wavelength in nanometres)
OSMIRIDIUM (Ir,Os) with Os-content <32 %. Cubic	1. Creamy white. ⟶ platinum, slightly darker. ⟶ osmium, yellowish. ⟶ iridosmium, creamy; iridosmium shows a distinct blue-grey tint. 2. + 3. Isotropic. 4. Not present.	Decreases with increasing Os-content. For osmiridium containing 10 % Os: 472 nm 75.8 556 80.0 582 78.4 Dispersion curve: [833].
IRIDIUM Ir Cubic	1. Similar to platinum. 2. + 3. Isotropic. 4. Not present.	472 nm 78.2 556 82.1 582 80.8 Dispersion curve: [833].

Hardness H: polishing hardness VHN: micro-indentation hardness in kg/mm^2	Miscellaneous
H: $\ll$ iridosmium. VHN: 297-645 [810, 1715].	For nomenclature of the osmiridium-iridosmium series see iridosmium. Osmiridium forms intergrowths with platinum and iridosmium. Encloses irregular or euhedral plates of iridosmium. Parag.: Pt-minerals. Ref.: 623, 742, 1485, 1680. Note: Aurosmiridium, (Ir,Os,Au), is cubic. Ass.: platinum, iridosmium. Ref.: 1738.
H: $\gg$ platinum. VHN: no data.	Only found as exsolution bodies in platinum. Various forms of the bodies: rounded or octahedral grains, lamellae $/\!/$ (100) of platinum, worm-like. These bodies, in turn, may contain exsolutions of platinum. Ref.: 833, 1237.

Name / Formula / Crystal System	1: Colour / 2: Bireflectance / 3: Anisotropy / 4: Internal reflections	Reflectance in % (Wavelength in nanometres)
BRAGGITE (Pt,Pd,Ni)S Tetr.	1. Creamy white with a faint violet tinge. ⟶ platinum, bluish or brownish grey. 2. Distinct (in oil): bluish tints. 3. Rather strong, esp. in oil: blue to brown, similar to arsenopyrite, but more subdued. 4. Not present.	589 nm 34.5-35.5 [1299a].
NIGGLIITE PtSn* Hex.	1. + 2. Strong and characteristic bireflectance: O bright pinkish cream, E pale cobalt-blue. 3. Extremely strong: fire-orange to greyish blue. 4. Not present.	About 25 to 65 % (due to bireflectance).
VYSOTSKITE (Pd,Ni,Pt)S Tetr.	1. Greyish white with a blue tint. 2. Only visible in oil: greyish blue to greyish lilac. 3. Moderate: bluish and brownish colour effects. 4. Not present.	About 45 %.
FROODITE $PdBi_2$ Mono.	1. Light grey. 2. Not indicated. 3. Strong, light to dark grey. 4. Not present.	About 50 %.
PALLADIUM BISMUTHIDE $PdBi_3$?	1. White with a creamy tint. 2. Weak. 3. Distinct. 4. Not present.	green 50.1 % [1725].

* The formula given by Groeneveld Meijer (1955), and by Gimpl et al. (1963) as PtTe, is not correct.

Hardness H: polishing hardness VHN: micro-indentation hardness in kg/mm^2	Miscellaneous
H: slightly > platinum, cooperite, stibiopalladinite, ≪ sperrylite. VHN$_2$: 742-1030 [1299a].	Occurs as idiomorphic tabular crystals. Twinning rarely observed. Forms alternating layers and myrmekitic intergrowths with platinum; also as inclusions in platinum. Other ass.: hematite, Pt-minerals. Ref.: 68, 348, 1237.
H: > galena, < tetrahedrite. VHN$_2$: 306-537 [1726].	Occurs as grains with six-sided or needle-like crystal sections in parkerite. Forms intergrowths with sperrylite and stannopalladinite. As inclusions in chalcopyrite and cubanite. Other ass.: hessite, other Te-minerals. Ref.: 765, 1209, 1342, 1680, 1726.
H: high. VHN: no data.	Occurs as well-formed prismatic crystals. Parag.: millerite, pyrite, chalcopyrite, linnaeite, cooperite. Ref.: 497, 1382.
H: no data. VHN: no data.	Parag.: occurs in the As- and Pb-Cu-rich ores of Sudbury. Ref.: 588, 589.
H: no data. VHN$_2$: 105-125 [1725].	Occurs often in close intergrowths with michenerite, niggliite, hessite and other tellurides, chalcopyrite. The mineral is similar to froodite, but the properties of froodite were shown to be similar to the synthetic phase PdBi$_2$. The grains of palladium bismuthide gave an X-ray powder pattern the last two lines of which only correspond to froodite. Ref.: 1725.

PLATINOID MINERALS, anisotropy distinct to strong

Name Formula Crystal System	1: Colour 2: Bireflectance 3: Anisotropy 4: Internal reflections	Reflectance in % (Wavelength in nanometres)

MONCHEITE
$(Pt,Pd)(Te,Bi)_2$
Hex.
	1. Air: bright greyish white.
	   Oil: light grey.
	   ⟶ ferroplatinum, more grey and less creamy.
	   ⟶ braggite, greyish white.
	   ⟶ chalcopyrite, bright white.
	2. Air: weak. Oil: distinct.
	3. Air: distinct to strong: light yellowish brown to dark brown.
	4. Not present.
		470 nm 53.0-56.8
		546 53.2-58.8
		589 52.9-58.1
		650 52.7-59.6
		
		Dispersion curves: [722].

KOTULSKITE $Pd(Te,Bi)_{1-2}$ Hex.	1. Air: cream or pale yellow. Oil: more yellow. Moncheite and merenskyite appear white, and chalcopyrite appears greenish yellow in comparison. 2. Air: distinct, light cream to a slightly darker greyish cream. Oil: more distinct. 3. Strong, grey or brownish to dark blue-grey. 4. Not present.	470 nm 53.0-57.3 546 58.7-64.4 589 61.7-65.6 650 64.1-67.9 Dispersion curves: [722].
MERENSKYITE $(Pd,Pt)(Te,Bi)_2$ Probably hex.	1. Air: white in comparison to the pale yellow kotulskite. ⟶ moncheite, lighter, and slightly more creamy. 2. Air: weak, white to greyish. Oil: more distinct, white with a slight creamy tint to light greyish white. 3. Distinct to strong, dark brown to light greenish grey. 4. Not present.	470 nm 60.9-62.2 546 63.2-65.2 589 64.4-67.0 650 64.3-67.4 Dispersion curves: [722].

Hardness H: polishing hardness VHN: micro-indentation hardness in kg/mm^2	Miscellaneous
H:>kotulskite, michenerite,~pentlandite, <chalcopyrite. VHN: no data.	Pt end-member of a series, the Pd end-member of which is merenskyite. Distinct cleavage $\parallel$ (0001). Generally occurs in the form of blebs or fine laths. Forms intergrowths with michenerite, kotulskite, ferroplatinum. Enclosed by chalcopyrite, pyrrhotite, violarite. Other ass.: braggite, magnetite. Ref.: 495, 722, 1382.
H:>chalcopyrite, ~michenerite,<pentlandite, moncheite, merenskyite. VHN$_{15}$: 236 [722].	Cleavage not observed. Forms intergrowths with moncheite, michenerite and merenskyite. Occurs as inclusions in chalcopyrite and pentlandite. Occasionally replaced by merenskyite. Ref.: 495, 722, 1615.
H:>chalcopyrite, kotulskite,<pentlandite. VHN: no data.	Pd end-member of a series, the Pt end-member of which is moncheite. Forms a solid solution series with melonite. Occurs intergrown with kotulskite, enclosed by or adjoining chalcopyrite and pentlandite. Other ass.: hollingworthite, millerite. Ref.: 722, 1301, 1302.

Name Formula Crystal System	1: Colour 2: Bireflectance 3: Anisotropy 4: Internal reflections	Reflectance in % (Wavelength in nanometres)
OSMIUM Os (Os-content>80 %). Hex.	1. Pure white, greyer with increasing Ir-content. ⟶ iridium, whiter. ⟶ osmiridium, whiter; osmiridium appears yellowish in contrast. ⟶ platinum, bluish grey. 2. Not present, even in oil. 3. Strong, vivid orange-red tints (similar to ilvaite). 4. Not present.	472 nm 64.7 -65.55 556 60.85-62.8 582 58.5 -59.9 Dispersion curve: [833].
IRIDOSMIUM (Os,Ir) with Os-content from 32 to 80 %. Hex.	1. White with a bluish grey tint. ⟶ platinum, darker. ⟶ osmiridium, distinct grey-blue tint. 2. Distinct, esp. in sections ∥ c-axis. 3. Weak to fairly strong; stronger with increasing Os-content; pale pinkish, reddish, bronze, and greyish to deep-blue tints. 4. Not present.	Decreases with increasing Os-content. For iridosmium with 74.5 % Os: 472 nm 68.2 556 65.15 582 61.5 For iridosmium with 49.5 % Os: 472 nm 71.7 556 72.0 582 69.3 Dispersion curves: [833].

Hardness H: polishing hardness VHN: micro-indentation hardness in kg/mm^2	Miscellaneous
H: very high. VHN: no data.	Forms elongated hexagonal crystals. Occurs as small lamellae in platinum. Parag.: platinum, osmiridium. Ref.: 833, 1485, 1680.
H: higher with increasing Os-content. ≫ osmiridium, platinum, < spinel. VHN: no data.	Note on the nomenclature: according to Hey (1963) osmiridium contains less than 32 % Os, and iridosmium more than 32 % Os. Iridosmium with more than 80 % Os should be called osmium. "Nevyanskite" is iridosmium with an Os-content between 32 and 50 %, and "sysertskite" is iridosmium with an Os-content higher than 50 %. Rutheniridosmium contains 40 % Os, 40 % Ir and 20 % Ru. It shows a perfect basal cleavage and is associated with gold. Ref.: 42. Iridosmium usually occurs as irregular or hexagonal flakes enclosed in platinum, irregularly distributed or orientated ∥ octahedral planes. A basal cleavage may be visible. Twinning not uncommon. Irregular or euhedral plates of iridosmium may be enclosed by osmiridium. Forms intergrowths with osmiridium. Parag.: other Pt-minerals. Ref.: 125, 623, 742, 1082, 1237, 1475, 1481, 1485, 1491, 1492, 1603, 1681. Note: An insufficiently defined sulphide of Os and Ir has been described by Ottemann and Augustithis (1967) and provisionally named "roseite". This proposal has not been submitted to the IMA-CNMMN; consequently, the name should not be used.

OXIDIC MANGANESE MINERALS

Name / Formula / Crystal System	1: Colour 2: Bireflectance 3: Anisotropy 4: Internal reflections	Reflectance in % (Wavelength in nanometres)
GAUDEFROYITE $Ca_4Mn_{3-x}[BO_3)_3/(CO_3)/(O,OH)_3]$ Hex.	1. + 2. Strong bireflectance, dark grey to a lighter brownish grey. 3. $\perp$ elongation, hexagonal grains: isotropic. $\parallel$ elongation: strong, grey-white tints. 4. Abundant, yellow-orange.	$\quad R_o \quad R_e$ 480 nm 10.3 13.5 520 10.2 13.1 600 10.1 12.5 [685].
RANCIEITE $(Fe,Mg)O.4MnO_2.4H_2O$ [1151]. $(Ca,Mn)O.4MnO_2.3H_2O$ [1266]. ?	1. Grey-white (for coarse aggregates); fine-grained masses show considerably lower refl. $\rightarrow$ todorokite, somewhat darker. 2. Distinct, grey-white to yellowish creamy grey. 3. Strong, stronger than for hausmannite; bluish grey to blue-black; undulatory extinction. 4. Not present (as distinct from todorokite).	green 12.5–15.0 [1151].
MANGANOSITE MnO Cubic	1. Grey with a greenish tint. Tarnishes rapidly in air. $\rightarrow$ sphalerite, distinctly darker. 2. + 3. Isotropic. 4. Always present; in fresh surfaces emerald-green, esp. well visible in oil; after some days a red int. refl. becomes more and more distinct, indicating beginning alteration to hausmannite.	470 nm 14.9 550 14.4 580 13.9 650 13.7 Dispersion curve: [1005].
HETAEROLITE $ZnMn_2O_4$ Tetr.	1. Dark grey. 2. Weak to distinct. 3. Strong (yellowish grey and brownish grey); in oil masked by the int. refl. 4. Reddish brown, common and abundant.	470 nm 14.2–18.6 550 13.4–17.5 580 13.2–17.2 650 12.2–16.0 Dispersion curves: [1005].

Hardness H: polishing hardness VHN: micro-indentation hardness in kg/mm^2	Miscellaneous
H: no data. VHN$_?$: 840 [685].	Forms acicular hexagonal prisms. A prismatic cleavage is present. Twinning not observed. May contain inclusions of hausmannite and braunite. Strongly veined by pyrolusite. Other ass.: cryptomelane, hematite, marokite, crednerite. Ref.: 531, 685.
H: < birnessite. VHN: no data.	Occurs as coarse- and fine-grained aggregates with fibrous texture. Forms rims around todorokite. Parag.: other Mn-oxides, Fe-hydroxides. Ref.: 403, 608, 1151, 1266, 1422.
H: no data. VHN: 314-325 [1004].	Commonly occurs as granular aggregates. Forms orientated intergrowths with zincite or periclase. Replaced by pyrochroite. Other ass.: hausmannite, manganite, jacobsite, other Mn- and Fe-oxides. Ref.: 362, 363, 441, 615, 1077, 1211.
H: ~hausmannite. VHN: 585-813 [1004, 1715].	Occurs as idiomorphic crystals and as polygonal aggregates; also forms lamellar aggregates often with radiated texture or concentric intergrowths with chalcophanite (alternating layers). Replaces manganite. The intimate, orientated intergrowth with franklinite is named zincian vredenburgite, franklinite forming the groundmass, hetaerolite the lamellar network; both minerals show a red int. refl., esp. in oil, which, however, is most marked in hetaerolite; the lightest position of hetaerolite (O) is whiter, the darker one (E) greyer than franklinite. Diff.: hydrohetaerolite has a more fibrous texture; hausmannite shows less int. refl. Ref.: 449, 451, 611, 882, 1077, 1178, 1241, 1255.

Name Formula Crystal System	1: Colour 2: Bireflectance 3: Anisotropy 4: Internal reflections	Reflectance in % (Wavelength in nanometres)
GROUTITE α-MnOOH (may contain some Sb) Ortho.	1. Greyish white to pale brownish grey. 2. Distinct; lightest position $\perp$ elongation. 3. Very strong, but no vivid colours; $\parallel$ elongation: dark violet-brown, $\perp$ elongation: pale brownish grey. 4. Abundant only in sections $\parallel$ (010) (in oil); rare in other sections; deep red-brown.	470 nm 12.6-20.8 550 12.4-20.0 580 12.2-20.0 650 11.6-18.9 Dispersion curves: [1005].
PYROCHROITE $Mn(OH)_2$ Hex.	1. Skye-blue in air, rapidly turning brown, and finally black by oxidation. Fresh polished surfaces show a low refl., lower than for sphalerite; the black oxidized material shows a much higher refl. 2. Weak (fresh material) to distinct (black material), esp. in oil. 3. Weak (fresh material) to distinct (black material), but here masked by the int. refl. 4. On oxidized black material: red.	470 nm 16.3-18.6 550 15.2-17.9 580 15.0-17.6 650 14.2-16.5 Dispersion curves: [1005].
HYDROHETAEROLITE $HZnMn^{3+}_{2-x}O_4$ Tetr.	1. Creamy grey. 2. Distinct. 3. Distinct to strong, black to white. 4. Abundant, in oil brown-red; rarer than for hausmannite and hetaerolite.	About 15-20 %.
MAROKITE $CaMn_2O_4$ Ortho.	1. Grey with a distinct brown tint. $\longrightarrow$ hausmannite and braunite, darker and browner. 2. $\parallel$ (100) and (010): yellowish grey to grey-brown. $\parallel$ (001): not apparent. 3. $\parallel$ (100) and (010): strong, intense and characteristic colours: yellowish green to greenish yellow. $\parallel$ (001): the colour (violet-red) remains intense on rotation of the stage. 4. Frequent, carmine-red.	480 nm 16.2-18.4 520 16.3-19.4 600 16.1-18.0 Dispersion curves: [471].

341

| Hardness
H: polishing hardness
VHN: micro-indentation hardness in kg/mm^2	Miscellaneous
H: no data.
VHN: 613-813 [1004].	Occurs as platy, tabular, wedge- or lens-shaped crystals, or as radiating aggregates of platy crystals. Cleavage $/\!/$ (010) and $/\!/$ (100).
Parag.: manganite, ramsdellite, franklinite, goethite, hematite, magnetite.
Very similar to manganite in polished section.
Ref.: 549, 576, 615, 724, 857, 1241, 1378. |
| H: no data.
VHN: 224-245 [1004]. | Forms lamellar aggregates commonly partly altered, as pyrochroite slowly alters to hausmannite by which process the mineral blackens, becomes opaque and shows a red int. refl. Perfect basal cleavage. When forming pseudomorphs after manganosite, pyrochroite shows bended or twisted fibres, indicating the mechanical deformation caused by volume increase due to hydration of manganosite.
Parag.: hausmannite, manganosite, psilomelane, alabandite, galaxite.
Ref.: 362, 363, 615, 1077, 1299, 1658.
Note:
An alteration product of pyrochroite has been described as "hydro-hausmannite" (Frondel, 1953, and Naganna, 1964). Bricker (1965) showed this alteration product to be an intergrowth of two minerals: hausmannite, and the new mineral feitknechtite, β-MnOOH, hexagonal. Feitknechtite forms very small hexagonal platelets within hausmannite.
Ref.: 167, 1299. |
| H:~hausmannite.
VHN: no data. | Very similar in all aspects to hetaerolite. Forms grains made up of mosaic aggregates each of which, when examined under crossed nicols, shows an internal spherulitic structure with very weak anisotropism. Twinning not observed.
Occurs as cavity filling as subhedral prismatic crystals in close association with chalcophanite. Alters to psilomelane.
Ref.: 789, 893, 1237, 1241, 1636. |
| H: no data.
VHN$_\varphi$: 800 [471]. | Forms well-developed prismatic crystals. Twinning not observed.
Parag.: hausmannite, braunite, cryptomelane, crednerite, pyrolusite, gaudefroyite.
Ref.: 471, 685, 826, 1611. |

Name Formula Crystal System	1: Colour 2: Bireflectance 3: Anisotropy 4: Internal reflections	Reflectance in % (Wavelength in nanometres)
MANGANITE γ-MnOOH Mono.	1. Grey to brownish grey. $\longrightarrow$ pyrolusite, dark grey, much darker. 2. Weak in cross sections; in parallel sections distinct to strong, especially in oil. $/\!/$ c (lightest) light greyish brown, $/\!/$ a dark grey with a brownish tint, $/\!/$ b (darkest) darker than $/\!/$ a, with olive tint. 3. Strong $/\!/$ elongation: yellowish, bluish grey, dark violet-grey; colours distinct from those shown by other Mn-minerals; cross sections show only dark colours and weak anisotr. 4. Blood-red, very common: especially well visible in oil and in sections $/\!/$ (010).	R_p R_m R_g 470 nm 15.0 17.5 21.9 550 14.8 17.0 21.4 580 14.7 17.0 20.7 650 13.7 15.7 19.4 Dispersion curves: [1005].
FRANKLINITE $(Zn,Fe,Mn)(Fe,Mn)_2O_4$ Cubic	1. Grey with a faint greenish tint. $\longrightarrow$ magnetite, grey-green tint; magnetite shows a more reddish tint. $\longrightarrow$ zincite, much lighter. $\longrightarrow$ sphalerite, lighter. $\longrightarrow$ hematite, much darker. 2. + 3. Isotropic. Due to tectonic deformation a very weak anomalous anisotropy may be visible: pinkish grey to grey-black. 4. Deep-red, abundant (on close examination in oil).	470 nm 18.9 550 18.4 580 18.2 650 17.1 Dispersion curve: [1005].

Hardness H: polishing hardness VHN: micro-indentation hardness in kg/mm^2	Miscellaneous
H: distinctly $<$ hausmannite, jacobsite, braunite, magnetite, $\ll$ pyrolusite. VHN: 367-803 [159, 194, 358, 1004, 1715]. VHN$_?$: 195-529 [510, 1298].	Forms prismatic crystals or lamellar crystal aggregates sometimes with radiated texture. Cleavage $/\!/$ (010) and $/\!/$ (110) often distinct, esp. in cross sections. Twinning lamellae very common. May be intergrown with psilomelane and pyrolusite. Alters to pyrolusite. Pyrolusite forms pseudomorphs after manganite. Replaced by pyrolusite, hetaerolite, psilomelane, coronadite. Other ass.: hausmannite, braunite, goethite. Diff.: hausmannite is harder and shows a much weaker birefl. and anisotr.; goethite shows a much weaker anisotr. and lower refl.; pyrolusite shows a much higher refl.; magnetite is isotropic and shows no int. refl.; braunite is much harder, shows no cleavage, a weaker birefl. and anisotr. Ref.: 320, 331, 358, 403, 510, 713, 747, 975, 1077, 1237, 1241, 1255, 1293, 1298, 1299.
H: $>$ zincite. VHN: 667-847 [194, 810, 1004, 1715].	Cleavage may be distinct. Twinning $/\!/$ (111) and $/\!/$ (100), and zoning occur. Forms orientated exsolution intergrowths with magnetite, hematite, gahnite or other spinels, or hetaerolite. Orientated intergrowths with hetaerolite are called "zincian vredenburgite": franklinite forms the groundmass, hetaerolite the anisotropic lamellar network. Other ass.: manganite, zincite. Diff.: magnetite does not show int. refl.; jacobsite is lighter and yellowish, and shows less int. refl. Ref.: 451, 726, 882, 1237, 1241, 1392.

Name Formula Crystal System	1: Colour 2: Bireflectance 3: Anisotropy 4: Internal reflections	Reflectance in % (Wavelength in nanometres)
LITHIOPHORITE $(Al,Li)MnO_2(OH)_2$ (Li-content may be as low as 0.2-0.3 %) Mono.	1. + 2. Very strong bireflectance (in oil): O white, E dark grey. 3. Extreme, black to white, sometimes with a steel-blue tint. 4. Not present (as distinct from chalcophanite).	About 10 to 20 %.
CHALCOPHANITE $(Zn,Mn,Fe)Mn_3O_7 \cdot 3H_2O$ Tricl.	1. + 2. Extremely strong and characteristic bireflectance, in oil much more distinct than in air; stronger than for molybdenite: O bright white, E dark grey. 3. Very strong, but without typical colours: white and grey-white tints. 4. Zn-rich chalcophanite: an intense carmine to deep-red int. refl. is very common, esp. in oil. Zn-deficient chalcophanite has no int. refl.	470 nm 10.6-32.2 550 9.6-27.3 580 9.4-25.7 650 9.1-23.5 Dispersion curves: [1005].
AURORITE $(Ag,Ba,Ca,Mn,...)Mn_3O_7 \cdot 3H_2O$ Tricl.	1. + 2. Strong bireflectance: creamy white to grey. 3. Strong: yellow-grey to brownish grey. 4. Absent, as distinct from chalcophanite.	About as for chalcophanite.

Hardness H: polishing hardness VHN: micro-indentation hardness in kg/mm^2	Miscellaneous
H: > cryptocrystalline cryptomelane, < pyrolusite. VHN: 60-100 [776b].	Usually occurs as fine-grained masses, sheet-like coatings, botryoidal crusts and colloform layers. Rarely as pseudohexagonal crystals. Mica-like cleavage. Replaces cryptomelane and nsutite. Forms pseudomorphs after garnet. Alters to pyrolusite. Other ass.: Mn-oxides. Very similar to chalcophanite and aurorite in polished section. Ref.: 402, 614, 615, 776b, 861, 925, 975, 1241, 1421, 1633.
H: low. VHN: 71-194 [159, 1004, 1715]. VHN$_{50}$: 107-246 [1715]. VHN$_{15}$: 72-86 [194].	Occurs as aggregates of tabular crystals and as radiating blades. As tiny prismatic crystals in secondary Mn-ores. Also as cryptocrystalline, colloform bands or layers which line or fill cavities in earlier Mn-oxides. Forms rhytmic banded textures with psilomelane or cryptomelane. Perfect basal cleavage always present (in crystals). Twinning-like intergrowths of tabular crystals not uncommon. Fills cracks in psilomelane, pyrolusite and other Mn-oxides. Replaces franklinite, manganite, pyrolusite. Other ass.: hetaerolite, hausmannite, todorokite, other Mn-oxides. Very similar to lithiophorite and aurorite in polished section. Ref.: 789, 796, 1077, 1178, 1198, 1211, 1241, 1392, 1627, 1635.
H: slightly < Ag-todorokite. VHN: no data.	Aurorite is the Ag-analogue of chalcophanite. Occurs intergrown and closely associated with Ag-bearing todorokite, cryptomelane, pyrolusite and birnessite (?). Very similar to lithiophorite and chalcophanite in polished section. Ref.: 1198.

Name Formula Crystal System	1: Colour 2: Bireflectance 3: Anisotropy 4: Internal reflections	Reflectance in % (Wavelength in nanometres)

HAUSMANNITE
$(Mn,Fe)Mn_2O_4$
Tetr.
	1. Bluish to brownish grey.
	$\longrightarrow$ jacobsite, somewhat greyer.
	$\longrightarrow$ bixbyite, much darker, no yellow tint.
	$\longrightarrow$ braunite, no brownish tint; otherwise similar.
	2. Very distinct in oil; in darkest position (in oil), a very typical dark lustre is shown, numerous fine scratches becoming visible; both lustre and scratches disappear on turning the stage.
	  O grey with a faint bluish tint,
	  E dark brownish grey.
	3. Strong, showing yellowish or yellow-brown, light grey or bluish grey colours; in lightest position numerous fine scratches appear. No definite extinction.
	4. Beautifully blood-red, not uncommon, especially in oil and in sections $\parallel$ (001).
		470 nm 18.1-21.4
		550 17.2-20.5
		580 16.8-19.8
		650 15.5-18.4
		Dispersion curves: [1005].

α-VREDENBURGITE $(Mn,Fe)_3O_4$ Tetr.	1. Grey. $\longrightarrow$ braunite, less grey. $\longrightarrow$ bixbyite, considerably duller; bixbyite is more yellowish. 2. Slight, even in oil: shades of grey. 3. Distinct: grey to black. Intensity of birefl. and anisotr. varies with composition; both increase with increasing Mn-content. 4. Not present.	About 18-20 %.

Hardness H: polishing hardness VHN: micro-indentation hardness in kg/mm^2	Miscellaneous
H:> manganite, pyrolusite, cryptomelane,< jacobsite, ≪ bixbyite. VHN: 466-724 [159, 194, 1004, 1715].	Usually forms coarse-grained aggregates often with mosaic texture or well-developed crystals; also as fine-grained veinlets replacing bixbyite. Irregular twinning very common and characteristic; lamellae often intersecting each other and of unequal width and hardness; untwinned spec. rare but may occur. Replaced by pyrolusite, psilomelane, cryptomelane, braunite. Replaces bixbyite, braunite. May contain inclusions of manganosite and braunite; psilomelane may be present in minute cracks. Alters to pyrolusite and psilomelane. The intimate exsolution intergrowth of lamellar hausmannite and jacobsite (groundmass) is called "β-vredenburgite". Hausmannite also occurs as exsolution lamellae in a groundmass of galaxite in a vredenburgite-type intergrowth. Forms orientated intergrowths with zincite. Diff.: braunite is weakly anisotropic and does not show compound twinning; manganite is more strongly anisotropic showing other colours. Ref.: 227, 320, 362, 363, 403, 415, 879, 880, 881, 1077, 1237, 1241, 1293, 1298, 1299, 1392, 1629, 1657. Note: An orthorhombic analogue of magnesian hausmannite has been reported by Fan-De-Lyan (1964) and named "rhombomagnojacobsite". It shows a light grey-white colour with a brownish yellow tint. Distinct bireflectance and anisotropy. VHN$_2$: 708. Refl. (green): 16.8-20.6 %. Polysynthetic twinning is present. The name has been disapproved by the IMA-CNMMN.
H:< bixbyite, braunite. VHN: no data.	Observed material was homogeneous, but in metastable state. Occurs as irregular masses, commonly as definite pseudomorphs after bixbyite. In larger crystals this replacement is more complete than in smaller ones. Marked lamellar parting, the result of orientated replacement of bixbyite. No cleavage. Ref.: 881. Notes: 1. β-Vredenburgite is the orientated, intimate intergrowth of lamellar hausmannite in a groundmass of jacobsite. Ref.: 881, 883, 1241. 2. A vredenburgite-like intergrowth of lamellar hausmannite in a groundmass of galaxite has been described by Watanabe and Kato (1966). 3. Zincian vredenburgite is a similar intergrowth of lamellar hetaerolite in a groundmass of franklinite. Ref.: 882.

Name Formula Crystal System	1: Colour 2: Bireflectance 3: Anisotropy 4: Internal reflections	Reflectance in % (Wavelength in nanometres)
QUENSELITE PbO.MnOOH Mono.	1. Bluish grey. With low magnification the colour is more creamy grey. 2. Distinct at twin boundaries (oil, high power); strongest in sections $/\!/$ (100) showing transverse cleavage. With low magnification the birefl. is very weak. $/\!/$ c lightest, bluish grey, $/\!/$ a and $/\!/$ b darker grey; between a and b no great difference. 3. Distinct at twin boundaries and in sections $/\!/$ (100); nicols not completely crossed: light grey to dark brown-grey. Random sections show very weak anisotr. with completely crossed nicols. 4. Usually visible, deep-red or yellowish brown; occasionally abundant.	546 nm 17.8-20.8 [194].
JACOBSITE $(Mn,Fe,Mg)(Fe,Mn)_2O_4$ Cubic	1. Varies with Mn-content: jacobsite with low Mn-content is rose-brown, and very similar to magnetite; jacobsite with very high Mn-content is brownish grey; jacobsite with intermediate Mn-content is grey with an olive tint. $\longrightarrow$ magnetite, distinctly olive. $\longrightarrow$ braunite, olive-green or yellowish tinge; braunite does not show olive tint. $\longrightarrow$ hausmannite, less grey. $\longrightarrow$ bixbyite, olive-grey. 2. + 3. Isotropic; occasionally very slightly anisotropic (dark greyish brown to light grey with a bluish tint). 4. Deep-red, may occur; more common with increasing Mn-content.	470 nm 18.8 550 19.6 580 19.6 650 19.0 Dispersion curve: [1005].

Hardness H: polishing hardness VHN: micro-indentation hardness in kg/mm^2	Miscellaneous
H: low; the tiny well-developed crystals must be prepared separately. VHN: 153-186 [194].	Occurs in minute crystals (<1 mm) grown on manganese ore (consisting of bixbyite, hausmannite, braunite, jacobsite, psilomelane) and formed in open fissures in this ore. Basal cleavage very pronounced and characteristic. Single twinning very common, often with irregular boundaries. Trilling and plagiolase-like lamellar twinning occur; lamellae of different width. Replaces pyrolusite and cryptomelane. Ref.: 1077, 1257.
H: ~magnetite, slightly <braunite. VHN: increases with increasing Mn-content VHN: 690-875 [159, 1004, 1306, 1715]. VHN$_?$: 575-724 [1298].	Forms polygonal grains, rounded idiomorphic crystals and fine-grained aggregates. Twinning and cleavage not observed. Often partly altered to secondary Mn- and Fe-minerals, such as goethite, pyrolusite, hematite. Occurs as inclusions in braunite, pyrolusite, psilomelane, bixbyite. Replaced by pyrolusite, psilomelane. Regular intergrowths with hausmannite are called "β-vredenburgite"; in these intergrowths jacobsite forms the groundmass and hausmannite the lamellar network. Other ass.: Mn-minerals, Fe-hydroxides, sulphides. Diff.: braunite only rarely shows int. refl. which are brown; magnetite shows other colours and no int. refl. Ref.: 320, 339, 403, 705, 726, 767, 880, 881, 901, 975, 1077, 1150, 1237, 1241, 1262, 1293, 1294, 1295, 1297, 1298, 1299, 1306, 1386, 1392, 1400, 1466.

Name Formula Crystal System	1: Colour 2: Bireflectance 3: Anisotropy 4: Internal reflections	Reflectance in % (Wavelength in nanometres)
BRAUNITE $MnMn_6[O_8/SiO_4]$ (Fe always replaces some Mn) Tetr.	1. Grey with a slight brownish tint. $\longrightarrow$ magnetite, brown tint less distinct, no reddish tinge. $\longrightarrow$ pyrolusite, much darker. psilomelane, darker. $\longrightarrow$ manganite, and hausmannite, similar colour but birefl. much weaker. $\longrightarrow$ bixbyite, dull grey, no yellow tint. $\longrightarrow$ jacobsite, greyer; jacobsite shows yellowish or olive tint. 2. Weak but distinct (in oil): shades of dark grey. 3. Weak but distinct (with strong illumination): brownish grey, slate-blue. Undulatory extinction often characteristic. 4. Rare, dark brown or deep-red; much rarer than in hausmannite, manganite or jacobsite.	470 nm 21.5-22.5 550 20.4-22.4 580 19.8-20.7 650 19.0-19.7 Dispersion curves: [1005].
TODOROKITE $(H_2O,...)_{\leqslant 2}(Mn,...)_{\leqslant 8}$ $(O,OH)_{16}$ Some varieties are Zn-, Ag-, or Sr-bearing. Ortho., or Mono. with β near 90°; may be isostructural with woodruffite.	1. Various shades of a characteristic pale grey. 2. Weak, shades of grey, with yellowish and brownish tints. 3. Strong, white to grey; undulatory extinction. 4. Not present.	525 nm 20-23 Dispersion curves: [776b].

Hardness H: polishing hardness VHN: micro-indentation hardness in kg/mm^2	Miscellaneous
H: slightly > magnetite, slightly < bixbyite, < most sections of hollandite. VHN: 280-1187 [159, 194, 776b, 1004, 1715]. VHN$_{50}$: 689-766 [610].	Usually forms compact or finely granular masses, hypidiomorphic crystals or well-developed crystals resembling octahedrons. Cleavage not observed; twinning rare. Zonal texture may occur. May show inclusions of jacobsite, hollandite or remnants of replaced bixbyite. Replaces bixbyite and hollandite along their cleavage planes, hematite; forms intergrowths with hematite, bixbyite, pyrolusite. May be replaced by pyrolusite, hausmannite, psilomelane. Other ass.: cryptomelane, magnetite, manganite, Mn-silicates. Diff.: hausmannite and manganite show much stronger anisotr., more often int. refl. and commonly lamellar twinning; magnetite and jacobsite are isotropic; bixbyite is distinctly yellowish. Ref.: 320, 331, 403, 415, 610, 767, 776b, 880, 901, 975, 1051, 1052, 1077, 1237, 1241, 1293, 1296, 1298, 1299, 1392, 1610. Note: "Braunite-II" is a more Fe-rich, ordered variety of braunite. Very similar to ordinary braunite, but the colour is yellow-brown, intermediate between that of ordinary braunite and bixbyite. Occurs intergrown with ordinary braunite along certain crystallographic directions, and as single crystals which are replaced by cryptomelane. Ref.: 1610.
H: no data. VHN: no data.	Occurs as columnar aggregates, fine fibrous and as irregular masses with botryoidal or layered structure. Also as fan- or sheave-like aggregates of radiating fibres or acicular crystallites. Cleavage perpendicular to the basal plane and parallel to the elongation. Replaces and replaced by pyrolusite. Rimmed by rancieite. Parag.: magnetite, goethite, pyrolusite, cryptomelane, hollandite, psilomelane, chalcophanite, franklinite, manganite, birnessite, nsutite, maghemite, hausmannite, hematite, rancieite, The fibrous texture is very common and typical. Ref.: 453, 776b, 788, 789, 804, 818, 829, 843, 982, 1178, 1196, 1198, 1299, 1422, 1468, 1714.

Name Formula Crystal System	1: Colour 2: Bireflectance 3: Anisotropy 4: Internal reflections	Reflectance in % (Wavelength in nanometres)
BIXBYITE $(Mn,Fe)_2O_3$ Cubic	1. Grey with a distinct creamy or yellow tint. $\longrightarrow$ braunite, lighter, yellowish. $\longrightarrow$ jacobsite, much lighter and yellowish. $\longrightarrow$ hausmannite, distinctly yellow, much lighter. $\longrightarrow$ hollandite, much darker, distinctly brownish; hollandite is white in contrast. $\longrightarrow$ hematite, brownish. 2. Usually not present; sometimes very weak (in oil). 3. Isotropic; sometimes weakly anomalously anisotropic (nicols not completely crossed). 4. Not present.	470 nm 22.2 550 22.7 580 22.4 650 21.2 Dispersion curve: [1005].
BIRNESSITE $(Ca,Mg,Na,K)_{\ll 1}$ $(Mn^{4+},Mn^{2+})(O,OH)_2$?	1. Grey. 2. Observable. 3. Weak to distinct, grey tints; undulatory extinction. 4. Not present.	About 25 %.
RAMSDELLITE γ-MnO_2 Ortho.	1. Yellowish white. $\longrightarrow$ pyrolusite, olive to brownish tint. 2. Distinct, stronger in oil: yellowish white to more greyish. 3. Strong: yellow-brown to dark grey. 4. Very common (in oil): deep violet-red.	548 nm 11.7-33.0 589 8.9-23.4 [1632]. 525 nm 39.2-41.0 [776b].
WOODRUFFITE $(Zn,H_2O)_{\leqq 2}(Mn,Zn,...)_{\leqq 8}$ $(O,OH)_{16}$ Tetr.	1. + 2. Distinct bireflectance: grey to yellowish grey. 3. Very distinct. 4. Not indicated.	About 26 %.

Hardness H: polishing hardness VHN: micro-indentation hardness in kg/mm^2	Miscellaneous
H:> hausmannite,≥ braunite,>or<hollandite, depending on the orientation of hollandite. VHN: 882-1168 [159, 194, 1004, 1715].	Occurs as well-developed idiomorphic crystals and as granular aggregates. Cleavage ∥ (111) may be distinct. Coarse lamellar twinning, sometimes forming a regular network, not uncommon (visible when nicols not completely crossed). Zoning occurs. May partly be altered to an intergrowth of hematite and braunite. Shows inclusions of pyrolusite, hollandite, lamellar braunite or hausmannite ∥ (100). Replaced by braunite, very fine-grained hausmannite (following crystallographic directions), hematite (forming network in bixbyite). Other ass.: other Mn-minerals; occasionally magnetite, cassiterite, pseudobrookite. Diff.: jacobsite may show int. refl., has a lower refl. and hardness. Ref.: 320, 437, 878, 880, 987, 1051, 1052, 1077, 1237, 1241, 1298, 1299, 1392, 1710. Note: Partridgeite is pure Mn_2O_3, cubic. Optical properties very similar to bixbyite. Shows no twinning. Ass.: braunite, bixbyite. Ref.: 1609.
H:>rancieite. VHN: no data.	Occurs as reniform to botryoidal masses. Only found as supergene or secondary mineral. Parag.: cryptomelane, pyrolusite, ramsdellite, chalcophanite, franklinite, nsutite, todorokite. Ref.: 452, 683, 789, 830, 1299, 1422.
H:< pyrolusite. VHN: 93 [358]. VHN: 300-450 [1632]. VHN: 1130-1200 [776b].	May show diamond-shaped cross-sections (pseudomorphs after groutite). Commonly as small patches of fine-grained fibrous and easily cleavable crystals. Cleavage ∥ (110) and ∥ (010). Replaced by pyrolusite, replaces groutite. Other ass.: manganite, braunite, other Mn-minerals, goethite. Ref.: 358, 404, 615, 728, 776b, 795, 857, 1237, 1241, 1299, 1351, 1632.
H: no data. VHN: 744 [976].	Occurs as very fine grains. Alteration product of franklinite. Parag.: chalcophanite, cryptomelane, pyrolusite, lithiophorite, ramsdellite, jacobsite, manganite, braunite, psilomelane, franklinite, hausmannite. Ref.: 444, 975, 976.

Name Formula Crystal System	1: Colour 2: Bireflectance 3: Anisotropy 4: Internal reflections	Reflectance in % (Wavelength in nanometres)
CRYPTOMELANE $A_{\leq 2}B_8O_{16}$ A = chiefly K, some Na and Ba. B = chiefly Mn^{4+}, some Mn^{2+}, Zn, Al, Cu, Co, Fe^{3+} K-free cryptomelane has been reported. Tetr. and Mono.	1. Varying with mode of occurrence: grey-white, light tan, bluish grey-white. $\longrightarrow$ psilomelane, very similar. 2. Distinct. 3. For crystalline material: fairly strong, shades of grey. Cryptocrystalline material is isotropic. 4. Not present.	470 nm 28.0 550 26.7 580 26.0 650 23.9 Dispersion curve: [1005].
PSILOMELANE $A_3X_6Mn_8O_{16}$* A = Ba, Mn, Al, Fe, Si, etc. X = $(O,OH)_6$ with OH = 5. Ortho.*	1. Bluish grey to greyish white; very fine-grained aggregates are slightly greyer than coarse crystals. $\longrightarrow$ pyrolusite and hollandite, much darker, without yellow tint. $\longrightarrow$ braunite, manganite, jacobsite, hausmannite and bixbyite, much lighter. $\longrightarrow$ cryptomelane, very similar. $\longrightarrow$ magnetite, bluish white; magnetite is distinctly brown. 2. Strong: $\parallel$ c almost white, even in oil, $\perp$ c much darker, dull grey or bluish grey. 3. Strong, white to grey, straight extinction. 4. Brown, occasionally visible.	Varying: about 15 to 30 %.
CESAROLITE $PbO \cdot 3MnO_2 \cdot H_2O$?	1. White. 2. + 3. Only visible on crystalline aggregates. 4. Not indicated.	About 28 %.

* Other authors (Wadsley, 1953, and Fleischer, 1960) give as formula $(Ba,H_2O)_2Mn_5O_{10}$, and a monoclinic symmetry.

Hardness H: polishing hardness VHN: micro-indentation hardness in kg/mm^2	Miscellaneous
H: varying, fibrous masses showing very low hardness. VHN: 525-1048 [776b, 1004, 1715].	With pyrolusite commonest of the Mn-minerals. Occasionally occurs as well-developed fibrous or acicular crystals. Usually forms very fine-grained masses, less commonly botryoidal masses; also as colloform layers concentric with, or alternating with, layers of pyrolusite or nsutite. Other ass.: hollandite, braunite, bixbyite, hausmannite, lithiophorite, magnetite, hematite. Very similar to psilomelane. Can only be identified with certainty with X-ray methods. Ref.: 132, 209, 358, 389, 403, 415, 548, 616, 713, 776b, 789, 885, 955, 962, 974, 975, 1251, 1252, 1265, 1293, 1299, 1421. Note: Manjiroite, $(Na,Ca,K,Ba)(Mn,Fe,Al,Mg)_8O_{16} \cdot 1,64H_2O$, tetragonal, is the Na-analogue of cryptomelane. It shows a weak bireflectance and a distinct anisotropy. VHN: 181. Parag.: pyrolusite, nsutite, birnessite, cryptomelane, goethite. Ref.: 983.
H: varying, finest aggregates being hardest; increases with the amount of admixed iron hydroxide; reduced by increasing porosity; commonly <coarse-grained Mn-minerals. VHN: 203-813 [159, 194, 358, 776b, 1004].	Common Mn-mineral; only pyrolusite and cryptomelane are encountered more frequently. Forms finely crystalline aggregates or minute acicular crystals not unlike fine-grained pyrolusite; most commonly botryoidal masses sometimes consisting of concentric layers, which layers may also contain pyrolusite or cryptomelane; also irregular or cellular masses, rarely long prismatic crystals resembling a common variety of pyrolusite. May show very fine felted ice-flower structure. Replaces hollandite, braunite. Other ass.: bixbyite, hausmannite, Fe-hydroxides. Diff.: may be very difficult to identify psilomelane, cryptomelane, hollandite, coronadite, todorokite or woodruffite under the microscope. X-ray powder analyses give most certain results; presence of Ba typical for psilomelane. Ref.: 320, 321, 358, 399, 403, 718, 776b, 955, 962, 963, 964, 975, 1237, 1251, 1255, 1293, 1299, 1598, 1634.
H: no data. VHN: no data.	Occurs as aggregates of needle-like crystals and in colloform masses. Isotropic cryptocrystalline centers may be bordered by tiny spherulitic aggregates consisting of the same material with the same refl. May contain cores of amorphous hydrated Fe-hydroxides. Ref.: 192, 615, 1077.

Name Formula Crystal System	1: Colour 2: Bireflectance 3: Anisotropy 4: Internal reflections	Reflectance in % (Wavelength in nanometres)
HOLLANDITE $Ba_{\leq 2}R_8O_{16}$ R = mainly Mn^{4+}, also Mn^{2+}, Fe and Co. Tetr. and Mono.	1. White with a faint yellowish tint. $\rightarrow$ hematite, yellowish, no bluish tint. $\rightarrow$ braunite and bixbyite, much lighter, no brownish or olive tint. $\rightarrow$ pyrolusite, very similar or darker depending on the polish of pyrolusite. Basal sections are much duller greyish white and similar to psilomelane. $\rightarrow$ bixbyite, much lighter, white. 2. Distinct (in oil): $\parallel$ c white, $\perp$ c light grey. Basal sections are darkest and show no birefl. 3. Strong: grey, yellowish, pinkish white, bluish, violet-grey; basal sections are practically isotropic; undulatory extinction not uncommon. 4. Not present.	470 nm 27.1-33.0 550 25.6-32.3 580 25.2-31.5 650 24.3-29.4 Dispersion curves: [1005].
CREDNERITE $CuMnO_2$ Mono.	1. Bright creamy white. 2. Very strong: white or yellowish white to grey. 3. Strong and vivid, comparable to stibnite: white, yellowish white, light grey, violet-grey. Sometimes with undulatory extinction. 4. Not present.	470 nm 24.9-36.8 550 23.6-35.0 580 23.1-34.0 650 21.3-30.2 Dispersion curves: [1005].
CORONADITE $Pb_{\leq 2}Mn_8O_{16}$ Tetr.	1. White to greyish white, sometimes like galena. 2. Distinct. 3. Strong: white, dark grey, dark brown; straight extinction. Cross sections appear isotropic. 4. Not present.	470 nm 28.3-34.2 550 26.7-32.3 580 26.0-31.5 650 24.7-28.7 Dispersion curves: [1005].

Hardness H: polishing hardness VHN: micro-indentation hardness in kg/mm^2	Miscellaneous
H: (all sections) slightly <bixbyite, most sections ≫ braunite; some sections slightly <braunite; sections ⊥ c show greatest hardness. VHN: 272-1048 [159, 194, 1004, 1715].	Forms well-developed prismatic or tabular crystals, fibrous aggregates of needle-like crystals (as distinct from braunite and hausmannite), or coarse-grained and fine-grained, compact or botryoidal masses. Cleavage commonly distinct; sections ⊥ c show no cleavage. Twinning may occur, similar to that of hausmannite or hematite; in oil, under crossed nicols, an extremely fine lattice-like, intersecting lamellar texture may be visible. Occurs as inclusions in psilomelane, bixbyite, jacobsite; may contain inclusions of braunite, hausmannite. May be replaced by and intergrown with braunite, bixbyite, hematite and psilomelane, often along cleavage planes. Other ass.: pyrolusite, manganite. Diff.: hematite shows other colour, weak birefl., and often deep-red int. refl. Ref.: 320, 400, 403, 449, 452, 548, 955, 986, 1051, 1052, 1077, 1237, 1296, 1298, 1299.
H: no data. VHN: 327-357 [1004]. VHN$_2$: 200 [472].	Occurs as platy crystals, arranged in sheaf-like aggregates. Cleavage often distinct. Twinning, polysynthetic and mosaic-like, often occurs. Forms intergrowths with hausmannite. Alters to a mixture of Mn-oxides. The twins of crednerite remain visible in the alteration products. Other ass.: secondary Mn-minerals, manganite, hausmannite, psilomelane. Ref.: 472, 745, 894, 1077, 1422.
H: > manganite. VHN: 359-813 [159, 194, 1004, 1715]. VHN$_2$: 410-840 [510, 1298].	Occurs as granular, lamellar or fibrous aggregates; also as irregular, cellular or botryoidal masses, or in concentric zones, the crystalline fibrous zones alternating with isotropic gel-zones (probably amorphous coronadite). Occasionally occurs as single crystals of extremely acicular habit. Developes arrow-head twins. Replaces manganite. Other ass.: chalcophanite, psilomelane, other Mn-minerals, Fe-hydroxides. Ref.: 209, 403, 449, 510, 548, 615, 796, 839, 1068, 1072, 1293, 1298.

Name Formula Crystal System	1: Colour 2: Bireflectance 3: Anisotropy 4: Internal reflections	Reflectance in % (Wavelength in nanometres)
PYROLUSITE β-MnO_2 Tetr.	1. White with a distinctly creamy tint (in oil). Brightest of all Mn-minerals. $\longrightarrow$ hollandite, coarse crystals are distinctly yellower. $\longrightarrow$ magnetite, yellowish white. $\longrightarrow$ hematite, yellowish; hematite is bluish. $\longrightarrow$ manganite, almost white. $\longrightarrow$ ramsdellite, very similar. 2. Distinct (not as strong as for manganite), for coarse-grained varieties in oil. $/\!/$ E yellowish white, $/\!/$ O distinctly darker, white-grey. 3. Very strong: yellowish, dark brown, greenish blue, slate-grey (nicols exactly crossed); most typical for coarsely crystalline spec.; straight extinction. Very fine-grained aggregates appear to be almost isotropic. 4. Not present.	550 nm 27.2-40.8 [610]. 548 29.5-35.5 589 30.0-36.3 [1632]. 525 39.0-47.0 [776b]. 470 33.8-35.0 550 33.8-35.4 580 33.5-35.0 650 32.0-33.2 Dispersion curves: [1005].
NSUTITE $Mn^{4+}_{1-x}Mn^{2+}_{x}O_{2-2x}(OH)_{2x}$ x = 0.06-0.07 x = 0.16 for Mn-nsutite. Hex.	1. White with a slight creamy tint. $\longrightarrow$ pyrolusite, slightly greyer. 2. Cryptocrystalline aggregates are isotropic. Coarser crystals: distinct to strong, white-grey to dark grey. 3. Coarser crystals: strong, light and dark grey. 4. Not present.	About 30 to 40 %. Mn-nsutite is much darker than nsutite.

Hardness H: polishing hardness VHN: micro-indentation hardness in kg/mm^2	Miscellaneous
H: depending on orientation and type of crystal aggregates; may be as high as braunite. VHN: 76-1500 [159, 194, 358, 776b, 1004, 1632, 1715]. VHN$_{50}$: 532-575 [194]. VHN$_{15}$: 320-570 [610].	Occurs as coarse-grained euhedral tabular to prismatic crystals. Commonly fine-grained showing intersecting irregular prisms; also massive or in radiated crystals. Banded texture not uncommon. Cleavage $/\!/$ (110) usually distinct in coarse crystals. Twinning, single and lamellar, may occur. Forms pseudomorphs after manganite and cryptomelane; lens-shaped crystals, sometimes showing zoning and transverse "cleavage" or parting planes; needle-shaped crystals, often enclosed in psilomelane or manganite. Cryptocrystalline material may be intimately intergrown with psilomelane, Fe-hydroxides or hematite. Replaces manganite, braunite, psilomelane, magnetite, todorokite, ramsdellite. Replaced by todorokite. Other ass.: bixbyite, other Mn-minerals. Ref.: 227, 320, 321, 331, 358, 403, 415, 610, 614, 713, 718, 747, 776b, 789, 795, 1077, 1237, 1255, 1293, 1299, 1392, 1421, 1469, 1632.
H:>pyrolusite. VHN: 1003-1288 [1004]. VHN$_?$: 350-1150 [1079, 1739].	Occurs as coarse-grained crystals resembling pyrolusite, or as cryptocrystalline aggregates. Nsutite shows shrinkage cracks in fine-grained colloform aggregates. These cracks are caused by the transition from Mn-nsutite to nsutite. May form fan-like aggregates of fine fibers. Forms alternating layers with cryptomelane, pyrolusite and birnessite. Other ass.: goethite, todorokite, other Mn-minerals. Ref.: 20, 388, 789, 1079, 1299, 1421, 1422, 1739.

REFERENCES

1 Adam, H.R., 1927. A note on a new palladium mineral from the Potgietersrust platinum fields. J. Chem. Metall. Mining Soc. S. Africa, 27, 249-250. Ref. in: Am. Mineralogist, 13, 201 (1928) and 15, 242 (1930).
2 Adam, H.R., 1930. Notes on platinum group minerals from Rustenburg and Potgietersrust Districts, Transvaal. Trans. Geol. Soc. S. Africa, 33, 103-109.
3 Agrinier, H. and Geffroy, J., 1967. Les minéraux séléniés du point uranifère de Liauzun-en-Olloix (Puy-de-Dôme): clausthalite, sélénium natif, sélénite de plomb et chalcoménite. Bull. Soc. Franç. Minéral. Crist., 90, 383-386.
4 Agrinier, H., Geffroy, J. and Pulou, R., 1969. Paragenèse filonienne à uranium-sélénium-cuivre à Prévinquières, près Entràygues (Aveyron). Bull. Soc. Franç. Minéral. Crist., 92, 232-234.
5 Ahlfeld, F., 1931. The tin ores of Uncia-Llallagua, Bolivia. Econ. Geol., 26, 241-257.
6 Ahlfeld, F., 1932a. Die Silbererzlagerstätte Colquijirca (Peru). Z. Prakt. Geol., 40, 81-87.
7 Ahlfeld, F., 1932b. Die Erzlagerstätten in der tertiären Magmaprovinz der bolivianischen Zentralanden. (Beiträge zur Geologie und Mineralogie Boliviens. Nr. 1.) Neues Jahrb. Mineral., Beil., 65A, 285-446.
8 Ahlfeld, F., 1933. Die Kupfererzlagerstätte Naukat. (Beiträge zur Lagerstättenkunde von Westturkestan. Nr. 1.) Neues Jahrb. Mineral., Beil., 67 A, 467-485.
9 Ahlfeld, F., 1934. Ueber Zinnkies. (Beiträge zur Geologie und Mineralogie Boliviens. Nr. 5.) Neues Jahrb. Mineral., Beil., 68 A, 268-286.
10 Ahlfeld, F.,1935a. Die Antimonit- und Zinnoberlagerstätten im Alai. (Beiträge zur Lagerstättenkunde von Westturkestan. Nr. 2.) Neues Jahrb. Mineral., Beil., 69 A, 255-275.
11 Ahlfeld, F., 1935b. Neue Beobachtungen am Cerro von Potosi. Z. Prakt. Geol., 43, 167-171.
12 Ahlfeld, F., 1936. The Bolivian tinbelt. Econ. Geol., 31, 48-72.
13 Ahlfeld, F., 1938a. Epithermale Wolframlagerstätten in Bolivien. (Beiträge zur Geologie und Mineralogie Boliviens. Nr. 11.) Neues Jahrb. Mineral., Beil., 74 A, 1-19.
14 Ahlfeld, F., 1938b. Die Silber-Zinnerzlagerstätten von Colquechaca. (Beiträge zur Geologie und Mineralogie Boliviens. Nr. 12.) Neues Jahrb. Mineral., Beil., 74 A, 466-492.
15 Ahlfeld, F., 1951. A new locality for greenockite crystals in Bolivia. Am. Mineralogist, 36, 165-166.
16 Ahlfeld, F. and Moritz, H., 1933. Beitrag zur Kenntnis der Sulfostannate Boliviens. (Beiträge zur Geologie und Mineralogie Boliviens. Nr. 3.) Neues Jahrb. Mineral., Beil., 66 A, 179-210.
17 Ahlfeld, F. and Reyes, J.M., 1939. Die Bodenschätze Boliviens. Berlin.
18 Aicard, P., Picot, P., Pierrot, R. and Poulain, P.A., 1968. Sur la présence de ménéghinite dans deux gîtes français. Bull. Soc. Franç. Minéral. Crist., 91, 497-499.
19 Aires-Barros, L., 1961a. Sur une occurrence de maghémite à Goa. Bol. Soc. Geol. Port., 14, 31-35.
20 Aires-Barros, L., 1961b. Sur une occurrence de γ-MnO_2 à Goa. Bol. Soc. Geol. Port., 14, 37-42.
21 Aires-Barros, L., 1961c. Mineralogia dos Filões da Concessão da Folha da Atalaia, Almeida. Bol. Soc. Geol. Port., 14, 105-119.
22 Akatsuka, K., 1961. On mineralogenesis appeared in the main lodes of the Nakase Mine, Hyôgo Prefecture, Japan. J. Sci. Hiroshima Univ., Ser. C, 4, 1-34.
23 Aksenov, V.S., Kosyak, Ye.A. Mergenov, Sh.K. and Rafikov, T.K., 1968. A new bismuth telluride, Bi_2Te_5. Dokl. Akad. Nauk SSSR, 181, 443-446. Translated in: Dokl. Acad. Sci. USSR, Earth Sci. Sect., 181, 113-115 (1969).
24 Aleksandrov, S.M., Akhmanova, M.V., and Karyakin, A.V., 1965. Investigation of the ludwigite-vonsenite group of borates by means of their infrared spectra. Geochem. Intern., 2, 822-827.

25 Alexandrov, A.I., 1955. On native tin from alluvial placers of the river Is (middle Urals). (in Russian). Zap. Vses. Mineralog. Obshchestva, 84, 462-464. Ref. in: Mineral. Abstr., 13, 160-161.
26 Aliyev, R.M., 1965. Cubic crystals of magnetite from the Dashkesan deposit. Dokl. Akad. Nauk SSSR, 162, 422-424. Translated in: Dokl. Acad. Sci. USSR, Earth. Sci. Sect., 162, 141-142 (1965).
27 Allman, R., Baumann, I., Kutoglu, A., Rösch, H., and Hellner, E., 1964. Die Kristalstruktur des Patronits. Naturwissenschaften, 51, 263-264. Ref. in: Mineral. Abstr., 17, 558.
28 Anderson, A.L., 1934. Some pseudo-eutectic ore textures. Econ. Geol., 29, 577-589.
29 Anderson, A.L., 1940. Aikinite and silver enrichment at the St. Louis mine, Butt County, Idaho. Econ. Geol., 35, 520-533.
30 Anderson, A.L., 1941. A copper deposit of the Ducktown type near the Coeur d'Alene District, Idaho. Econ. Geol., 36, 641-657.
31 Anderson, A.L., 1946. Lead-silver mineralization in the Clark Fork District, Bonner County, Idaho. Econ. Geol., 41, 105-123.
32 Anderson, A.L., 1947a. Cobalt mineralization in the Blackbird District, Lemhi County, Idaho. Econ. Geol., 42, 22-46.
33 Anderson, A.L., 1947b. Lead-silver mineralization in the Clark Fork District. Econ. Geol., 42, 305-306.
34 Anderson, A.L., 1948. Tungsten mineralization at the Ima mine, Blue Wing District, Lemhi County, Idaho. Econ. Geol., 43, 181-206.
35 Anderson, A.L., 1963. Silver mineralization in the Mineral District, Washington County, Idaho. Econ. Geol. 58, 1195-1217.
36 Anderson, A.L. and Rasor, A.C., 1934. Silver mineralization in the Banner District, Boise County, Idaho. Econ. Geol., 29, 371-387.
37 Anderson, R.J., 1940. Microscopic features of ore from the Sunshine mine. Econ. Geol., 35, 659-667.
38 Andronopoulos, B., 1961. Association de magnétite-chromite-pentlandite dans quelques gîtes de fer en Grèce. Bull. Soc. Franç. Minéral. Crist. 84, 345-349.
39 Angelelli, V., and Gordon, S.G., 1948. Sanmartinite, a new zinc tungstate from Argentina. Notulae Naturae Acad. Nat. Sci. Phila., Nr. 205, 7pp. Ref. in: Am. Mineralogist, 33, 653.
40 Ankinovitch, E.A., 1958. Sulvanit aus den Ton-Anthraxolith-Schiefern im Gebirge Karatau und Dsebaglinsk. (in Russian). Izv. Akad. Nauk Kaz. SSR, Ser. Geol., Nr. 1, 29-37. Ref. in: Zentr. Mineral. I, 1960, 463-464.
41 Antun, P., El Goresy, A., and Ramdohr, P., 1966. Ein neuartiger Typ hydrothermaler Cu-Ni-Lagerstätten. Mit Bemerkungen über die Mineralien: Valleriit, Mackinawit, Oregonit. Mineral. Deposita, 1, 113-132.
42 Aoyama, S., 1936. A new mineral "ruthenosmiridium". Sci. Rept. Tôhoku Univ., Ser. 1, K. Honda Anniv. Volume, 1936, 527-547. Ref. in: Mineral. Abstr., 7, 315-316.
43 Araya, R., 1968. Untersuchungen über das Reflexionsvermögen einiger Erzmineralien unter Berücksichtigung unterschiedlicher Versuchsbedingungen für seine Bestimmung. Ph.D. Thesis, University of Munich.
44 Armstrong, H.S., 1943. Gold ores of the Little Long Lac area, Ontario. Econ. Geol., 38, 204-252.
45 Arnold, R.G., 1966. Composition, structure and temperature of stability of natural and synthetic pyrrhotites. Can. Mineralogist, 8, 660-661.
46 Arnold, R.G., 1967. Range in composition and structure of 82 natural terrestrial pyrrhotites. Can. Mineralogist, 9, 31-50.
47 Aubert, G., Bariand, P., Burnol, L. and Geffroy, J., 1962. Présence de scheelite en association avec le mispickel à Tournebise (Puy-de-Dôme). Bull. Soc. Franç., Minéral. Crist. 85, 459-460.
48 Auger, P.E., 1941. Zoning and district variations of the minor elements in pyrite of Canadian gold deposits. Econ. Geol., 36, 401-423.
49 Augustithis, S.S., 1960. Alteration of chromite. Ore-microscopic observations on chromite-ores from Rodiani, Greece. Neues Jahrb. Mineral., Abhandl., 94, 890-904.
50 Augustithis, S.S., 1964. Geochemical and ore-microscopic studies of hydrothermal and pegmatitic primary uranium parageneses. Nova Acta Leopoldina, Neue Folge, Nr. 170, Band 28.
51 Augustithis, S.S., 1965. Mineralogical and geochemical studies of the platiniferous dunite-birbirite-pyroxenite complex of Yubdo/Birbir, W. Ethiopia. Chem. Erde, 24, 159-196.

52 Azaroff, L.V., and Buerger, M.J., 1955. Refinement of the structure of cubanite, $CuFe_2S_3$. Am. Mineralogist, 40, 213-225.
53 Babkin, P.V. and Kim, Ye.P., 1966. First find of native arsenic in a Chukotka mercury deposit. Dokl. Akad. Nauk SSSR, 169, 424-427. Translated in: Dokl. Acad. Sci. USSR, Earth Sci. Sect., 169, 120-122 (1967).
54 Bachmann, H.G., 1952. Ueber Pyrit mit vollkommener Teilbarkeit nach (111) von Persberg, Schweden. Neues Jahrb. Mineral., Monatsh., 1952, 103-108.
55 Bächtiger, K., 1960. Die Kupfermineralisation an der Mürtschenalp (Kanton Glarus, Schweiz). Neues Jahrb. Mineral., Abhandl., 94, 627-635.
56 Badalov, S.T. and Turesebekoc, A., 1966. Monoclinic pyrrhotite from anhydritized skarns at Almalyk (Uzbek, S.S.R.) (in Russian). Dokl. Akad. Nauk Uz. SSR. 12, 27-29. Ref. in: Mineral. Abstr., 19, 99.
57 Bader, H., 1934. Beitrag zur Kenntnis der Gesteine und Minerallagerstätten des Binnentals. Schweiz. Mineral. Petrog. Mitt., 14, 319-441.
58 Bagdasarov, Yu.A., Gaydukova, V.S., Kuznetsova, N.N. and Sidorenko, G.A., 1962. Lueshite from carbonatites of Siberia. Dokl. Akad. Nauk SSSR, 147, 1168-1171. Translated in: Dokl. Acad. Sci. USSR, Earth Sci. Sect., 147, 157-159 (1964).
59 Bailly, R., 1942. Propriétés optiques de la thoreaulite. Ann. Soc. Géol. Belg., 65, Bull., 169-171.
60 Baker, G., 1952. Opaque oxides in some rocks of the basement complex, Torricelli Mountains, New Guinea. Am. Mineralogist, 37, 567-577.
61 Baker, G., 1953. Ilvaite and prehnite in micropegmatitic diorite, Southeast Papua. Am. Mineralogist, 38, 840-844.
62 Baker, G., 1958. Tellurides and selenides in the Phantom Lode, Great Boulder mine, Kalgoorlie. Australasian Inst. Mining Met. Proc., Stillwell Ann. Volume, 15-40.
63 Baldock, J.W., 1968. Calzirtite and the mineralogy of residual soils from the Bukusu carbonatite complex, south-eastern Uganda. Mineral. Mag., 36, 770-774.
64 Ballmer, G.J., 1932. Native tellurium from northwest of Silver City, New Mexico. Am. Mineralogist, 17, 491-492.
65 Banás, M and Ottemann, J., 1967. Bohdanowiczyt - newy naturalny selenek srebra i bismutu z Kletna w Sudetach (in Polish). Przeglad Geol., 15, 240. Ref. in: Mineral. Abstr., 20, 148.
66 Bannister, F.A., 1941. The crystal structure of violarite. Mineral. Mag., 26, 16-18.
67 Bannister, F.A., Claringbull, G.F. and Hey, M.H., 1953. Crichtonite, a distinct species. Am. Mineralogist, 38, 734.
68 Bannister, F.A. and Hey, M.H., 1932. Determination of minerals in platinum concentrates from the Transvaal by X-ray methods. Mineral. Mag., 23, 188-206.
69 Bannister, F.A. and Hey, M.H., 1937. The identity of penroseite and blockite. Am. Mineralogist, 22, 319-324.
70 Bannister, F.A., Hey, M.H. and Stadler, H.P., 1947. Nigerite, a new tin mineral. Mineral. Mag., 28, 129-136.
71 Barabanov, V.F., 1957. Cosalite from the Bukuky deposit (in Russian). Dokl. Akad. Nauk SSSR, 112, 938-941. Ref. in: Mineral. Abstr., 13, 366-367.
72 Bariand, P., 1959. Deux nouveaux indices de plattnérite (PbO_2, tétragonal). Bull. Soc. Franç. Minéral. Crist., 82, 324.
73 Bariand, P., 1963. Contribution à la minéralogie de l'Iran. Bull. Soc. Franç. Minéral. Crist., 86, 17-64.
74 Bariand, P., Cesbron, F., Agrinier, H. Geffroy, J. and Issakhanian, V., 1968. La getchellite, $AsSbS_3$, de Zarehshuran, Afshar, Iran. Bull. Soc. Franç. Minéral. Crist., 91, 403-406.
75 Barić, L., 1958. Neuntersuchung des Lorandit-Vorkommens von Mazedonien und Vergleich der Mineralvergesellschaftungen in den beiden bisher bekannten Fundorten des Lorandits. Schweiz. Min.Petr. Mitt. 38, 247-253. Zentr. Min. I, 1960, 148.
76 Barrabé, L. and Orcel, J., 1953. Sur un nouveau gisement Pyrénéen d'ullmannite. Compt. Rend. 78me Congr. Soc. Savantes Paris Dep., Sect. Sci., 203-206.
77 Barsanov, G.P., Kumskova, N.M. and Chepizhnyi, K.I., 1964. On a new occurrence of tapiolite. (in Russian.) Mineraly Akad. Nauk SSSR, 15, 189-193. Ref. in: Mineral. Abstr., 18, 281.
78 Bartholomé, P., Duchesne, J.C. and Plas, L. van der 1968. Sur une forme monoclinique de l'ilvaite. Ann. Soc. Géol. Belg., 90, Bull., 779-788.
79 Bartikyan, P.M., 1966. Native lead and zinc in the rocks of Armenia. (in Russian.) Zap. Vses Mineralog Obshchestva, 95, 99-102. Ref. in: Mineral. Abstr., 18, 200.

80 Baryschnikov, E.K., Merlitsch, B.V. and Slavskaia, A.I., 1957. Metacinnabarit aus dem Transkarpaten-Gebiet (in Russian). Mineralog. Sb. L' vovsk Geol. Obshchestvo, 11, 342-346. Ref. in: Zentr. Mineral. I, 1961, 93.
81 Basta, E.Z., 1959. Some mineralogical relationships in the system Fe_2O_3-Fe_3O_4 and the composition of titanomaghemite. Econ. Geol. 54, 698-719.
82 Basta, E.Z., 1960. Natural and synthetic titanomagnetites (the system Fe_3O_4-Fe_2TiO_4-$FeTiO_3$). Neues Jahrb. Mineral., Abhandl., 94, 1017-1048.
83 Bastin, E.S., 1933. The chalcocite and native copper types of ore deposits. Econ. Geol., 28, 407-446.
84 Bastin, E.S., 1939. The nickel-cobalt-native silver ore type. Econ. Geol., 34, 1-40.
85 Bastin, E.S., 1941. Paragenetic relations in the silver ores of Zacatecas, Mexico. Econ. Geol., 36, 371-400.
86 Bastin, E.S., Graton, L.C., Lindgren, W., Newhouse, W.H., Schwartz, G.M. and Short, M.N., 1931. Criteria of age relations of minerals with especial reference to polished sections of ores. Econ. Geol., 26, 561-610.
87 Bateman, A.M. and Lasky, S.G., 1932. Covellite-chalcocite solid solution and exsolution. Econ. Geol., 27, 52-86.
88 Bauer, J. and Hřichová, R., 1962. Das Eisen von Teigarhorn, Island (in Czech). Sb. Vysoké Školy Chem. Technol. Mineral., (Praha), 6. 37-43. Ref. in: Zentr. Mineral. I, 1965, 100.
89 Baumann, I.H., 1964. Patronit, VS_4, und die Mineral-Paragenese der bituminösen Schiefer von Minasragra, Peru. Neues Jahrb. Mineral., Abhandl., 101, 97-108.
90 Baumann, I.H. and Amstutz, G.C., 1965. Natural X-ray amorphous lead-arsenic sulfides from Cerro de Pasco Mine, Peru. Naturwissenschaften. 52. 585-587.
91 Baumann, L., 1958. Tektonik und Genesis der Erzlagerstätte von Freiberg (Zentralteil). Freiberger Forschungsh., C 46, 208 pp.
92 Bautsch, H.J., 1960. Erzmikroskopische Beobachtungen an den Opakanteilen eines Metaserpentinits von Wurzbach (Thüringen). Neues Jahrb. Mineral., Abhandl., 94, 908-925.
93 Bayliss, P., 1968. The different crystal structures of gersdorffite (NiAsS). Can. Mineralogist, 9, 570.
94 Bayliss, P., 1969a. X-ray data, optical anisotropism, and thermal stability of cobaltite, gersdorffite, and ullmannite. Mineral. Mag., 37, 26-33.
95 Bayliss, P., 1969b. Isomorphous substitution in synthetic cobaltite and ullmannite. Am. Mineralogist, 54, 426-430.
96 Bayramgil, O., 1945. Mineralogische Untersuchung der Erzlagerstätte von Isikdag (Türkei). Schweiz. Mineral. Petrog. Mitt., 25, 23-112.
97 Bazhenov, I.K., Indukaev, Yu.V. and Yakhno, A.V., 1959. Native iron in gabbrodolerites of the Kureyka, Krasnoyarsk region (in Russian). Zap. Vses. Mineralog. Obshchestva., 88, 180-184. Ref. in: Mineral. Abstr., 14, 373-374.
97a Beeson, M.H. and Jackson, E.D., 1969. Chemical composition of altered chromites from the Stillwater Complex, Montana. Am. Mineralogist, 54, 1084-1100.
98 Benedicks, C., 1912. Le fer d'Ovifak. Compt. Rend. XI Congr. Géol. Intern., II, 885-890.
99 Benešová-Talandová, M., 1965. Find of clausthalite in Moravia, Czechoslovakia (in Czech). Časopis Mineral. Geol., 10, 83-86. Ref. in: Mineral. Abstr., 17, 722.
100 Bergeat, A., 1897. Mineralogische Mittheilungen über den Stromboli. Neues Jahrb. Mineral. Geol. Palaeontol., 1897, II. Band, 109-123.
101 Berman, H. and Gonyer, F.A., 1939. Re-examination of colusite. Am. Mineralogist, 24, 377-381.
102 Berman, H. and Harcourt, G.A., 1938. Natural amalgams. Am. Mineralogist, 23, 761-764.
103 Berman, H. and Wolfe, C.W., 1939. Crystallography of aramayoite. Mineral. Mag., 25, 466-473.
103a Berman, Yu.S. and Kazarinova, Ye.I., 1968. Aguilarite found for the first time in the USSR. Dokl. Akad. Nauk SSSR, 183, 1406-1409. Translated in: Dokl. Acad. Sci. USSR, Earth Sci. Sect., 183, 147-150 (1969).
104 Bernard, J.H., 1954. Gersdorffite from Rudňany (in Czech, Russian and English summaries). Sb. Ústřed. Ústavu Geol., Svázek 21, oddíl geol. - 2. díl, 805-812.
105 Bernard, J.H., 1961. Mineralogie und Geochemie der Siderit-Schwerspat-Gänge mit Sulfiden im Gebiet von Rudňany (Tschechoslowakei). Geol. Práce, 58, 222 pp.
106 Bernard, J.H., 1964. Cosalit aus einem Sideritgang in Prakovce, Zips-Gömörer Erzgebirge (in Czech, German summary). Geol. Práce, 33, 43-52. Ref. in: Zentr. Mineral. I, 1964, 150-151.

107 Bernard, J.H. and Padĕra, K., 1954. Bravoit aus dem Kladno-Rakonitzer Steinkohlenbecken. Geologie (Berlin), 3, 155-169.
108 Berner, R.A., 1962. Tetragonal iron sulfide. Science, 137, 669.
109 Berry, L.G., 1939. Studies of mineral sulpho-salts: I - Cosalite from Canada and Sweden. Univ. Toronto, Geol. Ser., 42, 23-30.
110 Berry, L.G., 1940a. Studies of mineral sulpho-salts: II - Jamesonite from Cornwall and Bolivia. Mineral. Mag., 25, 597-608.
111 Berry, L.G., 1940b. Studies of mineral sulpho-salts: III - Boulangerite and "epiboulangerite". Univ. Toronto, Geol. Ser., 44, 5-19.
112 Berry, L.G., 1940c. Studies of mineral sulpho-salts: IV - Galenobismutite and "lillianite". Am. Mineralogist, 25, 726-734.
113 Berry, L.G., 1943. Studies of mineral sulpho-salts: VII - A systematic arrangement on the basis of cell dimensions. Univ. Toronto, Geol. Ser., 48, 9-30.
114 Berry, L.G., 1946. Nagyagite. Univ. Toronto, Geol. Ser., 50, 35-48.
115 Berry, L.G., 1951. The unit cell of magnetoplumbite. Am. Mineralogist, 36, 512-514.
116 Berry, L.G., 1953. New data on lead sulpharsenides from Binnental, Switzerland. Am. Mineralogist, 38, 330.
117 Berry, L.G., 1954. The crystal structure of covellite, CuS, and klockmannite, CuSe. Am. Mineralogist, 39, 504-509.
118 Berry, L.G., 1963. The probable identity of laitakarite and selenjoseite. Can. Mineralogist, 7, 677-679.
119 Berry, L.G., 1966. Crystallography of hatchite. Can. Mineralogist, 8, 661.
120 Berry, L.G., Fahey, J.J. and Bailey, E.H., 1952. Robinsonite, a new lead antimony sulphide. Am. Mineralogist, 37, 438-446.
121 Berry, L.G. and Moddle, D.A., 1941. Studies of mineral sulpho-salts: V - Meneghinite from Ontario and Tuscany. Univ. Toronto, Geol. Ser., 46, 5-17.
122 Berry, L.G. and Thompson, R.M., 1962. X-ray powder data for ore minerals: the Peacock Atlas. Geol. Soc. Am., Mem. 85.
123 Besson, M., 1967. La teneur en geikiélite des ilménites des kimberlites. Bull. Soc. Franç. Minéral. Crist., 90, 192-201.
124 Betekhtin, A.G., 1941. The new mineral arsenosulvanite (in Russian). Zap. Vses. Mineralog. Obshchestva., 70, 161-164. Ref. in: Am. Mineralogist, 40, 368-369.
125 Betekhtin, A.G., 1961. Mikroskopische Untersuchungen an Platinerzen aus dem Ural. Neues Jahrb. Mineral., Abhandl., 97, 1-34.
126 Beugnies, A., 1967. Contribution à l'étude des wolframites. Ann. Soc. Géol. Belg., 90, Bull., 173-184.
127 Beugnies, A. and Mozafari, Ch., 1968. Contribution à l'étude des propriétés des columbotantalites et des tapiolites. Ann. Soc. Géol. Belg., 91, 35-91.
128 Beverly, B., Jr., 1934. Graphite deposits in Los Angeles County, California. Econ. Geol., 29, 346-355.
129 Bezmertnaya, M.S., and Soboleva, L.N., 1963. A new telluride of bismuth and silver, established by the newest micromethods (in Russian). Tr. Inst. Mineralog. Geokhim. Kristallokhim. Redkikh, Elementov, Akad. Nauk SSSR, 18, 70-84. Ref. in: Am. Mineralogist, 49, 818, and Geologie (Berlin), 15, 370.
130 Bezmertnaya, M.S. and Soboleva, L.N., 1965. Volynskite, a new telluride of bismuth and silver (in Russian). Akad. Nauk SSSR, Eksperim. Metody Issled. Rudn. Mineral., 1965, 129-141. Ref. in: Am. Mineralogist, 51, 531. Mineral. Abstr., 17, 696. Geologie (Berlin), 16, 350.
131 Bhaskara Rao, A., and Adusumilli, M.S., 1966. Bismuth minerals from Borborema region, Brazil. Mineral. Mag., 35, 785-787.
132 Bhattacharjee, S.B., and Bhattacharjee, S., 1963. Occurrence of cryptomelane in manganese ores, Balaghat District, India. Am. Mineralogist, 48, 1174-1176.
133 Bianconi, F. and Simonetti, A., 1967. La brannerite e la sua paragenesi nelle pegmatiti di Lodrino (Ct. Ticino). Schweiz. Mineral. Petrog. Mitt., 47, 887-934.
134 Bird, W.H., 1969. A note on the occurrence of violarite, Copper King Mine, Boulder County, Colorado. Econ. Geol., 64, 91-94.
135 Birks, L.S., Brooks, E.J., Adler, I. and Milton, C., 1959. Electron probe analysis of minute inclusions of a Cu-Fe mineral. Am. Mineralogist, 44, 974-978.
136 Bjørlykke, H., 1945. Inneholdet av kobolt i svovelkis fra norske nikkelmalmer. Norsk Geol. Tidsskr., 25, 11-15.
137 Bjørlykke, H., 1947. Flaat nickel mine. Norg. Geol. Undersøk., 168 b.
138 Blanchard, R. and Hall, G., 1937. Mount Isa ore deposition. Econ. Geol., 32, 1042-1057.
139 Block, H. and Ahlfeld, F., 1937. Die Selenerzlagerstätte Pacajake, Bolivia. Z. Prakt. Geol., 45, 9-14.

140 Blomstrand, C.W., 1870. Om några nya svenska mineralier samt om magnetkiesens sammansättning. Kgl. Vetenskapsakad., Förh., Ofversigt, 27, 19-27.
141 Bodechtel, J. and Klemm, D.D., 1966. Ueber die Lagerstättenkundliche Stellung und chemische Zusammensetzung der Bleiwismutspiessglanze. Geol. Rundschau, 55, 418-427.
142 Bolfa, J., Chevalier, R., de la Roche, H. and Kern, R., 1961. Contribution à l'étude des "ilménites" du sud-est de Madagascar et du Sénégal. Relations avec la nature de "l'arizonite". Bull. Soc. Franç. Minéral. Crist. 84, 33-39.
143 Boorman, R.S. and Abbott, D., 1967. Indium in co-existing minerals from the Mount Pleasant tin deposit. Can. Mineralogist, 9, 166-179.
144 Borchert, H., 1934. Ueber Entmischungen im System Cu-Fe-S und ihre Bedeutung als "geologische Thermometer". Chem. Erde, 9, 145-172.
145 Borchert, H., 1935a. Ueber das Golderzvorkommen des Wilhelmsstollens bei Cristior, unweit Brad, in Rumänien. Z. Prakt. Geol., 43, 115-122.
146 Borchert, H., 1935b. Neue Beobachtungen an Tellurerzen. Neues Jahrb. Mineral., Beil., 69 A, 460-477.
147 Borchert, W. and Schroeder, R., 1947. Kristallographische und röntgenographische Bestimmungen am "Wittichenit von Sadisdorf". Heidelberger Beitr. Mineral. Petrog., 1, 112-117.
148 Borgström, L.H., 1916. Algodonit och whitneyit. Geol. Fören. Stockholm Förhandl., 38, 95-100.
149 Born, L., and Hellner, E., 1960. A structural proposal for boulangerite. Am. Mineralogist, 45, 1266-1271.
150 Borodayev, Yu.S., Mozgova, N.N. and Senderova, V.M., 1968. Chalcostibite from Tereksay (Kirgiz SSR). Dokl. Akad. Nauk SSSR, 178, 675-678. Translated in: Dokl. Acad. Sci. USSR, Earth Sci. Sect., 178, 118-121 (1968).
151 Borodin, L.S., Bykova, A.V., Kapitonova, T.A. and Pyatenko, Yu.A., 1960. New data on zirconolite and its niobian variety. Dokl. Akad. Nauk SSSR, 134, 1188-1191. Translated in: Dokl. Acad. Sci. USSR, Earth Sci. Sect., 134, 1022-1024 (1961).
152 Borodin, L.S., Nazarenko, I.I. and Richter, T.L., 1956. The new mineral zirconolite - a complex oxide of the AB_3O_7 type (in Russian). Dokl. Akad. Nauk SSSR, 110, 845-848. Ref. in: Am. Mineralogist, 42, 581-582.
153 Borovskii, I.B., Deev, A.N. and Marchukova, I.D., 1959. Application of the method of local X-ray spectrographic analysis to the study of minerals of the platinum group (in Russian). Geol. Rudn. Mestorozhd., 6, 68-73. Ref. in: Am. Mineralogist, 46, 464.
154 Bose, M.K., 1958. Goethite-hematite relation - an ore microscope observation. Am. Mineralogist, 43, 989-990.
155 Bose, M.K. and Roy, A.K., 1966. Coexisting Fe-Ti oxide minerals in norites associated with anorthosites of Bengal, India. Econ. Geol., 61, 555-562.
156 Bouladon, J. and Picot, P., 1968. Sur les minéralisations en cuivre des ophiolites de Corse, des Alpes françaises et de Ligurie (pro parte). Bull. Bur. Rech. Géol. Minières, 1968, Deuxième Sér., Sect. II, no 1, 23-41.
157 Bourguignon, P. and Mélon, J., 1965. Wodginite du Rwanda. Ann. Soc. Géol. Belg., 88, Bull., 291-300.
158 Bourguignon, P. and Toussaint, J., 1955. Caractères minéralogiques d'hématites manganésifères d'Ardenne. Ann. Soc. Géol. Belg., 78, Bull., 419-426.
158a Bowie, S.H.U., 1967. Microscopy: Reflected light. In: J. Zussmann (Editor): Physical Methods in Determinative Mineralogy. Academy Press, London.
159 Bowie, S.H.U. and Taylor, K., 1958. A system of ore mineral identification. Mining Mag., 99, 265-277 and 337-345.
160 Boyer, F. and Picot, P., 1963. Sur la présence de maldonite (Au_2Bi) à Salsigne (Aude). Bull. Soc. Franç. Minéral. Crist., 86, 429.
161 Boyle, R.W., 1951. An occurrence of native gold in an ice lens, Giant Yellowknife Gold mines, Yellowknife, N.W.T. Econ. Geol., 46, 223-227.
162 Boyle, R.W., 1960. Occurrence and geochemistry of native silver in the Pb-Zn-Ag lodes of the Keno Hill-Galena Hill area, Yukon, Canada. Neues Jahrb. Mineral., Abhandl., 94, 280-297.
163 Boyle, R.W., 1961. Native zinc at Keno Hill., Can. Mineralogist, 6, 692-694.
164 Brandenberger, E., 1939. Die Kristallstruktur des Awaruit. Schweiz. Mineral. Petrog. Mitt., 19, 285-286.
165 Bray, J.M., 1939. Ilmenite-hematite-magnetite relations in some emery ores. Am. Mineralogist, 24, 162-170.

166 Brett, R. and Yund, R.A., 1964. Sulfur-rich bornites. Am. Mineralogist, 49, 1084-1098.
167 Bricker, O., 1965. Some stability relations in the system $Mn-O_2-H_2O$ at 25°C and one atmosphere total pressure. Am. Mineralogist, 50, 1296-1354.
168 Brixner, L.H., 1965. Note on schirmerite. Am. Mineralogist, 50, 259-260.
169 Brodin, B.V., 1958. Über endogenen Goethit in den Erzen von Ken-Schanyk (in Russian). Zap. Vses. Mineralog. Obshchestva., 87, 496-498. Ref. in: Zentr. Mineral. I, 1960, 162.
170 Brodin, B.V. and Dymkov, Yu.M., 1961. Montroseite from hydrothermal veins of Příbram (in Russian). Zap. Vses. Mineralog. Obshchestva., 90, 653-659. Ref. in: Mineral. Abstr., 17, 65.
171 Brovkin, A.A., Aleksandrov, S.M. and Nekrasov, I.Ya., 1963. An X-ray study of minerals of the ludwigite-vonsenite series (in Russian). Rentgenografiya mineralnogo syrya. Gosgeoltekhizdat, 3, 16-34. Ref. in: Mineral. Abstr., 19, 128.
172 Brown, J.C., 1934. Lagerstättliche und erzmikroskopische Untersuchung der Zinnerzgänge der East Pool mine bei Redruth in Cornwall. Neues Jahrb. Mineral., Beil., 68 A, 298-336.
173 Brown, J.S., 1959. Occurrence of jordanite at Balmat, New York. Econ. Geol. 54, 136-137.
174 Brownell, G.M., 1943. Chromite from Manitoba. Univ. Toronto, Geol. Ser., 48, 101-102.
174a Bückle, H., 1960. L'essai de microdureté et ses applications. Publications Scientifiques et Techniques du Ministère de l'Air, Note Technique No. 90.
175 Buddington, A.F., 1935. High temperature mineral associations at shallow to moderate depths. Econ. Geol., 30, 205-222.
176 Budko, I.A. and Kulagov, E.A., 1963. Natural cubic chalcopyrite. Dokl. Akad. Nauk SSSR, 152, 408-410. Translated in: Dokl. Acad. Sci. USSR, Earth Sci. Sect., 152, 135-137 (1965).
177 Budko, I.A. and Kulagov, E.A., 1966. Troilite in vein ores of Norilsk and Talnakh. Dokl. Akad. Nauk SSSR, 169, 428-429. Translated in: Dokl. Acad. Sci. USSR, Earth Sci. Sect., 169, 123-125 (1967).
178 Budzyńska, H., 1968a. Anwendung von Natriumazid NaN_3 zur Unterscheidung des Löllingits von Arsenopyrit. Ber. Deut. Ges. Geol. Wiss., B, 13, 367-371.
179 Budzyńska, H., 1968b. Einige physikalische und chemische Eigenschaften der Arsenminerale der Lagerstätte Zloty Stok, VR Polen. Ber. Deut. Ges. Geol. Wiss., B, 13, 391-398.
180 Buerger, M.J., 1936a. The symmetry and crystal structure of the minerals of the arsenopyrite group. Z. Krist., A 95, 83-113.
181 Buerger, M.J., 1936b. The probable non-existence of arsenoferrite. Am. Mineralogist, 21, 70-71.
182 Buerger, M.J., 1936c. Crystallographic data, unit cell and space group for berthierite ($FeSb_2S_4$). Am. Mineralogist, 21, 442-448.
183 Buerger, M.J., 1939. The crystal structure of gudmundite (FeSbS) and its bearing on the existence field of the arsenopyrite structural type. Z. Krist., A 101, 290-316.
184 Buerger, M.J. and Hahn, T., 1955. The crystal structure of berthierite, $FeSb_2S_4$. Am. Mineralogist, 40, 226-238.
185 Buerger, N.W., 1934. The unmixing of chalcopyrite from sphalerite. Am. Mineralogist, 19, 525-530.
186 Buerger, N.W., 1935. The copper ores of Orange County, Vermont. Econ. Geol., 30, 434-443.
187 Buerger, N.W., 1941. The chalcocite problem. Econ. Geol., 36, 19-44.
188 Bugge, C., 1935. Lead-bismuth ores in Bleka, Svartdal, Norway. Econ. Geol., 30, 792-799.
189 Buist, D.D., Gandalla, A.M.M. and White, J., 1966. Delafossite and the system Cu-Fe-O. Mineral. Mag., 35, 731-741.
190 Bulakh, A.G., Anastasenko, G.F. and Dakhiya, L.M., 1967. Calzirtite from carbonatites of Northern Siberia. Am. Mineralogist, 52, 1880-1885.
191 Burkart-Baumann, I., Ottemann, J. and Amstutz, G.C., 1966. Neue Beobachtungen an den röntgenamorphen Sulfiden von Cerro de Pasco, Peru. Neues Jahrb. Mineral. Monatsh., 1966, 353-361.
192 Burkart-Baumann, I., Ottemann, J. and Nicolini, P., 1967. Mineralogische Untersuchungen an Jordanit, Semseyit und Cesarolith von drei tunesischen Blei-Zink-Lagerstätten. Chem. Erde, 26, 256-270.

193 Burkart-Baumann, I., Ottemann, J. and Nuber, B., 1968. Gratonit aus der Lagerstätte Rio Tinto, Südspanien. Neues Jahrb. Mineral., Monatsh., 1968, 215-224.
194 Burke, E.A.J., 1967. In: Uytenbogaardt, W. (1967a); and many unpublished VHN and reflectance data.
195 Burke, E.A.J., Kieft, C., Felius, R.O. and Adusumilli, N.S., 1969. Staringite, a new Sn-Ta-mineral from northeastern Brazil. Mineral. Mag., 37, 447-452.
196 Burke, E.A.J., Kieft, C., Felius, R.O. and Adusumilli, M.S., 1970. Wodginite from northeastern Brazil. Geol. Mijnbouw, 49, 235-240.
197 Burnol, L., Picot, P. and Pierrot, R., 1965. La ménéghinite des gisements de Saki et de Gouskhalat (Iran). Bull. Soc. Franç. Minéral. Crist., 88, 290-293.
198 Burri, G., Graeser, S., Marumo, F. and Nowacki, W., 1965. Imhofit, ein neues Thallium-arsenosulfosalz aus dem Lengenbach (Binnatal, Kanton Wallis). Chimia, 19, 499-500.
199 Burschtejn, E.F. and Mukanov, K.M., 1958. Plattnerit in oxydischen Erzen von Alajgyr, Zentralkasachstan (in Russian). Zap. Vses. Mineral. Obshchestva., 87, 498-500. Ref. in: Zentr. Mineral. I, 1960, 157-158.
200 Buryanova, E.Z. and Komkov, A.I., 1955. A new mineral: ferroselite (in Russian). Dokl. Akad. Nauk SSSR, 105, 812-813. Ref. in: Am. Mineralogist, 41, 671.
201 Buryanova, E.Z., Kovalev, G.A. and Komkov, A.I., 1957. A new mineral: cadmoselite (in Russian). Zap. Vses. Mineralog. Obshchestva., 86, 626-628. Ref. in: Am. Mineralogist, 43, 623, and Mineral. Abstr., 14, 59.
202 Buseck, P.R., 1967. Contact metasomatism and ore deposition: Tem Piute, Nevada. Econ. Geol., 62, 331-353.
203 Butler, B.S. and Burbank, W.S., 1929. The copper deposits of Michigan. U.S., Geol. Surv., Profess. Papers, 144.
204 Butler, J.R. and Hall, R., 1960a. Chemical variations in members of the fergusoniteformanite series. Mineral. Mag., 32, 392-407.
205 Butler, J.R. and Hall, R., 1960b. Chemical characteristics of davidite. Econ. Geol. 55, 1541-1550.
206 Buttgenbach, H., 1933. Minéraux du Congo Belge. La thoreaulite, nouvelle espèce minérale. (Note préliminaire). Ann. Soc. Géol. Belg., 56, Bull., 327-328.
207 Byers, A.R., 1940. Geology of the Nighthawk peninsular gold mine. Econ. Geol., 35, 996-1011.
208 Byström, A., 1945. The structure of quenselite, $PbMnO_2OH$. Arkiv Kemi, Mineral., Geol., 19 A (35), 1-9.
209 Byström, A. and Byström, A.M., 1950. The crystal structure of hollandite, the related manganese oxide minerals, and α-MnO_2. Acta Cryst., 3, 146-154.
210 Cabri, L.J., 1965a. Phase relations in the Au-Ag-Te system and their mineralogical significance. Econ. Geol., 60, 1569-1606.
211 Cabri, L.J.P., 1965b. Discussion of "Empressite and stuetzite redefined" by R.M. Honea. Am. Mineralogist, 50, 795-801.
212 Cabri, L.J., 1967. A new copper-iron sulfide. Econ. Geol., 62, 910-925.
213 Cabri, L.J. and Rucklidge, J.C., 1968. Gold-silver tellurides: relation between composition and X-ray diffraction data. Can. Mineralogist, 9, 547-551.
214 Cabri, L.J. and Traill, R.J., 1966. New palladium minerals from Noril'sk, Western Siberia. Can. Mineralogist, 8, 541-550.
215 Çagatay, A., 1968. Erzmikroskopische Untersuchung des Weiss-Vorkommens bei Ergani Maden, Türkei, und genetische Deutung der Kupferlagerstätten von Ergani Maden. Neues Jahrb. Mineral., Abhandl., 109, 131-155.
216 Caillère, S., Avias, J. and Falgueirettes, J., 1961. Découverte en Nouvelle Calédonie d'une minéralisation arsénicale, sous forme d'un nouvel arséniure de nickel, Ni_2As. Bull. Soc. Franç., Minéral. Crist., 84, 9-12 (1961).
217 Callow, K.J., 1967. The geology of the Thanksgiving mine, Baguio District, Mountain Province, Philippines. Econ. Geol., 62, 472-481.
218 Callow, K.J. and Worley, B.W. Jr., 1965. The occurrence of telluride minerals at the Acupan Gold Mine, Mountain Province, Philippines. Econ. Geol., 60, 251-268.
219 Cambel, B., 1959. Hydrothermale Erzlagerstätten im Kristallinikum der Kleinen Karpathen (in Czech, Russian and German summaries). Acta Geol. Geograph. Univ. Comenianae, Geol., 3, 347 pp.
220 Cameron, E.N., 1961. Ore microscopy. Wiley and Sons, New York and London.
220a Cameron, E.N., Davis, J.H., Guilbert, J.M., Larson, L.T., Manusco, J.J. and Sorem, R.K., 1961. Rotation properties of certain anisotropic ore minerals. Econ. Geol., 56, 569-583.

221 Cameron, E.N. and Desborough, G.A., 1964. Origin of certain magnetite-bearing pegmatites in the eastern part of the Bushveld Complex, South Africa. Econ. Geol., 59, 197-225.
222 Cameron, E.N. and Threadgold, I.M., 1961. Vulcanite, a new copper telluride from Colorado, with notes on certain associated minerals. Am. Mineralogist, 46, 258-268.
223 Campbell, D.F., 1939. Geology of the Bonanza King mine, Humboldt Range, Pershing County, Nevada. Econ. Geol., 34, 96-112.
224 Campbell, D.F., 1942. The Oruro silver-tin district, Bolivia. Econ. Geol., 37, 87-115.
225 Campbell, D.F., 1947. Geology of the Colquiri tin mine, Bolivia. Econ. Geol., 42, 1-21.
226 Canavan, F. and Edwards, A.B., 1938. The iron ores of Yampi Sound, Western Australia. Australasian Inst. Mining Met. Proc., 110, 59-101.
227 Capdecomme, L., 1943. Caractères optiques nouveaux d'une série d'espèces minérales opaques. Bull. Soc. Franc. Minéral, Crist., 66, 79-104.
228 Carmichael, I.S.E., 1967. The mineralogy and petrology of the volcanic rocks from the Leucite Hills, Wyoming. Beitr. Mineral. Petrol., 15, 24-66.
228a Carpenter, R.H. and Cameron, E.N., 1963. Additional measurements of rotation properties of ore minerals. Econ. Geol., 58, 1309-1312.
229 Carpenter, R.H. and Desborough, G.A., 1964. Range in solid solution and structure of naturally occurring troilite and pyrrhotite. Am. Mineralogist, 49, 1350-1365.
230 Caye, R. and Cervelle, B.D., 1968. Détermination de l'indice de réfraction et du coefficient d'absorption des minéraux non transparents. Bull. Soc. Franç. Minéral. Crist., 91, 284-288.
231 Caye, R., Laurent, Y., Picot, P., Pierrot, R. and Lévy, C., 1968. La hocartite, Ag_2SnFeS_4, une nouvelle espèce minérale. Bull. Soc. Franç. Minéral. Crist., 91, 383-387.
232 Caye, R., Picot, P., Pierrot, R. and Permingeat, F., 1967. Nouvelles données sur la vrbaïte, sa teneur en mercure. Bull. Soc. Franç. Minéral. Crist., 90, 185-191.
233 Čech, F., Černý, P. and Povondra, P., 1964. Ilmenorutile from Udraž, near Pisek, and its desintegration products (in Czech, English summary). Acta Univ. Carolinae, Geol., 1964, 1-14.
234 Černý, P., 1957. The second occurrence of hawleyite, β-CdS (in Czech). Časopis Mineral. Geol., 2, 13-16. Ref. in: Zentr. Mineral. 1960, I, 329. Mineral Abstr., 15, 66.
235 Černý, P., Čech, F. and Povondra, P., 1964. Review of ilmenorutile-strüverite minerals. Neues Jahrb. Mineral., Abhandl., 101, 142-172.
236 Cervelle, B., 1967. Contribution à l'étude de la série ilménite-geikielite. Bull. Bur. Rech. Géol. Minières, 1967, no. 6, 1-26.
237 Cervelle, B., Lévy, C. and Caye, R., 1968. Sur l'inversion du signe de la biréflectance de certains minéraux absorbants: cas de la mawsonite. Bull. Soc. Franç. Minéral. Crist., 91, 468-478.
238 Cevales, G., 1961. I giacimenti minerari del Gran Paradiso. Rend. Soc. Mineral. Ital., 17, 193-217.
239 Chace, F.M., 1948. Tin-silver veins of Oruro, Bolivia. Econ. Geol., 43, 435-470.
240 Chakraborty, K.L., 1962. Note on the mineralogical characters of some Indian chromites. Mineral. Mag., 33, 68-70.
241 Chakrapani Naidu, M.G. and Rama Rao, G.V.S. 1965. A new occurrence of titanomaghemite from South India. Can. Mineralogist, 8, 334-338.
242 Challis, G.A. and Long, J.V.P., 1964. Wairauite, a new Co-Fe mineral. Mineral. Mag., 33, 942-948.
243 Chamberlain, J.A., 1966. Heazlewoodite and awaruite in serpentinites of the Eastern Townships, Quebec. Can. Mineralogist, 8, 519-522.
244 Chamberlain, J.A. and Delabio, R.N., 1965. Mackinawite and valleriite in the Muscox intrusion. Am. Mineralogist, 50, 682-695.
245 Chamberlain, J.A., McLeod, C.R., Traill, R.J. and Lachance, G.R., 1965. Native metals in the Muskox intrusion. Can. J. Earth Sci., 2, 188-215.
246 Chandy, K.C., 1965. An occurrence of wüstite. Mineral. Mag., 35, 664-666.
247 Chao, G.Y. and Hounslow, A.W., 1966. The first Canadian occurrence of geikielite. Can. Mineralogist, 8, 664.
248 Chatschaturian, E.A., 1958. Altait und Petzit in den Erzen der Lagerstätte Kafan (in Russian). Dokl. Akad. Nauk Arm. SSR, 26, 177-180. Ref. in: Zentr. Mineral., I, 1961, 94.

249 Chauris, L. and Geffroy, J., 1966. Cubanite dans la galène de la Chapelle-Saint-Maudé près Baud (Morbihan). Bull. Soc. Franç. Minéral. Crist., 89, 137-138.
250 Chernitsyn, V.B. and Apostolov, D.A., 1966. Native arsenic in north Caucasus mercury ore occurrences. Dokl. Akad. Nauk SSSR, 169, 199-200. Translated in: Dokl. Acad. Sci. USSR, Earth Sci. Sect., 169, 40-41 (1967).
251 Chesterman, C.W., 1959. Ludwigite from Fresno County, California. Bull. Geol. Soc. Am., 70, 1712-1713.
252 Chistyakova, M.B., Kazakova, M.E. and Ukhanov, E.V., 1964. A new occurrence of stibiotantalite (in Russian). Mineraly Akad. Nauk SSSR, 15, 251-255. Ref. in: Mineral. Abstr., 18, 281.
253 Choudari, R., Kosztolanyi, Ch. and Coppens, R., 1967. Etude d'une uraninite provenant de la pegmatite de Rajasthan (Inde). Bull. Soc. Franç. Minéral. Crist., 90, 77-81.
254 Christ, C.L. and Clark, J.R., 1955. The crystal structure of murdochite. Am. Mineralogist, 40, 907-916.
255 Christophe-Michel-Lévy, M. and Sandréa, A., 1953. La högbomite de Frain (Tchécoslovaquie). Bull. Soc. Franç. Minéral. Crist., 76, 430-433.
256 Chudoba, K.F., 1967. Handbuch der Mineralogie von Dr. Carl Hintze, Ergänzungsband III, Lieferung 4. Walter de Gruyter und Co., Berlin, 393-439.
257 Chukrov, F.V. and Bonstedt-Kupletskaja, E.M., 1967. Mineraly Spravotshnik, Tom II, III. Izdatelstvo Nauka, Moskva, p. 80.
258 Chukrov, F.V., Genkin, A.D., Soboleva, S.V. and Vasova, G.V., 1965. Smythite from the iron-ore deposits of the Kerch Peninsula. Litologiya Polezne Iskop., No. 2, 60-69. Translated in: Geochemistry Intern., 2, 372-381, 1965.
259 Cid, H., 1960. Etude aux rayons X de la cylindrite. Bull. Soc. Franç. Minéral. Crist., 83, 64-65.
260 Cissarz, A., 1930. Allopalladium und Clausthalit von Tilkerode im Harz und das Verhältnis von Allopalladium zu Potarit. Z. Krist., A 74, 501-510.
261 Cissarz, A., 1931. Beiträge zur Kenntnis der komplexen Indikatrix von Antimonglanz. Z. Krist., A 78, 445-461.
262 Clar, E., 1931. Zwei Erzentmischungen von Schneeberg in Tirol. Zentr. Mineral. Geol. Palaeontol., A, 147-153.
263 Claringbull, G.F. and Hey, M.H., 1957. Arsenopalladinite (Pd_3As), a new mineral from Itabira, Brazil. Mineral. Abstr., 13, 237.
264 Clark, A.H., 1965a. Rhombohedral molybdenite from the Minas da Panasqueira, Beira Baixa, Portugal. Mineral. Mag., 35, 69-71.
265 Clark, A.H., 1965b. The composition and conditions of formation of arsenopyrite and löllingite in the Ylöjärvi copper-tungsten deposit, Southwest Finland. Bull. Comm. Géol. Finlande, 217, 56 pp.
266 Clark, A.H. 1966a. Some comments on the composition and stability relations of mackinawite. Neues Jahrb. Mineral., Monatsh., 1966, 300-304.
267 Clark, A.H., 1966b. The mineralogy and geochemistry of the Ylöjärvi Cu-W deposit, Southwest Finland: mackinawite-pyrrhotite-troilite assemblages. Bull. Comm. Géol. Finlande, 222, 331-342.
268 Clark, A.H., 1967a. Note on zinckenite from Turhal antimony deposit, Turkey. Trans. Inst. Mining Met., 76, B 117-118.
269 Clark, A.H., 1967b. Note on zinckenite from Turhal antimony deposit, Turkey: an addendum. Trans. Inst. Mining Met., 76, B 167.
270 Clark, A.H., 1969a. Preliminary observations on chromian mackinawite and associated native iron, Mina do Abessedo, Vinhais, Portugal.Neues Jahrb. Mineral., Monatsh., 1969, 282-288.
270a Clark, A.H., 1969b. A second occurrence of berndtite: Lagares-do-Estanho, Queiriga, Viseu, Portugal. Neues Jahrb. Mineral., Monatsh., 1969, 426-430.
271 Clark, A.H. and Clark, A.M., 1968. Electron microprobe analysis of mackinawite from the Ylöjärvi deposit, Finland. Neues Jahrb. Mineral., Monatsh., 1968, 259-268.
271a Clark, A.H. and Moraga, A., 1969. Ternary solid solutions in the systems Cu-As-S, Mina El Guanaco, Taltal, Chile. Am. Mineralogist, 54, 1269-1273.
271b Clark, A.H. and Sillitoe, R.H., 1969. Supergene $CuMo_2S_5$ ("Castaingite"), Portrerillos, Atacama Province, Chile. Neues Jahrb. Mineral., Monatsh., 1969, 499-503.
272 Coats, R.R., 1936. Aguilarite from the Comstock Lode, Virginia City, Nevada. Am. Mineralogist, 21, 532-534.
273 Colbertaldo, D. Di, 1954. Marcasite idiomorfa nel giacimento di Raibl. Rend. Soc. Mineral. Ital., 10, 369-372.

274 Coleman, L.C., 1953. Mineralogy of the Yellowknife Bay area, N.W.T. Am. Mineralogist, 38, 506-527.
275 Coleman, L.C., 1957. Mineralogy of the Giant Yellowknife Gold mine, Yellowknife, N.W.T. Econ. Geol., 52, 400-425.
276 Coleman, R.G., 1959a. New occurrences of ferroselite. Geochim. Cosmochim. Acta, 16, 296-301.
277 Coleman, R.G., 1959b. The natural occurrence of galena-clausthalite solid solution series. Am. Mineralogist, 44, 166-175.
278 Coleman, R.G., and Delevaux, M., 1957. Occurrence of selenium in sulfides from some sedimentary rocks of the western United States. Econ. Geol., 52, 499-527.
279 Contag, B., and Strunz, H., 1961. Zur Kristallchemie des Uran-Minerals Davidit. Naturwissenschaften, 48, 597. Ref. in: Zentr. Mineral., I, 1962, 246-247.
280 Cooke, H.R., Jr., 1947. The original Sixteen to One gold-quartz vein. Alleghany, California. Econ. Geol., 42, 211-250.
281 Cooke, S.R.B. and Doan, D.J., 1935. The mineragraphy and X-ray analyses of stainierite from the Swansea mine, Goodsprings, Nevada. Am. Mineralogist, 20, 274-280.
282 Cooper, R.A., 1923. Mineral constituents of the Rand concentrates. J. Chem. Metall. Mining Soc. S. Africa, 24, 90-95. Ref. in: Mineral. Abstr., 3, 114-115.
283 Cornwall, H.R., 1951. Ilmenite, magnetite, hematite and copper in lavas of the Keweenawan Series. Econ. Geol., 46, 51-67.
284 Coulon, M., Heitz, F. and Le Bihan, M.Th., 1961. Contribution à l'étude structurale d'un sulfure de pomb, d'antimoine et d'étain: la frankeite. Bull. Soc. Franç. Minéral. Crist., 84, 350-353.
285 Craig, D.C. and Stephenson, N.C., 1965. The crystal structure of lautite. Acta Cryst., 19, 543-547.
286 Craig, J.R., 1967. Phase relations and mineral assemblages in the Ag-Bi-Pb-S system. Mineral. Deposita, 1, 278-306.
287 Cruys, A., Parfenoff, A. and Fauquier, D., 1964. Sur la présence de fergusonite et d'euxénite en Guyane. Bull. Soc. Franç. Minéral. Crist., 87, 625-626.
288 Cup, K.C. and Wensink, H., 1959. The lead-zinc ores of Yenefrito near Panticosa (Spanish Pyrenees). Geol. Mijnbouw, 21, 434-444.
289 Czamanske, G.K., 1969. The stability of argentopyrite and sternbergite. Econ. Geol., 64, 459-461.
289a Czamanske, G.K. and Larson, R.R., 1969. The chemical identity and formula of argentopyrite and sternbergite. Am. Mineralogist, 54, 1198-1201.
290 Dadson, A.S., 1937. A potential series of some minerals from the Temiskaming District, Ontario. Univ. Toronto, Geol. Ser., 40, 115-150.
291 Dana, E.S., 1932. A textbook of mineralogy, 4th Ed. (10th Impr. 1947). New York.
292 Darmon, R. and Wintenberger, M., 1966. Structure cristalline de $CoAs_2$. Bull. Soc. Franç. Minéral. Crist., 89, 213-215.
293 Darnley, A.G. and Killingworth, P.J., 1962. Identification of carrollite from Chibuluma by X-ray scanning microanalyser. Trans. Inst. Mining Met., 72, 165-168.
294 Darnley, A.G., Smith, G.H., Chandler, T.R.D. and Dance, D.F., 1962. The age of fergusonite from the Jos area, Northern Nigeria. Mineral. Mag., 33, 48-51.
295 Dasgupta, D.R., Poddar, B.C. and Sen Gupta, N.R., 1968. A note on the occurrence of geocronite and boulangerite in the Rajpura belt, Udaipur district, Rajasthan, India. Mineral. Mag., 36, 1174-1175.
296 Davidson, A., and Wyllie, P.J., 1965. Zoned magnetite and platy magnetite in Cornwall type ore deposits. Econ. Geol., 60, 766-771.
297 Davis, G.R., 1954. The origin of the Roan Antelope copper deposit of Northern Rhodesia. Econ. Geol., 49, 575-615.
298 Davydova, L.I. and Shaposhnikov, G.N., 1966. Geological conditions of davidite formation in an area of the USSR (in Russian). Zap. Vses. Mineralog. Obshchestva., 95, 474-476. Ref. in: Mineral. Abstr., 18, 122.
299 Dell'Anna, L. and Quagliarella, F., 1967. Jordanite nel marmo di Carrara. Periodico Mineral. (Roma) 36, 245-258. Ref. in: Mineral. Abstr., 19, 16.
300 Demfrsoy, S., 1968. Untersuchungen über den Einfluss der chemischen Zusammensetzung auf die spektralen Reflexionsfunktionen und Mikroeindruckhärten: - im Spinell-Dreistoffsystem unter besonderer Berücksichtigung der Chromspinelle, - im System FeS_2 -NiS_2 -CoS_2, an Zonen eines natürlichen Bravoit-Kristalls. Ph.D. Thesis, Aachen (Germany), 95 pp.

301 Demirsov, S., 1969. Untersuchungen über den Einfluss der chemischen Zusammensetzung auf die spektralen Reflexionsfunktionen und Mikroeindruckhärten im System $FeS_2-NiS_2-CoS_2$, an Zonen eines natürlichen Bravoit-Kristalls. Neues Jahrb. Mineral., Monatsh., 1969, 323-333.

301a Demirsoy, S., 1969b. Beitrag zu spektralen Reflexionsfunktionen von Millerit. Neues Jahrb. Mineral., Monatsh., 1969, 477-479.

302 Dennen, W.H., 1943. A nickel deposit near Dracut, Massachusetts. Econ. Geol., 38, 25-55.

303 Derriks, J.J. and Vaes, J.F., 1955. The Shinkolobwe uranium deposit: current status of our geological and metallogenic knowledge. Proc. U.N. Intern. Conf. Peaceful Uses At. Energy, 1st, Geneva, 1955, 6, 94-128.

304 Desborough, G.A. and Amos, D.H., 1961. Ilvaite: a late magmatic occurrence in gabbro of Missouri. Am. Mineralogist, 46, 1509-1511.

305 Desborough, G.A. and Carpenter, R.H., 1965. Phase relations of pyrrhotite. Econ. Geol., 60, 1431-1450.

306 Deudon, M., 1957. Présence de maghémite (Fe_2O_3-γ) dans le minerai de fer de Lorraine. Bull. Soc. Franç. Minéral. Crist., 80, 239-241.

307 Dickson, F.W. and Tunell, G., 1959. The stability relations of cinnabar and metacinnabar. Am. Mineralogist, 44, 471-487.

308 Dietrich, V., Huonder, N. and Rybach, L., 1967. Uranvererzungen im Druckstollen Ferrera-Val Niemet. Beitr. Geol. Schweiz, Geotech. Ser., 44, 27 pp.

309 Dimitrov, S., 1967. Jordisite and ilsemannite, new minerals for Bulgaria (in Bulgarian). Rev. Bulgar. Geol. Soc., 28, 345-542. Ref. in: Abstr. Bulgar. Sci. Literature, 1968, 2, p. 9.

310 Donnay, G., Donnay, J.D.H. and Kullerud, G., 1958. The crystal and twin structure of digenite, Cu_9S_5. Am. Mineralogist, 43, 228-242.

311 Donnay, G., Kracek, F.C. and Rowland, W.R., Jr., 1956. The chemical formula of empressite. Am. Mineralogist, 41, 722-723.

312 Donnay, J.D.H. and Donnay, G., 1954. Syntaxic intergrowths in the andorite series. Am. Mineralogist, 39, 161-171.

313 Dornberger-Schiff, K. and Höhne, E., 1959. Die Kristallstruktur des Betechtinits. Acta Cryst., 12, 646-651.

314 Dornberger-Schiff, K. and Höhne, E., 1962. Zur Symmetrie des Boulangerit. Chem. Erde, 22, 78-82.

315 Doucet, X.R., 1943. Etude du gîte de wolfram de Chatelus-le-Marcheix (Creuse). Bull. Soc. Franç. Minéral. Crist., 66, 238-250.

316 Douglass, R.M., Murphy, M.J. and Pabst, A., 1954. Geocronite. Am. Mineralogist, 39, 908-928.

317 Dreyer, R.M., 1940. The geochemistry of quicksilver mineralization. II. Econ. Geol., 35, 140-157.

317a Drummond, A.D., Trotter, J., Thompson, R.M. and Gower, J.A., 1969. Neyite, a new sulphosalt from Alice Arm, British Columbia. Can. Mineralogist, 10, 90-96.

318 Dunin-Barkovskaya, E.A., Lider, V.V. and Rozhanskii, V.N., 1968. Lead-bearing joseite from Ustarasai (in Russian). Zap. Vses. Mineralog. Obshchestva, 97, 332-341. Ref. in: Mineral. Abstr., 20, 55.

319 Dunn, J.A., 1934. Sulphide mineralization in Singhbhum. Trans. Mining Geol. Met. Inst. India, 29, 163-172.

320 Dunn, J.A., 1936. A study of some microscopical aspects of Indian manganese ores. Trans. Natl. Inst. Sci. India, I, 103-124.

321 Dunn, J.A., 1937a. The mineral deposits of Eastern Singhbhum and surrounding areas. Mem. Geol. Surv. India, 69, I.

322 Dunn, J.A., 1937b. A microscopical study of the Baldwin ores, Burma. Records Geol. Surv. India, 72, 333-359.

323 Dunn, J.A., 1938. Tin-tungsten mineralization at Mawchi, Karenni States, Burma. Records Geol. Surv. India, 73, 209-237.

324 Dunn, J.A. and Dey, A.K., 1937. Vanadium-bearing titaniferous iron ores in Singhbhum and Mayurbhanj, India. Trans. Mining Geol. Met. Inst. India, 31, 117-184.

325 Du Rietz, T., 1955. The content of chromium and nickel in the Caledonian ultrabasic rocks of Sweden. Geol. Fören. Stockholm Förhandl., 78, 233-300.

326 Eales, H.V., 1962. The occurrence of aurostibite in the Gwanda District, Southern Rhodesia. Trans. Geol. Soc. S. Africa, 65, 79-83.

327 Eales, H.V., 1964. Mineralogy and petrology of the Empress nickel-copper deposit, Southern Rhodesia. Trans. Geol. Soc. S. Africa, 67, 173-201.

327a Eales, H.V., 1967. Reflectivity of gold and gold-silver alloys. Econ. Geol., 62, 412-420.
328   Eales, H.V., 1969. The bases of a scheme for identification of ore minerals by optical methods. Mineral. Deposita, 4, 52-64.
329   Earley, J.W., 1949. Studies of natural and artificial selenides: I - Klockmannite. CuSe. Am. Mineralogist, 34, 435-440.
330   Earley, J.W., 1950. Description and synthesis of the selenide minerals. Am. Mineralogist, 35, 337-364.
331   Edwards, A.B., 1936. The iron ores of the Middleback Ranges, South Australia. Australasian Inst. Mining Met. Proc., 102, 155-207.
332   Edwards, A.B., 1938. Some ilmenite micro-structures and their interpretation. Australasian Inst. Mining Met. Proc., 110, 39-58.
333   Edwards, A.B., 1939. Some observations on the mineral composition of the Mount Lyell copper ores, Tasmania and their modes of occurrence. Australasian Inst. Mining Met. Proc., 114, 67-109.
334   Edwards, A.B., 1940a. On the mineral composition of the Mount Oxide copper ore, Queensland. Australasian Inst. Mining Met. Proc., 118, 83-99.
335   Edwards, A.B., 1940b. A note on some tantalum-niobium minerals from Western Australia. Australasian Inst. Mining Met. Proc., 120, 731-744.
336   Edwards, A.B., 1943a. The composition of the lead-zinc ore at Captain's Flat, N.S.W., Australasian Inst. Mining Met. Proc., 129, 23-40.
337   Edwards, A.B., 1943b. The copper deposits of Australia. Australasian Inst. Mining Met. Proc., 130, 105-171.
338   Edwards, A.B., 1946. Solid solution of tetrahedrite in chalcopyrite and bornite. Australasian Inst. Mining Met. Proc., 143, 141-155.
339   Edwards, A.B., 1947. Textures of the ore minerals and their significance. Melbourne.
340   Edwards, A.B., 1949. Natural ex-solution intergrowths of magnetite and hematite. Am. Mineralogist, 34, 759-761.
341   Edwards, A.B., 1951. Some occurrences of stannite in Australia. Australasian Inst. Mining Met. Proc., 160-161, 1-59.
342   Edwards, A.B., 1953. The mineral composition of the Yerranderie silver-lead ores. Australasian Inst. Mining Met. Proc., No. 170, 73-101.
342a Edwards, A.B., 1954. Textures of the ore minerals and their significance, second edition. Australasian Institute of Mining and Metallurgy.
343   Edwards, A.B., 1955a. The mineral composition of ore from the Hill 50 gold mine, Western Australia. Australasian Inst. Mining Met. Proc., 174, 33-42.
344   Edwards, A.B., 1955b. The composition of the Peko Copper orebody, Tennant Creek. Australasian Inst. Mining Met. Proc., 175, 55-82.
345   Edwards, A.B., 1956. Hypogene goethite at Peko mine, N.T., Australia. Am. Mineralogist, 41, 657-660.
346   Edwards, A.B., 1958. The mineral composition of the Maude and Yellow Girl gold ore, Glen Wills, Victoria. Australasian Inst. Mining Met. Proc., Stillwell Anniv. Volume, 105-132.
347   Edwards, A.B., 1960. Contrasting textures in the Ag-Pb-Zn ores of the Magnet mine, Tasmania. Neues Jahrb. Mineral., Abhandl., 94, 298-318.
348   Edwards, A.B., Anderson, J.S. and Hart, J.G., 1942. On the occurrence of platinum and palladium at the Tompson River Copper Mine, Victoria, with a note on the optical properties of braggite. Australasian Inst. Mining Met. Proc., 125, 61-69.
349   Edwards, A.B., Baker, G. and Callow, K.J., 1956. Metamorphism and metasomatism at King Island scheelite mine. J. Geol. Soc. Australia, 3, 55-98.
350   Ehrenberg, H., 1931. Der Aufbau der Schalenblenden der Aachener Bleizinkerzlagerstätten und der Einfluss ihres Eisengehaltes auf die Mineralbildung. Zugleich ein Beitrag zur mikroskopischen Diagnose von Wurtzit und Zinkblende. Neues Jahrb. Mineral., Beil., 64 A, 397-422.
351   Ehrenberg, H., 1932. Orientierte Verwachsungen von Magnetkies und Pentlandit. Z. Krist., A 82, 309-315.
352   Einaudi, M.T., 1968. Copper zoning in pyrite from Cerro de Pasco, Peru. Am. Mineralogist, 53, 1748-1752.
353   El Goresy, A., 1964. Neue Beobachtungen an der Nickel-Magnetkies-Lagerstätte von Abu Suwajel, Aegypten. Neues Jahrb. Mineral., Abhandl., 102, 107-113.
354   El Goresy, A. and Donnay, G., 1969. A new hexagonal form of carbon from the Ries Crater. Carnegie Inst., Wash., Ann. Rept. Geophys. Lab., 1967-1968, 215-216.

355 El Goresy, A. and Meixner, H., 1965. Brannerit aus den Eisenspatlagerstätten von Olsa bei Friesach, Kärnten. Neues Jahrb. Mineral., Abhandl., 103, 94-98.
356 Eliseev, E.N., 1955. On the composition and crystalline structure of pentlandite (in Russian). Zap. Vses. Mineralog. Obshchestva, 84, 53-62. Ref. in: Mineral. Abstr., 13, 13.
357 Ellsworth, H.V. and Osborne, F.F., 1934. Uraninite from Lac Pied des Monts, Saguenay District, Quebec. Am. Mineralogist, 19, 421-425.
358 El Shazly, E.M. and Saleeb, G.S., 1959. Contribution to the mineralogy of Egyptian manganese deposits. Econ. Geol., 54, 873-888.
359 Emmons, S.F., Irving, J.D. and Loughlin, G.F., 1927. Geology and ore deposits of the Leadville mining district, Colorado. U.S., Geol. Surv., Profess. Papers, 148.
360 Engel, P. and Nowacki, W., 1966. Die Verfeinerung der Kristallstruktur von Proustit und Pyrargyrit. Neues Jahrb. Mineral., Monatsh., 1966, 181-184.
361 Entwistle, L.P. and Gouin, L.O., 1955. The chalcocite-ore deposits at Corocoro, Bolivia. Econ. Geol., 50, 555-570.
362 Epprecht, W., 1946a. Die Eisen- und Manganerze des Gonzen. Beitr. Geol. Schweiz, Geotech. Ser., 24.
363 Epprecht, W., 1946b. Die Manganmineralien vom Gonzen und ihre Paragenese. Schweiz. Mineral. Petrog. Mitt., 26, 19-27.
364 Erd, R.C. and Evans, H.T., Jr., 1956. The compound Fe_3S_4 (smythite) found in nature. J. Am. Chem. Soc., 78, 2017. Ref. in: Am. Mineralogist;41, 815-816.
365 Erd, R.C., Evans, H.T., Jr., and Richter, D.H., 1957. Smythite, a new iron sulfide, and associated pyrrhotite from Indiana. Am. Mineralogist, 42, 309-333.
366 Ermilova, L.P. and Senderova, V.M., 1955. On the discovery of cosalite in Central Kazakhstan (in Russian). Dokl. Akad. Nauk SSSR, 105, 1325-1327. Ref. in: Mineral. Abstr., 13, 120.
367 Ervamaa, P., 1962. The Petolahti diabase and associated nickel-copper-pyrrhotite ore, Finland. Bull. Comm. Géol. Finlande, 199, 80 pp.
368 Euler, R. and Hellner, E., 1959. Die Struktur des Meneghinits. Fortschr. Mineral., 37, 57.
369 Evans, A.M., 1957. A tin-bearing ore from the Coal River area, Yukon Territory. Can. Mineralogist, 6, 119-127.
370 Evans, H.T., Jr. and Allman, R., 1967. Crystal chemistry of valleriite, a hybrid iron-copper sulfide, magnesium-aluminium hydroxide species. Geol. Soc. Am., Program Ann. Meeting, 1967, 61-62.
371 Evans, H.T., Jr. and Block, S., 1953. The crystal structure of montroseite, a vanadium member of the diaspore group. Am. Mineralogist, 38, 1242-1250.
372 Evans, H.T., Jr., Milton, C., Chao, E.C.T., Adler, I., Mead, C., Ingram, B. and Berner, R.A., 1964. Valleriite and the new iron sulfide, mackinawite. U.S., Geol. Surv., Profess. Papers, 475-D, 64-69.
373 Evans, H.T., Jr., and Mrose, M.E., 1955. A crystal chemical study of montroseite and paramontroseite. Am. Mineralogist, 40, 861-875.
374 Evans, H.T., Jr.,and Mrose, M.E., 1958. The crystal structures of three new vanadium oxide minerals. Acta Cryst., 11, 56-58.
375 Evans, H.T., Jr., and Mrose, M.E., 1960. A crystal chemical study of the vanadium oxide minerals, häggite and doloresite. Am. Mineralogist, 45, 1144-1166.
376 Evrard, P., 1945. Minor elements in sphalerites from Belgium. Econ. Geol., 40, 568-574.
377 Eyde, T.H., 1958. The Potosi tungsten district, Madison Country, Idaho. State of Montana, Bureau of Mines and Geology, Informal circular 21.
378 Faber, W., 1933a. Zur Kenntnis des Rotnickelkieses. Z. Krist., A 84, 408-435.
379 Faber, W., 1933b. Das Reflexionsvermögen und die Bireflexion des Rotnickelkieses. Z. Krist., A 85, 223-231.
380 Fabregat Guinchard, F.J., 1966. Los minerales Mexicanos. 5. Livingstonita. Univ. Nac. Autónoma Méx., Inst. Geol., Bol., 79, 84 pp.
380a Fabregat Guinchard, F.J., 1967. Los minerales Mexicanos. 6. Jalpaita. Univ. Nac. Autónoma Méx., Inst. Geol., Bol., 83, 52 pp.
381 Faddegon, J.M., 1961. Some mineragraphic data on copper-magnetite ores from the Ertsberg, Carstensz Mountains, Neth. New Guinea. Oost-Borneo Maatschappij N.V., report, 19 pp.
382 Faessler, C. and Schwartz, G.M., 1941. Titaniferous magnetite deposits of Sept-Iles, Quebec. Econ. Geol., 36, 712-728.
383 Fahey, J.J., 1955. Murdochite, a new copper lead oxide mineral. Am. Mineralogist, 40, 905-906.

384 Fairbanks, E.E., 1946. The punched card identification of ore minerals. Econ. Geol., 41, 761-768.
385 Fan' De-Lyan (Fan Tieh-Lin), 1964. The new mineral, rhombomagnojacobsite (in Chinese). Acta Geol. Sinica, 44, 343-350. Ref. in: Mineral. Abstr., 17, 398.
386 Fander, H.W., 1967. Selenium at Peko and Cobar. Australian Mineral. Develop. Lab., Bull., no 2, 73-75. Ref. in: Mineral. Abstr., 18, 281.
387 Fastré, P., 1933. Mesure des pouvoirs réflecteurs de quelques tellurures naturels par la méthode photo-électrique. Compt. Rend. Acad. Sci. Paris, 196, 630-632.
388 Faulring, G.M., 1965. Unit cell determination and thermal transformation of nsutite. Am. Mineralogist, 50, 170-179.
389 Faulring, G.M., Zwicker, W.K. and Forgeng, W.D., 1960. Thermal transformation and properties of cryptomelane. Am. Mineralogist, 45, 946-959.
390 Fediuk, F. and Kušnír, I., 1967. Groupe de gîtes polymétalliques de Cho Diên en République Démocratique du Vietnam. Acta Univ. Carolinae, Geol., 29-58.
391 Ferguson, R.B., 1947. The unit cell of glaucodot. Am. Mineralogist, 32, 199.
392 Ferguson, R.B., 1957. The crystallography of synthetic $YTaO_4$ and fused fergusonite. Can. Mineralogist, 6, 72-77.
393 Ferrari, A. and Curti, R., 1934. I solfoarseniti di piombo. Periodico Mineral. (Roma), 5, 155-174.
394 Filipenko, Y.S., 1958. In: Polikarpova and Ambartsumian: New data on uranium minerals in the USSR. Coffinite. Proc. U.N. Intern. Conf. Peaceful Uses At. Energy, 2nd, Geneva, 1955, 2, 302 and 304.
395 Fiorentini, M. and Minutti, L., 1965. Nuove osservazioni nel argirodite e canfieldite. Rend. Soc. Mineral. Ital., 21, 355.
396 Firsov, L.V., 1963. Pentagonal to dodecahedral microcrystals of gold in contact-metamorphosed recrystallized veins in northeast USSR. Dokl. Akad. Nauk SSSR, 148, 681-683. Translated in: Dokl. Acad. Sci. USSR, Earth Sci. Sect., 148, 104-106 (1964).
397 Fleet, M.E. and MacRae, N., 1969. Two-phase hexagonal pyrrhotites. Can. Mineralogist, 9, 699-705.
398 Fleischer, M., 1955. Minor elements in some sulfide minerals. Econ. Geol., 50th Anniversary Volume, 970-1024.
399 Fleischer, M., 1960. Studies of the manganese oxide minerals. III: Psilomelane. Am. Mineralogist, 45, 176-187.
400 Fleischer, M., 1964. Manganese oxide minerals. VIII: Hollandite. Advan. Frontiers Geol. Geophys., Hyderabad, 221-232. Ref. in: Mineral. Abstr., 17, 389.
401 Fleischer, M., 1966. Index of new mineral names, discredited minerals, and changes of mineralogical nomenclature in Volumes 1-50 of The American Mineralogist. Am. Mineralogist, 51, 1248-1357.
402 Fleischer, M. and Faust, G.T., 1963. Studies on manganese oxide minerals. VII. Lithiophorite. Schweiz. Mineral. Petrog. Mitt., 43, 197-216.
403 Fleischer, M. and Richmond, W.E., 1943. The manganese oxide minerals: A preliminary report. Econ. Geol., 38, 269-286.
404 Fleischer, M., Richmond, W.E. and Evans, H.T., Jr., 1962. Studies of the manganese oxides. V: Ramsdellite, MnO_2, an orthorhombic dimorph of pyrolusite. Am. Mineralogist, 47, 47-58.
405 Flinter, B.H., 1959a. Re-examination of "strüverite" from Salak North, Malaya. Am. Mineralogist, 44, 620-632.
406 Flinter, B.H., 1959b. The alteration of Malayan ilmenite grains and the question of "arizonite". Econ. Geol., 54, 720-729.
407 Flinter, B.H., 1963. A note on ferroan gahnite from Malaya and its bearing on the published data for hercynite. Am. Mineralogist, 48, 194-199.
408 Flinter, B.H., 1964. Re-examination of "strüverite", a further note. Am. Mineralogist, 49, 792-794.
409 Folinsbee, R.E., 1949. Determination of reflectivity of the ore minerals. Econ. Geol., 44, 425-436.
410 Forman, S.A. and Peacock, M.A., 1949. Crystal structure of rickardite, $Cu_{4-x}Te_2$. Am. Mineralogist, 34, 441-451.
411 Forster, I.F., 1960. Beobachtungen an einem primären Goldvorkommen in ultra-basischen Gesteinen des Lowvelds (Nordost Transvaal, S.A.). Neues Jahrb. Mineral., Abhandl., 94, 228-266.
412 Foslie, S., 1946. Melkedalen grube i Ofoten. Norg. Geol. Undersøk., 169.
413 Foslie, S. and Høst, M.J., 1932. Platina i sulfidisk Nikkelmalm. Norg. Geol. Undersøk., 137.

414 Francotte, J., Moreau, J., Ottenburgs, R. and Lévy, C., 1965. La briartite, une nouvelle espèce minérale. Bull. Soc. Franç. Minéral. Crist., 88, 432-437.
415 Frankel, J.J., 1958. Manganese ores from the Kurumam District, Cape Province, South Africa. Econ. Geol., 53, 577-597.
417 Frebold, G., 1927. Beiträge zur Kenntnis der Erzlagerstätten des Harzes. II. Ueber einige Selenerze und ihre Paragenesen im Harz. Zentr. Mineral. Geol. Palaeontol., A, 16-32.
418 Frebold, G., 1928. Ueber einige Mineralien der Enargitgruppe und ihre paragenetischen Verhältnisse in der Kupfererzlagerstätte von Mancayan auf Luzon (Philippinen). Neues Jahrb. Mineral., Beil., 56 A, 316-333.
418a Fredriksson, K. and Andersen, C.A., 1964. Electron probe analyses of Cu in meneghinite. Am. Mineralogist, 49, 1467-1469.
419 Frenzel, G., 1953. Die Erzparagenese des Katzenbuchels im Odenwald. Heidelberger Beitr. Mineral. Petrog., 3, 409-444.
420 Frenzel, G., 1954a. Erzmikroskopische Beobachtungen an natürlich erhitzten insbesondere pseudobrookitführende Vulkaniten. Heidelberger Beitr. Mineral. Petrog., 4, 343-376.
421 Frenzel, G., 1954b. Ueber eines ungewöhnlichen Hochtemperaturmagnetkies vom Katzelbuchel im Odenwald. Heidelberger Beitr. Mineral. Petrog., 4, 377-378.
422 Frenzel, G., 1955. Zur Kenntnis von Bismutotantalit. Neues Jahrb. Mineral., Monatsh., 1955, 241-251.
423 Frenzel, G., 1957. Deszendente Umbildungsstrukturen von Magnetkies. Fortschr. Mineral. 35, 23-25.
424 Frenzel, G., 1959a. Idait und "blaubleibender Covellin". Neues Jahrb. Mineral., Abhandl., 93, 87-132.
425 Frenzel, G., 1959b. Ein neues Mineral: Idait. Neues Jahrb. Mineral., Monatsh., 1959, 142.
426 Frenzel, G., 1961a. Zerfallserscheinungen bei Rotnickelkies. Schweiz. Mineral. Petrog. Mitt., 41, 395-406.
427 Frenzel, G., 1961b. Ueber ein neues, dem Högbomit verwandtes Nb-haltiges Fe-Ti Mineral. Fortschr. Mineral., 39, 82.
428 Frenzel, G., 1961c. Ein neues Mineral: Freudenbergit. Neues Jahrb. Mineral., Monatsh., 1961, 12-22.
429 Frenzel, G., 1963. On crystal data for idaite. Am. Mineralogist, 48, 676-677.
430 Frenzel, G. and Bloss, F.D., 1967. Cleavage in pyrite. Am. Mineralogist, 52, 994-1002.
431 Frenzel, G. and Ottemann, J., 1967. Eine Sulfidparagenese mit kupferhaltigem Zonarpyrit von Nukundamu/Fiji. Mineral. Deposita, 1, 307-316.
432 Frenzel, G. and Ottemann, J., 1968. Ueber ein neues Ni-As-Mineral und eine bemerkenswerte Uranmineralisation von der Anna-Procopi-Grube bei Přibram. Neues Jahrb. Mineral., Monatsh., 1968, 420-429.
432a Freund, H. (Editor), 1966. Applied Ore Microscopy, Theory and Technique. The MacMillan Company, New York and London.
433 Fricke, G., 1966. Molybdänglanzfund in Artenberg bei Steinach i.K., Schwarzwald. Aufschluss, 17, 110. Ref. in: Zentr. Mineral., II, 1967, 22.
434 Friedenreich, O., 1956. Die Chrom-Nickelvererzungen des Peridotitstockes von Finero-Centovalli. Schweiz. Mineral. Petrog. Mitt., 36, 227-243.
435 Friedman, G.M., 1952. Study of hoegbomite. Am. Mineralogist, 37, 600-608.
436 Friedrich, G., 1960. Petrographische und erzmikroskopische Beobachtungen an der Goldlagerstätte Rodalquir, Almeria, Spanien. Neues Jahrb. Mineral., Abhandl., 94, 208-227.
437 Fries, C., Jr., Schaller, W.T. and Glass, J.J., 1942. Bixbyite and pseudobrookite from the tin-bearing rhyolite of the Black Range, New Mexico. Am. Mineralogist, 27, 305-322.
438 Frohberg, M.H., 1933. Beiträge zur Kenntnis der turmalinführenden Goldquartzgänge des Michipicoten-Distriktes, Ontario. Tschermaks Mineral. Petrog. Mitt., 44, 349-409.
439 Frohberg, M.H. and Nuffield, E.W., 1960. Ueber ein ungewöhnliches Vorkommen von Millerit auf Temagami Island, Temagami District, Ontario. Neues Jahrb. Mineral., Abhandl., 94, 1183-1186.
440 Frondel, C., 1940a. Redefinition of tellurobismuthite and vandiestite. Am. J. Sci., 238, 880-888.

441 Frondel, C., 1940b. Exsolution growths of zincite in manganosite and of manganosite in periclase. Am. Mineralogist, 25, 534-538.
442 Frondel, C., 1941a. Unit cell and space group of vrbaite (Tl(As.Sb)$_3$S$_5$). seligmannite (CuPbAsS$_3$) and samsonite (Ag$_4$MnSb$_2$S$_6$). Am. Mineralogist, 26, 25-28.
443 Frondel, C., 1941b. Paramelaconite, a tetragonal oxide of copper. Am. Mineralogist, 26, 657-672.
444 Frondel, C., 1953. New manganese oxides: hydrohausmannite and woodruffite. Am. Mineralogist, 38, 761-769.
445 Frondel, C., 1958. Systematic mineralogy of uranium and thorium. U.S., Geol. Surv., Bull. 1064, 400 pp.
446 Frondel, C., 1962. Non-existence of native tantalium. Am. Mineralogist, 47, 786-787.
447 Frondel, C., 1963. Isodimorphism of the polybasite and pearceite series. Am. Mineralogist, 48, 565-572.
448 Frondel, C., 1967. Voltzite. Am. Mineralogist, 52, 617-634.
449 Frondel, C. and Heinrich, E.W., 1942. New data on hetaerolite, hydrohetaerolite, coronadite and hollandite. Am. Mineralogist, 27, 48-56.
450 Frondel, C. and Honea, R.M., 1968. Billingsleyite, a new silver sulfosalt. Am. Mineralogist, 53, 1791-1798.
451 Frondel, C. and Klein, C. Jr., 1965. Exsolution in franklinite. Am. Mineralogist, 50, 1670-1680.
452 Frondel, C., Marvin, U.B. and Ito, J., 1960a. New data on birnessite and hollandite. Am. Mineralogist, 45, 871-875.
453 Frondel, C., Marvin, U.B. and Ito, J., 1960b. New occurrences of todorokite. Am. Mineralogist, 45, 1167-1173.
454 Frueh, A.J., Jr., 1959. The crystallography of petzite, Ag$_3$AuTe$_2$. Am. Mineralogist, 44, 693-701.
455 Frueh, A.J., Jr., 1961. The use of zone theory in problems of sulfide mineralogy, Part III; Polymorphs of Ag$_2$Te and Ag$_2$S. Am. Mineralogist, 46, 654-660.
456 Fryklund, V.C., Jr. and Fletcher, J.D., 1956. Geochemistry of sphalerite from the Star Mine, Coeur d'Alène District, Idaho. Econ. Geol., 51, 223-247.
457 Fryklund, V.C., Jr. and Hutchinson, M.W., 1954. The occurrence of Co and Ni in the Silver Summit mine, Coeur d'Alène District, Idaho. Econ. Geol. 49, 753-758.
458 Furnival, G.M., 1939. A silver-pitchblende deposit at Contact Lake, Great Bear Lake area, Canada. Econ. Geol., 34, 739-776.
459 Gabrielson, O., 1945. Studier över elementfördelningen i zinkbländen från svenska fyndorter. Sveriges Geol. Undersökn., Ser. C, 468, Årsbok 39 (1).
459a Gahm, J., 1969. Einige Probleme der Mikrohärtemessung. Zeiss Mitt., 5, 40-80.
460 Gaines, R.V., 1957. Luzonite, famatinite, and some related minerals. Am. Mineralogist, 42, 766-779.
461 Gaines, R.V., 1965. Mineralización de telurio en la mina La Moctezuma, cerca de Moctezuma, Sonora. Univ. Nac. Autónoma Méx., Inst. Geol., Bol., 75, 1-15.
462 Galbraith, F.W., 1937. A microscopic study of goethite and hematite in the brown iron ores of East Texas. Am. Mineralogist, 22, 1007-1015.
463 Galbraith, F.W., 1940. Identification of the commoner tellurides. Am. Mineralogist, 25, 368-371.
464 Galbraith, F.W., 1941. Ore minerals of the La Plata Mountains, Colorado, compared with other telluride districts. Econ. Geol., 36, 324-334.
465 Galkin, M.A., 1966. Mineral associations in mercury deposits of northeastern Yakutia. Dokl. Akad. Nauk SSSR, 169, 438-440. Translated in: Dokl. Acad. Sci. USSR, Earth Sci, Sect., 169, 165-167 (1967).
466 Gallagher, D., 1941. A microscopic study of some ores of the Lupa goldfield, Tanganyika Territory, East Africa. Econ. Geol., 36, 306-323.
467 Galopin, R., 1936. Différenciation chimique des minéraux métalliques par la méthode des empreintes. Schweiz. Mineral. Petrog. Mitt., 16, 1-18.
468 Gammon, J.B., 1966. Some observations on minerals in the system CoAsS-FeAsS. Norsk Geol. Tidsskr., 46, 405-426.
469 Gamyanin, G.N., 1968. Bismuth sulfotellurides from northern Yakutia. Dokl. Akad. Nauk SSSR, 178, 679-682. Translated in: Dokl. Acad. Sci. USSR, Earth Sci. Sect., 178, 121-124 (1968).
470 Garrido, J. and Feo, R., 1938. Sur les sulfotellurures de bismuth. Bull. Soc. Franç. Minéral. Crist., 61, 196-204.
471 Gaudefroy, C., Jouravsky, G. and Permingeat, F., 1963. La marokite, CaMn$_2$O$_4$, une nouvelle espèce minérale. Bull. Soc. Franç. Minéral. Crist., 86, 359-367.

472 Gaudefroy, C., Dietrich, J., Permingeat, F. and Picot, P., 1966. La crednérite, sa composition chimique et sa signification génétique. Bull. Soc. Franç. Minéral. Crist., 89, 80-88.
473 Gavelin, S., 1936. Auftreten und Paragenese der Antimonminerale in zwei Sulfidvorkommen im Skelleftefelde, Nordschweden. Sveriges Geol. Undersökn., Ser. C., 404, Årsbok 30 (11).
474 Gavelin, S., 1939. Geology and ores of the Malånäs District, Västerbotten, Sweden. Sveriges Geol. Undersökn., Ser. C, 424, Årsbok 33 (4).
475 Gavelin, S., 1945. Arsenic-cobalt-nickel-silver veins in the Lindsköld Copper mine, N. Sweden. Sveriges Geol. Undersökn., Ser. C, 469, Årsbok 39 (2).
476 Gavelin, S., 1954. A telluride assemblage in the Rudtjebäcken pyrite ore, Vesterbotten, N. Sweden. Sveriges Geol. Undersökn., Årsbok 48 (1), N: o 536.
477 Geffroy, J., 1958. La berthiérite du gisement aurifère du Châtelet (Creuse). Bull. Soc. Franç. Minéral Crist., 81, 162.
478 Geffroy, J., 1963. La brannerite du filon aurifère de La Gardette (Isère) et sa signification métallogénique. Bull. Soc. Franç. Minéral. Crist., 86, 129-132.
479 Geffroy, J., Cesbron, F. and Lafforgue, P., 1964. Données preliminaires sur les constituants profonds des minerais uranifères de Mouana (Gabon). C.R. Acad. Sci., Paris, 259, 601-603.
480 Geffroy, J. and Lenfant, M., 1963. Présence d'une minéralisation microscopique à bismuth, nickel et cobalt dans le district du Kaymar (Aveyron). Bull. Soc. Franç. Minéral. Crist., 86, 201-203.
481 Geffroy, J. and Lissillour, J., 1963. Présence d'énargite en association avec la pechblende de la Crouzville (Hautte-Vienne). Bull. Soc. Franç., Minéral. Crist., 86, 14-16.
482 Geffroy, J., Sarcia, J. and Sarcia, J.-A., 1957. Présence de jordanite dans le district du Djebel-Hallouf (Tunisie). Bull. Soc. Franç. Minéral. Crist., 80, 99-100.
483 Gehlen, K. von, 1964. Anomaler Bornit und seine Umbildung zu Idait und "Chalcopyrit" in deszendenten Kupfererzen von Sommerkahl (Spessart). Fortschr. Mineral., 41, 163.
484 Gehlen, K. von, and Hausmann, K., 1969. New data on the optical properties and micro-hardness of galena, bornite, pyrite, chalcopyrite, and magnetite. Third Annual Regional Conference, Copenhagen. Medd. Dansk Geol. Foren., 19, 327-328.
485 Gehlen, K. von and Piller, H., 1964. Zur Optik von Covellin. Beitr. Mineral. Petrog., 10, 94-110.
486 Gehlen, K. von and Piller, H., 1965a. Zur Optik von Hämatit und Ilmenit. Neues Jahrb. Mineral., Monatsh., 1965, 97-108.
487 Gehlen, K. von and Piller, H., 1965b. Optics of hexagonal pyrrhotite ($\sim Fe_9S_{10}$). Mineral. Mag., 35, 335-346.
488 Geier, B., 1933. Die Kupferwismuterze von Neubulach in Schwarzwald. Z. Prakt. Geol., 41, 137-146.
489 Geijer, P., 1924. On cubanite and "chalcopyrrhotite" from Kaveltorp. Geol. Fören. Stockholm Förhandl., 46, 354-355.
490 Geller, S. and Wernick, J.H., 1959. Ternary semiconducting compounds with sodium chloride-like structure: $AgSbS_2$, $AgSbTe_2$, $AgBiS_2$, $AgBiSe_2$. Acta Cryst., 12, 46-54.
491 Genkin, A.D., 1959. Conditions of occurrence and features of the composition of minerals of the platinum group in the Norilsk deposits (in Russian). Geol. Rudn. Mestorozhd. 1959, 74-84. Ref. in:Am. Mineralogist, 46, 464.
492 Genkin, A.D., Filimonova, A.A., Shadlun, T.N., Soboleva, S.V. and Troneva, N.V., 1966. On cubic cubanite and cubic chalcopyrite. Geol. Rudn. Mestorozhd., 1966, 51-54. Translated in: Geochemistry Intern., 2, 766-781 (1965).
493 Genkin, A.D. and Muraveva, I.V., 1963. Indite and dzhalindite, new indium minerals (in Russian). Zap. Vses. Mineralog. Obshchestva, 92, 445-457. Ref. in: Mineral. Abstr., 16, 457 and Am. Mineralogist, 49, 439-440.
494 Genkin, A.D., Muraveva, I.V. and Troneva, N.V., 1966. Zvyagintsevite, a natural intermetallic compound of palladium, platinum, lead and tin (in Russian). Geol. Rudn. Mestorozhd., 1966, 94-100. Ref. in: Am. Mineralogist, 52, 299.
495 Genkin, A.D., Zhuravlev, N.N. and Smirnova, E.M., 1963. Moncheite and kotulskite, new minerals, and the composition of michenerite (in Russian). Zap. Vses. Mineralog. Obshchestva, 92, 33-50. Ref. in: Am. Mineralogist 48, 1181 and Mineral. Abstr., 16, 283.

496 Genkin, A.D., Zhuravlev, N.N., Troneva, N.V. and Muraveva, I.V., 1966. Irarsite, a new sulfoarsenide of iridium, rhodium, ruthenium and platinum (in Russian). Zap. Vses. Mineralog. Obshchestva, 95, 700-712. Ref. in: Am. Mineralogist, 52, 1580 and Mineral. Abstr., 18, 283.
497 Genkin, A.D. and Zvyagintsev, O.E., 1962. Vysotskite, a new sulfide of palladium and nickel (in Russian). Zap. Vses. Mineralog. Obshchestva, 91, 718-725. Ref. in: Mineral. Abstr., 16, 180 and Am. Mineralogist 48, 708.
498 Gianella, V.P., 1938. Epithermal hübnerite from the Monitor District, Alpine County, California. Econ. Geol., 33, 339-348.
499 Gibbons, G.S., 1967. Optical anisotropy in pyrite. Am. Mineralogist, 52, 359-370.
500 Gies, H.J., 1969. Relations between reflectivity, micro-hardness, and element-contents regarding gersdorffite and ullmannite. Third Annual Regional Conference, Copenhagen. Medd. Dansk Geol. Foren., 19, 339.
501 Giese, R.F., Jr. and Kerr, P.F., 1965. The crystal structure of ordered and disordered cobaltite. Am. Mineralogist, 50, 1002-1014.
502 Gimpl, M.L., Nelson, C.E. and Fuschillo, N., 1963. Some properties of platinum monotelluride (PtTe). Am. Mineralogist, 48, 689-691.
503 Ginzburg, A.I., Nazarova, A.S. and Sukhomazova, L.L., 1963. Nigerite from Siberian pegmatites. New Data on Rare Element Mineralogy, Edited by A.I. Ginzburg, pp. 42-46, Consultants Bureau Enterprises, Inc., New York.
504 Giraud, R., Picot, P., Grammont, X. de and Tollon, F., 1968. Sur la présence de millérite dans la région de Gèdre (Hautes-Pyrénées). Bull. Soc. Franç. Minéral. Crist., 91, 279-283.
505 Giusça, D., 1930. Die Erze der Lagerstätte vom Lengenbach im Binnental (Wallis). Schweiz. Mineral. Petrog. Mitt., 10, 152-177.
506 Gjelsvik, T., 1957. Pitchblende mineralization in the Precambrian plateau of Finnmarksvidda, Northern Norway. Geol. Fören. Stockholm Förhandl., 79, 572-580.
507 Glass, J.J., Koschmann, A.H. and Vhay, J.S., 1958. Minerals of the cassiterite-bearing veins at Irish Creek, Virginia, and their paragenetic relations. Econ. Geol., 53, 65-84.
508 Godovikov, A.A. and Ferantschitsch, F.A., 1960. Ueber den Fund Laitakarits, eines seltenen Bi-Selenids in der Sowjet-Union (in Russian). Geol. Geofys. (USSR), 10, 19-26. Ref. in: Zentr. Mineral. I, 1962, 237.
509 Godovikov, A.A., Fedorova, Zh.N. and Bogdanova, V.I., 1966. An artificial subtelluride of bismuth similar to hedleyite. Dokl. Akad. Nauk SSSR, 169, 925-928. Translated in: Dokl. Acad. Sci. USSR, Earth Sci. Sect., 169, (1967), 142-144.
510 Golovanov, I.M., 1960. Coronadite from the zone of oxidation of the lead-zinc deposits of Kurgashinkan (Uzbek. SSR). Dokl. Akad. Nauk SSSR, 130, 843-845. Translated in: Dokl. Acad. Sci. USSR, Earth Sci. Sect., 130, 148-150 (1961).
511 Goñi, J. and Picot, P., 1965. Certaines particularités mineralogiques des tactites à scheelite du nord-est du Brésil. Bull. Soc. Franç. Minéral. Crist., 88, 11-16.
512 Gorbunov, G.I. and Kornilov, N.A., 1954. On a new discovery of ilvaite in copper-nickel sulphide ores (in Russian). Dokl. Akad. Nauk SSSR, 94, 323-325. Ref. in: Mineral. Abstr., 12, 496-497.
513 Gordon, S.G., 1926. Penroseite and trudellite: two new minerals. Proc. Acad. Nat. Sci., Phila., 77, 317-324. Ref. in: Mineral. Abstr., 3 (1926), 112.
514 Gordon, S.G., 1939. Greenockite from Llallagua, Bolivia. Notulae Naturae Acad. Nat. Sci. Phila., 1, 1-16.
515 Gordon, S.G. and Shannon, E.V., 1928. Chromrutile, a new mineral from California. Am. Mineralogist, 13, 69.
516 Gorman, D.H., 1951. An X-ray study of the mineral livingstonite. Am. Mineralogist, 36, 480-483.
517 Gorzhevskaya, S.A., Sidorenko, G.A. and Smorchkov, I.E., 1963. A new modification of fergusonite, β-fergusonite. New Data on Rare Element Mineralogy, Edited by A.I. Ginzburg, pp. 16-17. Consultants Bureau Enterprises, Inc., New York.
518 Gorzhevsky, D.I., 1955. A discovery of a rare mineral - zinkenite (in Russian). Mineralog. Sb. L'vovsk. Geo. Obshchestvo, 9, 313-314. Ref. in: Mineral. Abstr., 13, 288.
519 Gotman, Ya.D., Polyakova, V.M. and Miguta, A.K., 1968. Another variety of brannerite. Dokl. Akad. Nauk SSSR, 179, 429-430. Translated in: Dokl. Acad. Sci. USSR, Earth Sci. Sect., 179, 124-125 (1968).
520 Graeser, S., 1963. Giessenit, ein neues Pb-Bi-sulfosalz aus dem Dolomit des Binnatales. Schweiz. Mineral. Petrog. Mitt., 43, 471-478.

521 Graeser, S., 1964. Ueber Funde der neues rhomboedrischen MoS_2-Modification (Molybdänit-3R) und von Tungstenit in den Alpen. Schweiz. Mineral. Petrog. Mitt., 44, 121-128.

522 Graeser, S., 1965. Die Mineralfundstellen im Dolomit des Binnatales. Schweiz. Mineral. Petrog. Mitt., 45, 597-795.

523 Graeser, S., 1967. Ein Vorkommen von Lorandit ($TlAsS_2$) in der Schweiz. Beitr. Mineral. Petrog., 16, 45-50.

524 Grafenauer, S., 1960. Seltene natürliche Bleioxyde von Mežica (Mies), Jugoslawien. Neues Jahrb. Mineral., Abhandl., 94, 1187-1190.

524a Grafenauer, S., Gorenc, B., Marinković, V. and Strmole, D., 1969. Physical properties and the chemical composition of sphalerites from Yugoslavia. Mineral. Deposita, 4, 275-282.

525 Grafenauer, S., Ottemann, J. and Strmole, D., 1968. Ueber Descloizit und Wulfenit von Mezica (Mies), Jugoslavien. Neues Jahrb. Mineral., Abhandl., 109, 25-32.

526 Graham, A.R., 1951. Matildite, aramayoite, miargyrite. Am. Mineralogist, 36, 436-449.

527 Graham, A.R. and Kaiman, S., 1952. Aurostibite, $AuSb_2$; a new mineral in the pyrite group. Am. Mineralogist, 37, 461-469.

528 Graham, A.R., Thompson, R.M. and Berry, L.G., 1953. Studies of mineral sulphosalts: XVII: Cannizzarite. Am. Mineralogist, 38, 536-544.

529 Granger, H.C., 1966. Ferroselite in a roll-type uranium deposit, Powder River Basin, Wyoming. U.S., Geol. Surv., Profess. Papers, 550-C, 133-137.

530 Granger, H.C. and Ingram, B.L., 1966. Occurrence and identification of jordisite at Ambrosia Lake, New Mexico. U.S., Geol. Surv., Profess. Papers, 550-B, 120-124.

531 Granger, M.M., and Protas, J., 1965. Détermination de la structure de la gaudefroyite. Compt. Rend. Acad. Sci., Paris, 260, 4553-4555.

532 Grasselly, J., 1948. Analyses of some bismuth minerals. Publ. Mineral. Petrog. Inst. Univ. Szeged, II, 24-30.

533 Graton, L.C., 1933. The depth-zones in ore deposition. Econ. Geol., 28, 513-555.

534 Graton, L.C., 1940 Nature of the ore-forming fluid. Econ. Geol., 35, 197-358.

535 Graton, L.C. and Bowditch, S.I., 1936. Alkaline and acid solutions in hypogene zoning at Cerro de Pasco. Econ. Geol., 31, 651-698.

536 Gray, A., 1932. The Mufulira copper deposits, Northern Rhodesia. Econ. Geol., 27, 315-343.

537 Grigoriev, Iv.F. and Dolomanova, E.I., 1955. Joseite from a tin-ore deposit of the Central Transbaikal region (in Russian). Tr. Mineralog. Muzeya, Akad. Nauk SSSR, 7. 154-157. Ref. in: Mineral. Abstr., 13, 160.

538 Grimmer, A., 1962. Mineralogische und paragenetische Untersuchungen an einigen Sulfiden des Kobalts und Nickels. Bergakademie, 14, 296-302.

539 Grip, E., 1942. Nickel förekomsten Lainejaur. Geol. Fören. Stockholm Förhandl., 64, 273-276.

540 Grip, E., 1951. Tungsten and molybdenum in sulphide ores in Northern Sweden. Geol. Fören. Stockholm Förhandl., 73, 455-472.

541 Grip, E. and Odman, O.H., 1942. The telluride-bearing andalusite-sericite rocks of Mångfallberget at Boliden, N. Sweden. Sveriges Geol. Undersökn., Ser. C, 447, Årsbok 36 (4).

542 Groeneveld Meijer, W.O.J., 1955. Niggliite, a monotelluride of platinum? Am. Mineralogist, 40, 693-696.

543 Grondijs, H.F. and Schouten, C., 1937. A study of the Mount Isa ores. Econ. Geol., 32, 407-450.

544 Gross, E.B., 1956. Mineralogy and paragenesis of the uranium ore, Mi Vida mine, San Juan County, Utah. Econ. Geol., 51, 632-648.

545 Gross, E.B., 1965. A unique occurrence of U-minerals, Marshall Pass, Saguache County, Colorado. Am. Mineralogist, 50, 909-923.

546 Grubb, P.L.C., 1967. Solid solution relationships between wolframite and scheelite. Am. Mineralogist, 52, 418-426.

547 Grünenfelder, M., 1957. Erzmikroskopische Beobachtungen an den Goldquarzgängen von Gondo (Simplon, Wallis) und Alpe Formazzolo (Val Calneggia, Tessin). Schweiz. Mineral. Petrog. Mitt., 37, 1-8.

548 Gruner, J.W., 1943. The chemical relationship of cryptomelane (psilomelane), hollandite and coronadite. Am. Mineralogist, 28, 497-506.

549 Gruner, J.W., 1947. Groutite, $HMnO_2$, a new mineral of the diaspore-goethite group. Am. Mineralogist, 32, 654-659.

550 Gruner, J.W., Fetzer, W.G. and Rapaport, I., 1955. The uranium deposits near Marysvale, Piute County, Utah. Econ. Geol., 50, 243-251.
551 Grybeck, D. and Finney, J.J., 1968. New occurrences and data for jalpaite. Am. Mineralogist, 53, 1530-1542.
552 Guilbert, F., Picot, P. and Schubnel, H.J., 1968. La gersdorffite des Ait-Ahmane, Maroc. Bull. Soc. Franç. Minéral. Crist., 91, 412-413.
553 Guild, F.N., 1932. The microstructure and paragenesis of stromeyerite. Schweiz. Mineral. Petrog. Mitt., 12, 222-233.
554 Guild, F.N., 1934. Microscopic relations of magnetite, hematite, pyrite and chalcopyrite. Econ. Geol., 29, 107-120.
555 Guild, P.W., 1953. Iron deposits of Minas Gerais, Brasil. Econ. Geol., 48, 639-676.
556 Guillemin, C. and Lévy, C., 1957. Sur les minéraux du sondage de Petitchet (Isère). Bull. Soc. Franç. Minéral. Crist., 80, 237-238.
557 Haan, W. de, Schouten, C. and Matthijsen, P.M., 1933. Monografie van de ertsafzettingen te Mangani (Sumatra). Verhandel. Geol. Mijnbouwk. Genoot. Ned. Kol., Mijnbouwk. Ser., III, 1-212.
558 Háber, M., 1965. Beitrag zum Studium der Mikrohärte von Pyrit. Geol. Sborník (Bratislava), 16, 113-127.
558a Háber, M., 1969. Unpublished values, measured at the Institute for Earth Sciences, Free University, Amsterdam.
558b Háber, M. and Streško, V., 1969. Ein neues Kobellit-Vorkommen in dem Zips-Gömörer Erzgebirge. Geol. Sborník (Bratislava), 20, 133-151.
559 Hackl, O., 1921. Ein neues Nickel-Arsen-Mineral. Verhandl. Geol. Staatsanstalt (Wien), 1921, 107-108.
560 Hagni, R.D., 1968. Titanium occurrence and distribution in the magnetite-hematite deposit at Benson mines, New York. Econ. Geol., 63, 151-155.
561 Hak, J., 1958. Zinckenit und seine Mineralbegleitung von Husárhy in Nízké Tatry, Slovakei (in Czech). Časopis Mineral. Geol., 3, 397-406. Ref. in: Zentr. Mineral., I, 1960, 332.
562 Hak, J. and Kupka, F., 1958. Identifikation von Kobellit aus Hummel in Zips-Gömörer Erzgebirge (in Czech). Časopis Mineral. Geol., 3, 16-20. Ref. in: Zentr. Mineral., I, 1960, 332-333.
563 Hak, J., Kvaček, M. and Johan, Z., 1964. Chemical-mineralogical investigation of chalcostibite from the Nízhé Tatry Mountains (in Czech). Sb. Národ. Musea Praze (Acta Musei Nationalis Pragae), XX B, 241-256. Ref. in: Zentr. Mineral., I, 1966, 122.
564 Häkli, T.A., Vuorelainen, Y. and Sahama, Th.G., 1965. Kitkaite (NiTeSe), a new mineral from Kuusamo, NE Finland. Am. Mineralogist, 50, 581-586.
565 Halahyjová-Andrusovová, G. and Šamajová, E., 1966. Vorkommen von Cu-Arseniden in der Verwitterungszone in Šankovce (in Czech, German summary). Acta Geol. Geograph. Univ. Comenianae, Geol., 11, 125-131.
566 Hall, H.T., 1967. The pearceite and polybasite series. Am. Mineralogist, 52, 1311-1321.
567 Halls, C., Clark, A.M. and Stumpfl, E.F., 1967. Some observations on silver-antimony phases from Silverfields mine, Ontario, Canada. Trans. Inst. Mining Met., 76, B 19-24.
568 Hannay, J.B., 1877. On some new minerals from the collection in the University of Glasgow. Mineral. Mag., 1, 149-153.
569 Hansen, M. and Anderko, K., 1958. Constitution of binary alloys, 2nd edition. McGraw Hill, New York.
570 Hanson, A.W., 1958. The crystal structure of nolanite. Acta Cryst., 11, 703-709.
571 Harada, Z., 1948. Chemical analyses of Japanese minerals (II). J. Fac. Sci., Hokkaidô Univ., Ser. IV, 7, 143-210.
572 Harada, Z., 1951. Star-shaped trillings of enargite from the Kamikita mine, Aomori Prefecture. J. Fac. Sci., Hokkaidô Univ., Ser. IV, 7, 319-323.
573 Haranczyk, C., 1958. Thallium jordanite. Bull. Acad. Polon. Sci., Sér. Sci., Chim. Géol. Géograph., 6, 201-208. Ref. in: Zentr. Mineral., I, 1959, 361.
574 Haranczyk, C., and Skiba, M., 1961. Gahnite from the Sn-bearing zone of Krobica-Gierczyn-Przecznica in Lower Silesia. Bull. Acad. Polon. Sci., Sér. Sci., Chim. Géol. Géograph., 9, No. 3, 149-154.
575 Harcourt, G.A., 1937. The distinction between enargite and famatinite (luzonite). Am. Mineralogist, 22, 517-525.

576 Hariya, Y., 1959. A new find of groutite, $HMnO_2$, in Japan. J. Fac. Sci., Hokkaidô Univ., Ser. IV, 10, 255-262.
577 Harker, D., 1934. The crystal structure of the mineral tetradymite, Bi_2Te_2S. Z. Krist., A 89, 175-181.
578 Harrington, B.J., 1907. Isomorphism as illustrated by certain varieties of magnetite. Mineral. Mag., 14, 373-377.
579 Harris, D.C., 1965. Zinckenite. Can. Mineralogist, 8, 381-382.
580 Harris, D.C., Jambor, J.L., Lachance, G.R. and Thorpe, R.I., 1968. Tintinaite, the antimony analogue of kobellite. Can. Mineralogist, 9, 371-382.
581 Harris, D.C., Nuffield, E.W. and Frohberg, M.H., 1965. Studies of mineral sulphosalts: XIX: Selenian polybasite. Can. Mineralogist, 8, 172-184.
582 Harris, D.C. and Thorpe, R.I., 1969. New observations on matildite. Can. Mineralogist, 9, 655-662.
583 Harrison, J.B. and Bourne, C.L.C., 1925. The occurrence of palladium amalgam - palladium mercuride - in British Guiana. Official Gazette, British Guiana, Febr. 27, no. 71, 3 pp. Mineral. Abstr., 3, 4-5.
584 Harry, W.T. and Oen Ing Soen, 1964. The Pre-Cambrian basement of Alángorssuaq. South Greenland, ant its copper mineralization at Josvaminen. Medd. Grønland, Bd 179, Nr. 1, 72 pp.
585 Harwood, H.F., 1951. The greenockite locality at Bishopton, Scotland. Am. Mineralogist, 36, 630.
586 Hawley, J.E., 1939. The association of gold, tungsten and tin at Outpost Islands, Great Slave Lake. Univ. Toronto, Geol. Ser., 42, 53-66.
587 Hawley, J.E., 1941. Boulangerite from Montgay Township, Abitibi County, Quebec. Univ. Toronto, Geol. Ser., 46, 25-32.
588 Hawley, J.E., 1962. The Sudbury ores: their mineralogy and origin. Can. Mineralogist, 7, 1-207.
589 Hawley, J.E. and Berry, L.G., 1958. Michenerite and froodite, palladium bismuthide minerals. Can. Mineralogist, 6, 200-209.
590 Hawley, J.E. and Haw, V.A., 1957. Intergrowths of pentlandite and pyrrhotite. Econ. Geol., 52, 132-139.
591 Hawley, J.E. and Hewitt, D.H., 1948. Pseudo-eutectic and pseudo-exsolution intergrowths of nickel-arsenides due to heat effects. Econ. Geol., 43, 273-279.
592 Hawley, J.E., Lewis, C.L. and Wark, W.J., 1951. Spectrographic study of Pt and Pd in common sulphides and arsenides of the Sudbury District, Ontario. Econ. Geol., 46, 149-162.
593 Hawley, J.E. and Rimsaite, Y., 1953. Platinum metals in some Canadian uranium and sulphide ores. Am. Mineralogist, 38, 463-475.
594 Hawley, J.E., Stanton, R.L. and Smith, A.Y., 1961. Pseudo-eutectic intergrowths in arsenical ores from Sudbury. Can. Mineralogist, 6, 555-575.
595 Hayase, K., 1955. Minerals of the bismuthinite-stibnite series with special reference to horobetsuite from the Horobetsu mine, Hokkaido, Japan. Mineral. J. (Tokyo), 1, 189-197. Ref. in: Mineral. Abstr., 15, 293, and Am. Mineralogist, 43, 623-624.
596 Hayashi, S., 1961. Cosalite from the Hagidaira mine, Gun'Ma Prefecture, Japan. Mineral. J. (Tokyo), 3, 148-155. Ref. in: Zentr. Mineral., I, 1962, 239-240.
597 Hayton, J.D., 1960. The constitution of davidite. Econ. Geol., 55, 1030-1038.
598 Head, R.E. and Loofbourow, R.W., 1934. A microchemical method for the determination of bournonite. Econ. Geol., 29, 301-305.
599 Heier, K., 1953. Clausthalite and selenium-bearing galena in Norway. Norsk Geol. Tidsskr., 32, 228-231.
600 Heinrich, E.W., 1960. Stibiotantalite from the Brown Derby No. 1 pegmatite, Colorado. Am. Mineralogist, 45, 728-731.
601 Heinrich, E.W., 1962. Radioactive columbite. Am. Mineralogist, 47, 1363-1379.
602 Helke, A., 1933. Beiträge zur Kenntnis der Golderzgänge am Ungarberge und am Fericel bei Stanija im Siebenbürgischen Erzgebirge, Rumänien. Tschermaks Mineral. Petrog. Mitt., 44, 265-326.
603 Helke, A., 1934. Die Goldtellurerzlagerstätten von Sacaramb (Nagyag) in Rumänien. Neues Jahrb. Mineral., Beil., 68 A, 19-85.
604 Helke, A., 1938. Die jungvulkanischen Gold-Silber-Erzlagerstätten des Karpathenbogens unter besonderer Berücksichtigung der Genesis und Paragenesis des gediegenen Goldes. Arch. Lagerstättenforschung, 66, 1-175.
605 Hellner, E., 1958. Ueber komplex zusammengesetzte Spiessglanze. III: Zur Struktur des Diaphorits. Z. Krist., 110, 169-174. Ref. in: Zentr. Mineral., I, 1959, 217.

606 Hellner, E., 1959. An intergrowth between galena and gratonite ($Pb_9As_4S_{15}$). J. Geol., 67, 473-475.
607 Hellner, E. and Burzlaff, H., 1964. Die Struktur des Smithits. Naturwissenschaften, 51, 35-36. Ref. in: Mineral. Abstr., 17, 558.
608 Hendricks, T.A. and Laird, W.M., 1943. The manganese deposits of the Turtle Mountains, North Dakota. Econ. Geol., 38, 591-602.
609 Henriques, Å., 1957. The Vickers hardness of zinc blende. Arkiv Mineral. Geol., 2, nr. 15, 283-297.
610 Herbosch, A., 1967. La viridine et la braunite de Salm-Château. Bull. Soc. Belge Géol. Paléontol. Hydrol., 76, 183-202.
611 Herzenberg, R., 1945. Blockita versus penroseita en el nuevo "Dana's System". Inst. Boliviano Ing. Mineria Geol., Publ. Tecnol. (La Paz), 5, 1-4.
612 Herzenberg, R., 1949. Montesita, nuevo mineral de estaño en Bolivia. Mineria Bolivia, 6, no. 44, 5-7. Ref. in: Am. Mineralogist, 35, 334-335.
613 Herzenberg, R. and Ahlfeld, F., 1935. Blockit, ein neues Selenerz aus Bolivien. Zentr. Mineral, A, 277-279.
614 Hewett, D.F., Cornwall, H.R. and Erd, R.C., 1968. Hypogene veins of gibbsite, pyrolusite, and lithiophorite in Nye County, Nevada. Econ. Geol., 63, 360-371.
615 Hewett, D.F. and Fleischer, M., 1960. Deposits of the manganese oxides. Econ. Geol. 55, 1-55.
616 Hewett, D.F. and Olivares, R.S., 1968. High-potassium cryptomelane from Tarapaca Province, Chile. Am. Mineralogist, 53, 1551-1557.
617 Hewett, D.F. and Rove, O.N., 1930. Occurrence and relations of alabandite. Econ. Geol., 25, 36-56.
618 Hewett, D.F., Stone, J. and Levine, H., 1957. Brannerite from San Bernardino County, California. Am. Mineralogist, 42, 30-38.
619 Hewitt, D.F., 1948. A partial study of the NiAs-NiSb system. Econ. Geol., 43, 408-417.
620 Hewitt, R.L., 1938. Experiments, bearing on the relation of pyrrhotite to other sulphides. Econ. Geol., 33, 305-338.
621 Hey, M.H., 1962a. A new analysis of villamaninite. Mineral. Mag., 33, 169-170.
622 Hey, M.H., 1962b. Cobaltic hydroxide in nature. Mineral. Mag., 33, 253-259.
623 Hey, M.H., 1963. The nomenclature of the natural alloys of osmium and iridium. Mineral. Mag., 33, 712-717.
624 Hey, M.H., 1968. On the composition of natural delafossite. Mineral. Mag., 36, 651-653.
624a Hey, M.H., Embrey, P.G. and Fejer, E.E., 1969. Crichtonite, a distinct species. Mineral. Mag., 37, 349-356.
625 Heyl, A.V., 1964. Enargite in the Zn-Pb deposits of the Upper Mississippi Valley District. Am. Mineralogist, 49, 1458-1461.
626 Heyl, A.V., Milton, C. and Axelrod, J.M., 1959. Nickel minerals from Near Linden, Iowa County, Wisconsin. Am. Mineralogist, 44, 995-1009.
627 Hickok, W.O., 1933. The iron ore deposits at Cornwall, Pennsylvania. Econ. Geol., 28, 193-255.
628 Hiller, J.E., 1938. Röntgenographische Bestimmungsmethoden und Untersuchung der Bleispiessglanze. Z. Krist., A 100, 128-156.
629 Hiller, J.E. and Walenta, K., 1960. Zwei Mineralien der Andorit-Ramdohrit-Fizelyit-Gruppe aus dem Schwarzwald. Neues Jahrb. Mineral. Abhandl., 94, 1160-1168.
630 Hiller, T., 1937. Sur l'application de la méthode des empreintes à la détermination des minéraux opaques en section polie. Schweiz. Mineral. Petrog. Mitt., 17, 88-145.
631 Hjelmqvist, S., 1949. The titaniferous iron-ore deposit of Taberg in the South of Sweden. Sveriges Geol. Undersökn., Årsbok 43, Ser. C: N:o 512.
632 Hoehne, K., 1936. Ueber einige Arsen-, Nickel-, Kobalt-, Silber-, Wismut- und Uranerzführende Kalkspatgänge der Grube Bergfreiheit zu Oberschmiedeberg im Riesengebirge. Chem. Erde, 10, 432-474.
633 Hofmann, W., 1935. Ergebnisse der Strukturbestimmung komplexer Sulfide. I. Die Struktur von Zinnsulfür, SnS, und Teallit, $PbSnS_2$. Z. Krist., A 92, 161-173.
634 Höhne, E., 1957. Die Kristallstruktur des Betechtinits. Fortschr. Mineral. 35, 50.
635 Höhne, E. and Kulpe, S., 1959. Zur Struktur des Lautits. Monatsber. Deut. Akad. Wiss. Berlin, 1, 283-285. Ref. in: Zentr. Mineral., I., 1959, 300-301.
636 Hollander, N.B., 1968. Electron microprobe analyses of spinels and their alteration products from Månsarp and Taberg, Sweden. Am. Mineralogist, 53, 1918-1928.
637 Holmes, R.J., 1936. An X-ray study of allemontite. Am. Mineralogist, 21, 202.

638 Holmes, R.J., 1947. Higher mineral arsenides of cobalt, nickel and iron. Bull. Geol. Soc. Am., 58, 299-391.
639 Honea, R.M., 1964. Empressite and stuetzite redefined. Am. Mineralogist, 49, 325-338.
640 Hornor, A.P., 1939. Magnetite and hematite veins in Triassic lavas of Nova Scotia. Econ. Geol., 34, 921-930.
641 Hubaux, A., 1960. Les gisements de fer titané de la région d'Egersund, Norvège. Neues Jahrb. Mineral., Abhandl., 94, 926-992.
642 Hugel, E., 1912. Magnoferrit von Schelingen (im Kaiserstuhl). Inaug.-Diss. Freiburg i. Br., 50-53. Ref. in: Neues Jahrb. Mineral. Geol. Paläontol., 1913, I. Band, 200-201.
643 Hugi, E., 1931. Ueber ein schweizerisches Cosalit-Vorkommen. Schweiz. Mineral. Petrog. Mitt., 11, 163-171.
644 Hügi, Th., Köppel, V., de Quervain, F. and Rickenbach, E., 1967. Die Uranvererzungen bei Isérables (Wallis). Beitr. Geol. Schweiz, Geotech. Ser., 42, 88 pp.
645 Hurlbut, C.S., Jr., 1957a. Bismutotantalite from Brazil. Am. Mineralogist, 42, 178-183.
646 Hurlbut, C.S., Jr., 1957b. The wurtzite-greenockite series. Am. Mineralogist, 42, 184-190.
647 Hutchinson, R.W., 1955. Preliminary report on investigations of minerals of columbium and tantalium and of certain associated minerals. Am. Mineralogist, 40, 432-452.
648 Huttenlocher, H.F., 1936. Zur Mangan-Zinn-Silber-Lagerstätte aus dem Wasserstollen des Amsteger Kraftwerkes. Schweiz. Mineral. Petrog. Mitt., 16, 406-408.
649 Ibach, R., 1939. Zur Entstehungsgeschichte der Kieslagerstätte von Kupferberg in Oberfranken. Z. Angew. Mineral., II, 114-152.
650 Iitaka, Y. and Nowacki, W., 1962. A redetermination of the crystal structure of galenobismutite. Acta Cryst., 15, 691-698.
651 Indolev, L.N., 1962. The first occurrence of diaphorite in the USSR. Dokl. Akad. Nauk SSSR, 142, 1387-1390. Translated in: Dokl. Acad. Sci. USSR, Earth Sci. Sect., 142, 127-130 (1964).
652 Indolev, L.N., 1964. Owyheeite from deposits of the south Verkhoyansk region. Dokl. Akad. Nauk SSSR, 154, 1351-1354. Translated in: Dokl. Acad. Sci. USSR, Earth Sci. Sect., 154, 122-124 (1964).
653 Indolev, L.N., Flerov, B.L., Zhdanov, Yu.Ya. and Brovkin, A.A., 1964. Herzenbergite from the Deputat deposit. Dokl. Akad. Nauk SSSR, 159, 1044-1047. Translated in: Dokl. Acad. Sci. USSR, Earth Sci. Sect., 159, 90-93 (1965).
654 Ishibasi, M., 1952. A Sn-Te-Bi-Sb paragenesis in ores from the Suttsu mine, Hokkaidô. (Study of the minor minerals of ore. Rept. 3). J. Fac. Sci., Hokkaidô Univ., Ser. IV, 8, 97-106.
655 Itô, T.-I. and Muraoka, H., 1960. Nakaséite, an andorite-like new mineral. Z. Krist., 113, 94-98. Ref. in: Am. Mineralogist, 45, 1314-1315. Mineral. Abstr., 15, 44.
656 Ivanov, V.V., 1963. Mineralogical-geochemical features and some physico-chemical characteristics of the formation of cassiterite-silicate-sulfide deposits in northern Yakutiya. Geochemistry (USSR; English transl.), 1963, 860-873.
657 Ivanov, V.V. and Pyatenko, Yu.A., 1959. On the so-called kusterite (in Russian). Zap. Vses. Mineralog. Obshchestva, 88, 165-168. Ref. in: Am. Mineralogist, 44, 1329, and Mineral. Abstr., 14, 280.
658 Jackson, G.C.A., 1932. The ores of the N'Changa mine and extensions, Northern Rhodesia. Econ. Geol., 27, 247-280.
659 Jacobsen, W., 1965. Untersuchungen über den Silbergehalt der Kupfererze von Mangula, Südrhodesien. Neues Jahrb. Mineral., Abhandl., 104, 1-28.
660 Jacobson, R. and Webb, J.S., 1947. The occurrence of nigerite, a new tin mineral in quarz-sillimanite rocks from Nigeria. Mineral. Mag., 28, 118-128.
661 Jambor, J.L., 1967a. New lead sulfantimonides from Madoc, Ontario - Part 1. Can. Mineralogist, 9, 7-24.
662 Jambor, J.L., 1967b. New lead sulfantimonides from Madoc, Ontario. Part 2 - Mineral descriptions. Can. Mineralogist, 9, 191-213.
663 Jambor, J.L., 1968. New lead sulfantimonides from Madoc, Ontario. Part 3 - Syntheses, Paragenesis, origin. Can. Mineralogist, 9, 505-521.
663a Jambor, J.L., 1969a. Dadsonite (minerals Q and QM), a new lead sulphantimonide. Mineral. Mag., 37, 437-441.
663b Jambor, J.L., 1969b. Sulphosalts of the plagionite group. Mineral. Mag., 37, 442-446.
664 Jambor, J.L. and Lachance, G.R., 1968. Bismuthian robinsonite. Can. Mineralogist, 9, 426-428.

665 Janković, S., 1955. Ueber entmischugsartige Strukturen zwischen Pyrit und Zinkblende. Neues Jahrb. Mineral., Monatsh., 1955, 224-232.
666 Janković, S., 1960. Allgemeine Charakteristika der Antimon-Erzlagerstätten Jugoslawiens. Neues Jahrb. Mineral., Abhandl., 94, 506-538.
667 Jaskólski, S., 1960. Beitrag zur Kenntis über die Herkunft der Zinnlagerstätten von Gierczyn (Giehren) im Iser-Gebirge, Niederschlesien. Neues Jahrb. Mineral., Abhandl., 94, 181-190.
668 Jaskólski, S. and Banaš, M., 1958. Vorkommen und mikroskopische Beobachtungen an Jordanit oberschlesischer Pb-Zn-Lagerstätten (in Polish). Arch. Mineral., 22, 5-16. Ref. in: Zentr. Mineral., I, 1961, 447-448.
669 Jedwab, J., 1967. Minéralisation en greigite de débris végétaux d'une vase récente (Grote Geul). Bull. Soc. Belge Géol. Paléontol. Hydrol., 76, 27-38.
670 Jenness, S.E., 1959. "Magnetic" chromite from Shoal Pond, Northeastern Newfoundland. Econ. Geol., 54, 1298-1301.
671 Jicha, H.L., Jr., 1951. Alpine lead-zinc ores of Europe. Econ. Geol., 46, 707-730.
672 Jicha, H.L., Jr., 1954. Paragenesis of the ores of the Palomas (Hermosa) District, Southwestern New Mexico. Econ. Geol., 49, 759-778.
673 Johan, Z., 1958. Koutekite, a new mineral. Nature, 181, 1553-1554.
674 Johan, Z., 1959a. Arsenolamprit - die rhombische Modifikation des Arsens aus Černý Důl (Schwarzental) im Riesengebirge. Chem. Erde, 20, 71-80.
675 Johan, Z., 1959b. Koutekit - Cu_2As, ein neues Mineral. Chem. Erde, 20, 217-226.
676 Johan, Z., 1960. A vanadium-containing cuprite from Popelky near Lomnice on Popelka (in Czech, Engl. summary). Acta Univ. Carolinae, Geol., 51-59.
677 Johan, Z., 1961. Paxite, Cu_2As_3, a new Cu-arsenide from Černý Důl in the Giant Mts. (Krkonoše) (in Czech, Engl. summary). Acta Univ. Carolinae, Geol., 77-86.
678 Johan, Z., 1967. Etude de la jalpaite, $Ag_{1.55}Cu_{0.45}S$. Acta Univ. Carolina, Geol., 113-122.
679 Johan, Z. and Hak, J., 1959. Novákit - $(Cu,Ag)_4As_3$, ein neues Mineral. Chem. Erde, 20, 49-50.
680 Johan, Z. and Hak, J., 1961. Novákite, $(Cu, Ag)_4As_3$, a new mineral. Am. Mineralogist, 46, 885-891.
681 Johan, Z., Picot, P. and Pierrot, R., 1969. Nouvelles données sur la raguinite. Bull. Soc. Franç. Minéral. Crist., 92, 237.
682 Johnson, R.F., 1955. Geology of the Atacocha mine, Department of Pasco, Peru. Econ. Geol., 50, 249-270.
683 Jones, L.H.P. and Milne, A.A., 1956. Birnessite, a new manganese oxide mineral from Aberdeenshire, Scotland. Mineral. Mag., 31, 283-288.
684 Joralemon, P., 1951. The occurrence of gold at the Getchell mine, Nevada. Econ. Geol., 46, 267-310.
685 Jouravsky, G. and Permingeat, F., 1964. La gaudefroyite, une nouvelle espèce minérale. Bull. Soc. Franç. Minéral. Crist., 87, 216-229.
686 Jurković, I., 1960. Quecksilberfahlerz vom Mačkaragang bei Gornji Vakuf in Bosnien (Jugoslawien). Neues Jahrb. Mineral., Abhandl., 94, 539-558.
687 Kaaden, G. van der, 1958. On the genesis and mineralization of the tungsten deposit Uludag. Bull. Mineral Res. Exploration Inst. Turkey, Foreign Ed., 50, 33-42.
688 Kaemmel, T., 1961. Geologie, Petrographie und Geochemie der Zinnlagerstätte "Tannenberg" (Vogtland). Geologie (Berlin), 10, Beih. 30, 1-105.
689 Kaiman, S., 1947. The crystal structure of rammelsbergite, $NiAs_2$. Univ. Toronto, Geol. Ser., 51, 49-58.
690 Kaiman, S., 1959. Synthesis of brannerite. Can. Mineralogist, 6, 389-390.
691 Kalenov, A.D., 1962. Cosalite from eastern Mongolia. Dokl. Akad. Nauk SSSR, 142, 443-444. Translated in: Dokl. Acad. Sci. USSR, Earth Sci. Sect., 142, 116-118 (1964).
692 Kalenov, A.D., 1964. Cosalite from the Tumen-Tsogto deposit and the specific composition of cosalite in the Transbaikal-Mongolian ore province. Dokl. Akad. Nauk SSSR, 157, 1376-1378. Translated in: Dokl. Acad. Sci. USSR, Earth Sci. Sect., 157, 69-72 (1965).
693 Kalenov, A.D., Anikeyeva, V.I. and Sokova, K.P., 1963. A case of the complex replacement of loparite. Dokl. Akad. Nauk SSSR, 152, 183-186. Translated in: Dokl. Acad. Sci. USSR, Earth Sci. Sect., 152, 132-134 (1965).
694 Kanehira, K., 1966. A note on the texture of pyrrhotite in some rocks of the Muskox intrusion. Can. Mineralogist, 8, 531-536.
695 Kanehira, K., Banno, S. and Hashimoto, M., 1964. Notes on rock-forming minerals (28). Finding of awaruite (native nickel-iron) from serpentinite near the city of Kôti, Sikoku. J. Geol. Soc. Japan, 70, 272-277.

696 Karamjan, K.A., 1958. Alabandin aus der Cu-Mo Lagerstätte Dastakertskoe, Armenien, Transkaukasien (in Russian). Izv. Akad. Nauk Arm. SSR, 1958. Ref. in: Zentr. Mineral., I, 1959, 359.
697 Karamjan, K.A., 1959. Germanit und renierit in den Erzen der Cu-Mo Lagerstätte Dastakertsk (in Russian). Zap. Arm. Otd. Vses. Mineralog. Obshchestva, 1, 101-108. Ref. in: Zentr. Mineral., I, 1961, 271-272.
698 Karpova, Ch.N., Konkova, E.A., Larchin, E.D. and Savelev, V.F., 1958. Avicennite, a new thallium mineral (in Russian). Dokl. Akad. Nauk Uz. SSR, 2, 23-25. Ref. in: Am. Mineralogist, 44, 1324-1325. Zentr. Mineral., I, 1959, 362.
699 Karup-Møller, S., 1966. Berryite from Greenland. Can. Mineralogist, 8, 414-423.
700 Karup-Møller, S., 1969. Secondary violarite and bravoite, English Lake, Manitoba. Can. Mineralogist, 9, 629-643.
701 Kašpar, P., 1966. Morphologische Kristallographie einiger Proustitkristalle von Jachymov. Acta Univ. Carolinae, Geol., 1-15.
702 Kato, A., 1959. Ikunolite, a new bismuth mineral from the Ikuno mine, Japan. Mineral. J. (Tokyo), 2, 397-407. Ref. in: Am. Mineralogist, 45, 477-478. Mineral. Abstr., 15, 43. Zentr. Mineral., I, 1960, 462.
703 Kato, A., 1965. Sakuraiite, a new mineral (in Japanese). Chigaku Kenkyu (Earth Sci. Stud.), Sakurai Vol., 1-5. Ref. in: Am. Mineralogist, 53, 1421.
703a Kato, A., 1969. Stannoidite, $Cu_5(Fe, Zn)_2SnS_8$, a new stannite-like mineral from the Konjo Mine, Okayama Prefecture, Japan. Bull. Nat. Sci. Mus. (Tokyo), 12, 165-172. Ref. in: Am. Mineralogist, 54, 1495-1496.
703b Kato, A. and Shinohara, K., 1968. The occurrence of roquesite from the Akenobe mine, Hyogo, Prefecture, Japan. Mineral J. (Tokyo), 5, 276. Ref. in: Am. Mineralogist, 54, 1203.
704 Katsura, T. and Kushiro, I., 1961. Titanomaghemite in igneous rocks. Am. Mineralogist, 46, 134-145.
705 Katz, G., 1960. Jacobsite from the Negev, Israel. Am. Mineralogist, 45, 734-739.
706 Kay, H.F. and Bailey, P.C., 1957. Structure and properties of $CaTiO_3$. Acta Cryst., 10, 219-226.
707 Keil, K., 1929. Beiträge zur Kenntnis der Kobalt-Nickel-Wismut-Silber-Erzgänge. Diss. Freiberg.
708 Keil, K., 1933. Ueber die Ursachen der charakteristischen Paragenesenbildung von gediegen Silber und gediegen Wismut. Neues Jahrb. Mineral., Beil., 66 A, 407-424.
709 Kelley, V.C. and Silver, C., 1946. Stages and epochs of mineralization in the San Juan Mountains, Colorado, as shown at the Dunmore mine, Ouray County. Colorado. Econ. Geol., 41, 139-159.
710 Kerr, P.F., 1945. Cattierite and vaesite: new Co-Ni minerals from the Belgian Congo. Am. Mineralogist, 30, 483-497.
711 Kerr, P.F., and Holland, H.D., 1951. Differential thermal analysis of davidite. Am. Mineralogist, 36, 563-572.
712 Keys, M., 1940. Paragenesis in the Hollinger veins. Econ. Geol., 35, 611-628.
713 Khan, M.A., 1964. On the occurrence and mineralogy of the manganese ore bodies of Eklingji, Udaipur District, Rajasthan. Proc. Indian Acad. Sci., 60, 155-161.
714 Khurshudyan, E.Kh., 1966. Mode of origin of the rhombohedral modification of molybdenite. Dokl. Akad. Nauk SSSR, 171, 186-189. Translated in: Dokl. Acad. Sci. USSR, Earth Sci. Sect., 171, 149-151 (1967).
715 Khvostova, V.A. and Maksimova, N.V., 1963. New discovery of ixiolite. Dokl. Akad. Nauk SSSR, 148, 424-426. Translated in: Dokl. Acad. Sci. USSR, Earth Sci. Sect., 148, 101-103 (1964).
716 Khvostova, V.A., Pavlova, V.N., Aleksandrov, V.B. and Maksimova, N.V., 1966. The first find of wodginite in the U.S.S.R. Dokl. Akad. Nauk SSSR, 167, 1135-1138. Translated in: Dokl. Acad. Sci. USSR, Earth Sci. Sect., 167, 109-112 (1966).
717 Kidd, D.F., 1932. A pitchblende-silver deposit, Great Bear Lake, Canada. Econ. Geol., 27, 145-159.
718 Kidd, D.F. and Haycock, M.H., 1935. Mineragraphy of the ores of Great Bear Lake. Bull. Geol. Soc. Am., 46, 879-960.
719 King, A.G., 1957. Pyrite-uraninite polycrystal. Am. Mineralogist, 42, 648-656.
720 Kingsbury, A.W.G., 1965. Tellurbismuth and meneghinite, two minerals new to Britain. Mineral. Mag., 35, 424-426.
721 Kingsbury, A.W.G. and Hartley, J., 1956. Cosalite and other lead sulpho-salts at Grainsgill, Carrock Fell, Caldbech, Cumberland. Mineral. Mag., 31, 296-300.
722 Kingston, G.A., 1966. The occurrence of platinoid bismuthotellurides in the Merensky Reef at Rustenburg platinum mine in the Western Bushveld. Mineral. Mag., 35, 815-834.

723 Kingston, P.W., 1968. Studies of mineral sulphosalts: XXI - Nuffieldite, a new species. Can. Mineralogist, 9, 439-452.
724 Klein, C., Jr. and Frondel, C., 1967. Antimonian groutite. Am. Mineralogist. 52, 858-860.
725 Klemm, D.D., 1962a. Untersuchungen über die Mischkristallbildung im Dreieckdiagramm FeS_2-CoS_2-NiS_2 und ihre Beziehungen zum Aufbau der natürlichen "Bravoite". Neues Jahrb. Mineral., Monatsh., 1962, 76-91.
726 Klemm, D.D., 1962b. Anisotropie Effekte bei kubischen Erzmineralien. Neues Jahrb. Mineral., Abhandl., 97, 337-356.
727 Klemm, D.D. and Weiser, T., 1965. Hochtemperierte Glanzkobalt-Gersdorffite-Mischkristalle von Outokumpu, Finland. Neues Jahrb. Mineral., Monatsh., 1965, 236-241.
728 Klingsberg, C. and Roy, R., 1957. Ramsdellite, newly observed in Minnesota. Econ. Geol., 52, 574-577.
729 Klingsberg, C. and Roy, R., 1959. Stability and inconvertibility of phases in the system Mn-O-OH. Am. Mineralogist, 44, 819-838.
730 Kluth, C., 1959. Mikrohärtemessungen an Zinkblende. Neues Jahrb. Mineral., Monatsh., 1959, 25-33.
731 Kluth, C., 1965. Ueber das Vorkommen des Coffinits im mittlerem Schwarzwald. Jahrb. Geol. Landesamt Baden-Württemberg, 7, 45-53.
732 Kluth, C., 1967. Ueber ein Vorkommen von Emplektit zu Bieber in Hessen. Aufschluss, 18, 9-12. Ref. in: Zentr. Mineral. I, 1966, 262.
733 Klyakhin, V.A., 1964. Synthesis of ferroselite under hydrothermal conditions. Dokl. Akad. Nauk SSSR, 155, 346-348. Translated in: Dokl. Acad. Sci. USSR, Earth Sci. Sect., 155, 116-117 (1965).
734 Klyakhin, V.A. and Dmitriyeva, M.T., 1968. More information about synthetic and natural lillianite. Dokl. Akad. Nauk SSSR, 178, 173-175. Translated in: Dokl. Acad. Sci. USSR, Earth Sci. Sect., 178, 106-108 (1968).
735 Knorring, O. von and Cox, K.G., 1961. Kennedyite, a new mineral of the pseudobrookite series. Mineral. Mag., 32, 676-682.
736 Knorring, O. von and Hornung, G., 1963. Simpsonite and stibiotantalite from Benson pegmatite mine, Mtoko, Southern Rhodesia. Mineral. Mag., 33, 458-466.
737 Knowles, C.R., 1964. A redetermination of the structure of miargyrite. Acta Cryst., 17, 847-851.
738 Koch, A., 1943. Das Bleierzvorkommen auf dem Szárhegy im Komitat Fejér. Acta Univ. Szeged, Acta Mineral. Petrog., I, 7-12.
739 Koch, A., 1948. Bismuth minerals in the Carpathian basin. Acta Univ. Szeged, Acta Mineral. Petrog., II, 17-23.
740 Koch, S., 1960. Ludwigite from Ocna de Fer (Vaskö, Banat, Rumania). Acta Univ. Szeged, Acta Mineral. Petrog., 13, 9-16. Ref. in: Zentr. Mineral., I, 1961, 284-285, and Zentr. Mineral., II, 1962, 192-193.
741 Koch, S., Grasselly, G. and Padĕra, K., 1960. Contribution to the jamesonite problem. Acta Univ. Szeged, Acta Mineral. Petrog., 13, 17-32. Ref.in: Zentr. Mineral., I, 1961, 273.
742 Koen, G.M., 1964. Rounded platinoid grains in the Witwatersrand Banket. Trans. Geol. Soc. S. Africa, 67, 139-147.
743 Komkov, A.I., 1957. On fergusonite (in Russian). Zap. Vses. Mineralog. Obshchestva, 86, 432-444. Ref. in: Mineral. Abstr., 14, 52.
744 Komkov, A.I., 1959. The structure of natural fergusonite, and of a polymorphic modification (in Russian). Kristallografiya (USSR), 4, 836-841. Ref. in: Mineral. Abstr., 15, 264.
745 Kondrashev, Yu.D., 1958. The crystal structure and composition of crednerite (in Russian). Kristallografiya (USSR), 3, 696-699. Ref. in: Mineral. Abstr., 15, 261.
746 Koning, L.P.G., 1941. On gersdorffite in the Falconbridge ore deposit, Sudbury District, Ontario, Canada. Koninkl. Ned. Akad. Wetenschap., Proc., 44, 93-101.
747 Koning, L.P.G., 1947. On manganite. Koninkl. Ned. Akad. Wetenschap., Proc., 50, 1348-1352.
748 Konkova, E.A. and Savelev, V.F., 1960. A new thallium mineral, avicennite (in Russian). Zap. Vses. Mineralog. Obshchestva, 89, 316-320. Ref. in: Mineral. Abstr., 15, 290-291.
749 Korich, D., 1964. Erzmikroskopische Beobachtungen an Serpentiniten des südwestlichen sächsischen Granulitgebirges. Geologie (Berlin), 13, 26-36.

750 Kornetsova, V.A. and Kazakova, M.Ye., 1964. Discovery of formanite in the USSR. Dokl. Akad. Nauk SSSR, 154, 359-362. Translated in: Dokl. Acad. Sci. USSR, Earth Sci. Sect., 154, 86-89 (1964).
751 Koslov, M.T., 1957. Ueber den Fund von Minium auf einer Blei-Lagerstätte in West-Tuwa (in Russian). Tr. Voronezhsk. Gos. Univ., 60, 31-36. Ref. in: Zentr. Mineral., I, 1960, 154.
752 Kostov, I., 1958. Bonchevite, $PbBi_4S_7$, a new mineral. Mineral. Mag., 31, 821-828.
753 Kouvo, O., Huhma, M. and Vuorelainen, Y., 1959. A natural cobalt analogue of pentlandite. Am. Mineralogist, 44, 897-900.
754 Kouvo, O. and Vuorelainen, Y., 1958. Eskolaite, a new chromium mineral. Am. Mineralogist, 43, 1098-1106.
755 Kouvo, O. and Vuorelainen, Y., 1959. Valleriitista (in Finnish). Geologi (Helsinki), 14, 162.
756 Kouvo, O. and Vuorelainen, Y., 1962. Magneettiküsun koostumuksesta ja rakenteesta. (in Finnish). Geologi (Helsinki), 1962, 79-82. Ref. in: Zentr. Mineral, I, 1963, 390.
757 Kouvo, O., Vuorelainen, Y. and Long, J.V.P., 1963. A tetragonal iron sulfide. Am. Mineralogist, 48, 511-524.
758 Kranck, E.H., 1945. The molybdenum deposit at Mätäsvaara in Carelia (E.Finland). Geol. Fören. Stockholm Förhandl., 67, 325-350.
759 Krause, H., 1960. Ueber Lievrit aus dem Huttal bei Clausthal. Neues Jahrb. Mineral., Abhandl., 94, 1277-1283.
760 Krause, H., 1965. Contributions to the mineralogy of Norway. No. 33. Idaite, Cu_5FeS_6, from Konnerud near Drammen. Norsk Geol. Tidsskr., 45, 417.
760a Kräutner, H.G. and Medeşan, A., 1969. Ilvait in den polymetamorphen Eisenerzen der Piana Ruscă (Rumänien). Tschermaks Mineral. Petrog. Mitt., 13, 157-164.
761 Krieger, P., 1932. An association of gold and uraninite from Chihuahua, Mexico. Econ. Geol., 27, 651-660.
762 Krieger, P., 1935a. Primary silver mineralization at Sabinal, Mexico. Econ. Geol., 30, 242-259.
763 Krieger, P., 1935b. Primary native silver ores at Batopilas, Mexico, and Bullard's Peak, New Mexico. Mineral. Mag., 20, 715-723.
764 Krieger, P., 1940. Bornite-klapritholite relations at Conception del Oro, Mexico. Econ. Geol., 35, 687-697.
765 Krieger, P. and Hagner, A.F., 1943. Gold-nickel mineralization at Alistos, Sinalva, Mexico. Am. Mineralogist, 28, 257-271.
766 Krishna Rao, J.S.R., 1962. Erzmikroskopische und experimentelle Studien der Paragenese von Awaruit. Chem. Erde, 21, 398-412.
767 Krishna Rao, J.S.R., 1963. Microscopic examination of manganese ores of Srikakulam and Visakhapatnam (Vizagapatam) Districts, Andhra Pradesh, India. Econ. Geol., 58, 434-440.
768 Krishnarao, J.S.R., 1964. Native nickel-iron alloy, its mode of occurrence, distribution and origin. Econ. Geol., 59, 443-448.
769 Krishnarao, J.S.R. and Malleswararao, V., 1965. Occurrence and origin of graphite in parts of Eastern Ghats, South India. Econ. Geol., 60, 1046-1051.
770 Kukharenko, A.A., Kondrateva, V.V. and Kovyazina, V.M., 1959. Cafetite, a new hydrous titanate of calcium and iron (in Russian). Zap. Vses. Mineralog. Obshchestva, 88, 444-453. Ref. in: Am. Mineralogist, 45, 476.
771 Kulagov, E.A., Izoitko, V.M. and Mitenkov, G.A., 1967. Heazlewoodite in the Talnakh copper-nickel sulfide ores. Dokl. Akad. Nauk SSSR, 176, 900-902. Translated in: Dokl. Acad. Sci. USSR, Earth Sci. Sect., 176 (1968), 134-136.
772 Kullerud, G. and Donnay, G., 1958. Natural and synthetic ferroselite. A roentgenographic mimesis of rammelsbergite. Geochim. Cosmochim. Acta, 15, 73-79.
773 Kullerud, G., Donnay, G. and Donnay, J.D.H., 1960. A second find of euhedral bornite crystals on barite. Am. Mineralogist, 45, 1062-1068.
774 Kullerud, G. and Yund, R.A., 1962. The Ni-S system and related minerals. J. Petrol., 3, 126-175.
774a Kupčik, V., Franc, L. and Makovický, E., 1969. Mineralogical data on a sulphosalt from the Rhodope Mountains, Bulgaria. Tschermaks Mineral. Petrog. Mitt., 13, 149-156.
775 Kupčik, V. and Matherny, M., 1960. Rutil aus dem Gebiete von Rožňava (in Czech, Russian and German summaries). Acta Geol. Geograph. Univ. Comenianae, Geol., 4, 7-30.

776 Kupčik, V., Matherny, M. and Varček, C., 1961. Contribution to the problem of the structure of the mineral "lillianite". (in Czech, English summary). Geol. Sborník (Bratislava), 12, 103-113.

776a Kupčik, V., Schneider, A. and Varček, C., 1969. Chemismus von einigen Bi-sulfosalzen aus dem Zips-Gömörer Erzgebirge (CSSR). Neues Jahrb. Mineral., Monatsh., 1969, 445-454.

776b Kusnaeny, K., 1968. Die Manganerzvorkommen in West-Kalimantan (Indonesien) und Orissa (Indien). Geol. Jahrb., 86, 655-692.

777 Kutina, J., 1953. Selektive Verdrängung der inneren Teile der Arsenkieskristalle usw. aus Horní Malá Upa (Ober-Kleinaupa), Riesengebirge. Neues Jahrb. Mineral. Abhandl., 86, 86-102.

778 Kutoglu, A., 1968. Die Struktur des Pyrostilpnits (Feuerblende) Ag_3SbS_3. Neues Jahrb. Mineral., Monatsh., 1968, 145-160.

779 Kutoglu, A., 1969. Röntgenographische und thermische Untersuchungen im quasibinären System $PbS-As_2S_3$. Neues Jahrb. Mineral., Monatsh., 1969, 68-72.

780 Kuznetzov, K.F., 1957. Geocronite in the ores of the Ekaterino-Blagodatsky deposit (in Russian). Dokl. Akad. Nauk SSSR, 114, 880-883. Ref. in: Mineral. Abstr., 13, 587-588.

781 Kuznetzov, K.F. and Wolfson, F.I., 1957. Fund des Geokronits im Erz der Ekaterino-Blagodaskoe-Lagerstätte (in Russian). Sb. Nauchn. Tr. Mosk. Inst. Tsvetn. Metal. Zolota, 27, 397-405. Ref. in: Zentr. Mineral., I, 1959, 218.

782 Kvaček, M., 1967. Ein neues Selenid von Bukov in Mähren. (in Czech). Časopis Národního Muzea (Prague), 136, 10-12. Ref. in: Zentr. Mineral. I, 1966, 263.

783 Labhart, T.P., 1967. Die Uranvererzungen am Südrand des Aarmassivs bei Naters (Kt. Wallis, Schweiz). Beitr. Geol. Schweiz. Geotech. Ser., 43, 30 pp.

784 Lacroix, A., 1892. Sur la magnésioferrite du Roc de Cuzau (Mont-Dore). Bull. Soc. Franç. Minéral. Crist., 15, 11-13.

785 Lambot, H., 1950. Sur la reniérite. Ann. Soc. Géol. Belg., 73, Bull., 183-186.

786 Landon, R.E. and Mogilnor, A.H., 1933. Colusite, a new mineral of the sphalerite group. Am. Mineralogist, 18, 528-533.

787 Lapin, A.V. and Kazakova, M.Ye., 1966. Titanium lueshite from the Kovdor massif and isomorphism in the perovskite group. Dokl. Akad. Nauk SSSR, 171, 956-959. Translated in: Dokl. Acad. Sci. USSR, Earth Sci. Sect., 171, 160-163 (1967).

788 Larson, L.T., 1962. Zinc-bearing todorokite from Philipsburg, Montana. Am. Mineralogist, 47, 59-66.

789 Larson, L.T., 1964. Geology and mineralogy of certain manganese oxide deposits. Econ. Geol., 59, 54-78.

790 Laurent, Y., Picot, P., Pierrot, R., Permingeat, F. and T. Ivanov., 1969. La raguinite, $TlFeS_2$, une nouvelle espèce minérale et le problème de l'allcharite. Bull. Soc. Franç. Minéral. Crist., 92, 38-48.

791 Lausen, C., 1930. Graphic intergrowths of niccolite and chalcopyrite, Worthington mine, Sudbury. Econ. Geol., 25, 356-364.

792 Lauzac, F., 1965. Découverte de reniérite dans une blende germanifère de Sardaigne. Bull. Soc. Franç. Minéral. Crist., 88, 347-348.

793 Lawrence, L.J., 1951. Note on an occurrence of native tin at Emmaville, N.S.W. Australian J. Sci., 14, 82-84.

794 Lawrence, L.J., 1957a. Native lead from the Redcap mine, Chillagoe. Proc. Roy. Soc. Queensland, 68, 21-23.

795 Lawrence, L.J., 1957b. An unusual cassiterite paragenesis and its genetic implications. Mineral. Mag., 31, 402-406.

796 Lawrence, L.J., 1961a. Notes on some additional minerals from the oxidized portion of the Broken Hill Lode, N.S.W., with observations on crystals of coronadite. J. Proc. Roy. Soc. N.S. Wales, 95, 13-16.

797 Lawrence, L.J., 1961b. The ore minerals of Webb's Silver mine, Emmaville, N.S.W. Australasian Inst. Mining Met. Proc., 199, 113-131.

798 Lawrence, L.J., 1962a. Owyheeite from Rivertree, N.S.W. Mineral. Mag., 33, 315-319.

799 Lawrence, L.J., 1962b. The mineral composition of the sulphide ores of the Drake and Rivertree mining fields, N.S.W., Australasian Inst. Mining Met. Proc., 201, 15-42.

800 Lawrence, L.J., 1963a. Multiple solid solution on sulphides in sphalerite. Nature, 197, 171-172.

801 Lawrence, L.J., 1963b. Textures of some copper-bismuth sulphide ores from Mount Elliston, N.T. Australasian Inst. Mining Met. Proc., 205, 121-129.
802 Lawrence, L.J., 1967a. Mineralogy and textures of Thackeringa sulphide ore. Australasian Inst. Mining Met. Proc., 222, 85-94.
803 Lawrence, L.J., 1967b. A mineragraphic study of Mount Morgan copper-gold ore. Australasian Inst. Mining Met. Proc., 223, 29-47.
804 Lawrence, L.J., Bayliss, P. and Tonkin, P., 1968. An occurrence of todorokite in the deuteric stage of a basalt. Mineral. Mag., 36, 757-760.
805 Lawrence, L.J. and Chand, F., 1962. Ore minerals of the Rockvale mine, N.S.W., with special reference to unmixing textures. Australasian Inst. Mining Met. Proc., 204, 161-183.
806 Lawrence, L.J. and Markham, N.L., 1962. A contribution to the study of the molybdenite pipes of Kingsgate, N.S.W., with special reference to ore mineralogy. Australasian Inst. Mining Met. Proc., 203, 67-94.
807 Lawrence, L.J. and Markham, N.L., 1963. The petrology and mineralogy of the pegmatite complex at Bismuth, Torrington, N.S.W. J. Geol. Soc. Australia, 10, 343-364.
808 Lawrence, L.J., See, G.T., Mc Bride, F. and Hofer, H., 1957. Davidites from the Mt. Isa-Cloncurry District, Queensland. Econ. Geol., 52, 140-148.
809 Layton, W. and Bagley, A.S., 1965. The nature of rutile and estimation of titanium in rutile concentrates. Australasian Inst. Mining Met. Proc., 215, 37-52. Ref. in: Mineral. Abstr., 18, 92.
810 Lebedeeva, S.I., 1963. The determination of the microhardness of minerals (in Russian). Izdatelstvo Nauka S.S.S.R., Moskva, 122 pp.
811 Lebedev, L.M., 1967. Contemporary deposition of native lead from the Cheleken thermal brines. Dokl. Akad. Nauk SSSR, 174, 197-200. Translated in: Dokl. Acad. Sci. USSR, Earth Sci. Sect., 174, 173-176 (1967).
812 Le Bihan, M.-Th., 1961a. Contribution à l'étude structurale des sulfures d'arsenic et de plomb. Structure de la baumhauerite. Acta Cryst., 14, 1210-1211.
813 Le Bihan, M.-Th. 1961b. Structure de la rathite II. Comparaison entre les differentes structures connues de sulfures d'arsenic et de plomb. Acta Cryst., 14, 1211-1212.
814 Le Bihan, M.-Th., 1962. Etude structurale de quelques sulfures de plomb et d'arsenic naturales du gisement de Binn. Bull. Soc. Franç. Minéral. Crist., 85, 15-47.
815 Lee, D.E., 1955. Occurrence of pyrophanite in Japan. Am. Mineralogist, 40, 32-40.
816 Leenheer, L. de, 1938. Neue Beobachtungen an Stainieriten aus Katanga. Zentr. Mineral., A, 281-288.
818 Lemaire, B., 1965. Étude géologique de l'île Erromango (Nouvelles-Hébrides). Métallogénie locale du manganèse. Mém. Bur. Rech. Géol. Minières, 38, 183 pp.
819 Leonard, B.F., 1960. Reflectivity measurements with a Hallimond visual microphotometer. Econ. Geol., 55, 1306-1312.
820 Leonard, B.F., 1969. Microidentation hardness of members of the ludwigite-vonsenite series. U.S., Geol. Surv., Profess. Papers, 650-B, 47-52.
820a Leonard, B.F., Desborough, G.A. and Page, N.J., 1969. Ore microscopy and chemical composition of some laurites. Am. Mineralogist, 54, 1330-1346.
821 Leonard, B.F. and Vlisidis, A.C., 1960. Vonsenite from St. Lawrence County, Northwest Adirondacks, New York. Am. Mineralogist, 45, 439-442.
822 Leonard, B.F. and Vlisidis, A.C., 1961. Vonsenite at the Jayville magnetite deposit, St. Lawrence County, New York. Am. Mineralogist, 46, 786-811.
823 Leonard, B.F., Hildebrand, F.A. and Vlisidis, A.C., 1962. Members of the ludwigite-vonsenite series and their distinction from ilvaite. Geol. Soc. Am., Petrologic Studies: A Volume to Honor A.F. Buddington, 523-568.
824 Leonard, B.F., Mead, C.M. and Conklin, N., 1968. Silver-rich disseminated sulfides from a tungsten-bearing quartz lode, Big Creek District, Central Idaho. U.S., Geol. Surv., Profess. Papers, 594-C, 1-24.
825 Leow, J.H., 1966. Reflectivity measurements on molybdenite. Econ. Geol., 61, 598-612.
826 Lepicard, G. and Protas, J., 1966. Etude structurale de l'oxyde double de manganèse et de calcium orthorhombique $CaMn_2O_4$ (marokite). Bull. Soc. Franç. Minéral. Crist., 89, 318-324.
827 Lerz, H., 1968. Ueber eine hydrothermale Paragenese von Anatas, Brookit und Rutil vom Dorfer Keesfleck, Prägraten Osttirol. Neues Jahrb. Mineral., Monatsh., 1968, 414-420.
828 Leube, A. and Stumpfl, E.F., 1963. The Rooiberg and Leeuwpoort tin mines, Transvaal, South Africa. Part II: Petrology, mineralogy and geochemistry. Econ. Geol., 58, 527-557.

829 Levinson, A.A., 1960. Second occurrence of todorokite. Am. Mineralogist, 45, 802-807.
830 Levinson, A.A., 1962. Birnessite from Mexico. Am. Mineralogist, 47, 790-791.
831 Lévy, C., 1956. La stannite jaune du gisement de Vaulry (Haute-Vienne). Bull. Soc. Franç. Minéral. Crist., 79, 383-391.
832 Lévy, C., 1967. Contribution à la minéralogie des sulfures de cuivre du type Cu_3XS_4. Mém. Bur. Rech. Géol. Minières, 54, 178 pp.
833 Lévy, C. and Picot, P., 1961. Nouvelles données sur les composés iridium-osmium. Existence de l'osmium natif. Bull. Soc. Franç. Minéral. Crist., 84, 312-317.
834 Lévy, C. and Prouvost, J., 1957. Rapport entre la chalcopyrite, la stannite et la reniérite. Bull. Soc. Franç. Minéral. Crist., 80, 59-66.
835 Lewis, R.W., Jr., 1956. The geology and ore deposits of the Quiruvilca District, Peru. Econ. Geol., 51, 41-63.
836 Li, A.F., 1957. Tellurium minerals in the north-eastern Sub-Baikal region (in Russian). Zap. Vses. Mineralog. Obshchestva, 86, 40-47. Ref. in: Mineral. Abstr., 13, 366.
837 Lietz, J., 1939. Mikroskopische und chemische Untersuchungen an Kongsberger Silbererzen. Z. Angew. Mineral., II, 65-113.
838 Lima de Faria, J. and Quadrado, R., 1966. Ilmenorútilo e struverite de Nampoça, Alto Ligonha, Moçambique. Garcia de Orta, 14, 305-310. Ref. in: Mineral. Abstr., 20, 143-144.
839 Lindgren, W., 1933. Coronadite "redivivus". Am. Mineralogist, 18, 548-550.
840 Lindgren, W., 1935. The silver mine at Colquijirca, Peru. Econ. Geol., 30, 331-346.
841 Lindgren, W. and Abbott, A.C., 1931. The silver-tin deposits of Oruro, Bolivia. Econ. Geol., 26, 453-479.
842 Lingen, G.J. van der, 1960. The arsene-bearing copper ores of Canfranc Estación (Central Spanish Pyrenees). Geol. Mijnbouw, 22, 729-736.
843 Ljunggren, P., 1960. Todorokite and pyrolusite from Vermlands Taberg, Sweden. Am. Mineralogist, 45, 235-238.
844 Llambías, E.J. and Malvicini, L., 1969. The geology and genesis of the Bi-Cu mineralized brecciapipe, San Francisco de los Andes, San Juan, Argentina. Econ. Geol., 64, 271-286.
845 Löfquist, H. and Benedicks, C., 1941. Det stora Nordenskiöldska järnblocket från Ovifak: mikrostruktur och bildningssätt. Kgl. Svenska Vetenskapsakad., Handl., Ser. 3, 19 (3).
846 Long, J.V.P., Vuorelainen, Y. and Kouvo, O., 1963. Karelianite, a new vanadium mineral. Am. Mineralogist, 48, 33-41.
847 Loomis, F.B., Jr., 1937. Boulder County tungsten ores. Econ. Geol., 32, 952-963.
848 López, V.M., 1939. The primary mineralization at Chuquicamata, Chile, S.A. Econ. Geol., 34, 674-711.
849 López-Soler, A. and Bosch-Figueroa, J.M., 1969. Optical characteristics of some opaque materials. Third Annual Regional Conference, Copenhagen. Medd. Dansk. Geol. Foren., 19, 330-332.
850 Lowell, W.R., 1942. The paragenesis of some gold and copper ores of Southwestern Oregon. Econ. Geol., 37, 557-595.
851 Luke Li-Yu Chang, 1963. Dimorphic relation in Ag_3SbS_3. Am. Mineralogist, 48, 429-432.
852 Lukesh, J.S., 1940. Optical evidence of polysynthetic twinning in arsenopyrite. Am. Mineralogist, 25, 619-621.
853 Luna, J., 1965. Wodginite and columbite-tantalite from a pegmatite at Krasonice, Western Moravia. (in Czech, Engl. summary). Acta Univ. Carolinae, Geol., 157-162.
854 Lyakhnitskaya, I.V. and Shumskaya, N.I., 1966. A new variety of gersdorffite in the Berikulsk deposit (in Russian). Zap. Vses Mineralog. Obshchestva, 95, 567-570. Ref. in: Min. Abstr., 18, 206.
855 Machairas, G. and Blais, R., 1966. La transformation de l'hédenbergite manganésifère en ilvaite dans les sulfures de cuivre et de zinc de la région de Noranda (Québec). Bull. Soc. Franç. Minéral. Crist., 89, 372-376.
856 Mackay, A.L., 1962. β-Ferric oxyhydroxide - akaganéite. Mineral. Mag., 33, 270-280.
857 Majmundar, H.H., 1969. Ramsdellite and groutite from Nova Scotia. Can. Mineralogist, 9, 718-720.
858 Maksimova, N.V. and Ilyukhin, V.V., 1967. Crystal structure of thoreaulite, $Sn(Ta,Nb)_2O_7$. Soviet Phys. Crist., 12, 105-106. Ref. in: Mineral. Abstr., 19, 15.

859 Malakhov, A.A., Nazirova, R. and Likhoidova, I.I., 1968. Lillianite from quartz-rare metal veins of the Chavata ore manifestation (in Russian). Dokl. Akad. Nauk Uzb. SSR, 25, 42-44. Chem. Abstr., 69, 9227.
860 Malvicini, L., 1962. Algodonita en la paragénesis mineralógica de la mina Kokito II, Provincia de Neuquen. Rev. Asoc. Geol. Arg., 17, 85-95. Ref. in: Mineral. Abstr., 16, 425.
861 Marchandise, H., 1958. Le gisement et les minerais de manganèse de Kisenge (Congo Belge). Bull. Soc. Belge Géol. Paléontol. Hydrol., 67, 187-210.
862 Marić, L., 1953. Scheelit aus Ost-Serbien (Jugoslawien). Neues Jahrb. Mineral., Monatsh., 1953, 180-185.
863 Markham, N.L., 1959. Occurrence of jordanite in the Otavi Mountains, South West Africa. Am. Mineralogist, 44, 682-685.
864 Markham, N.L., 1960. Synthetic and natural phases in the system Au-Ag-Te. Econ. Geol., 55, 1148-1178, and 1460-1477.
865 Markham, N.L., 1962. Plumbian ikunolite from Kingsgate, New South Wales. Am. Mineralogist, 47, 1431-1434.
866 Markham, N.L. and Lawrence, L.J., 1962. Primary ore minerals of the Consols Lode, Broken Hill, New South Wales. Australasian Inst. Mining Met. Proc., 201, 43-80.
867 Markham, N.L. and Lawrence, L.J., 1965. Mawsonite, a new Cu-Fe-Sn sulfide from Mt. Lyell, Tasmania and Tingha, New South Wales. Am. Mineralogist, 50, 900-908.
868 Markham, N.L. and Ottemann, J., 1968. Betekhtinite from Mt. Lyell, Tasmania. Mineral Deposita, 3, 171-173.
869 Markova, E.A., 1961. An occurrence of wehrlite in the Chalkuyryuk-Akdzhilginsk deposit. Dokl. Akad. Nauk SSSR, 141, 713-714. Translated in: Dokl. Acad. Sci. USSR, Earth Sci. Sect., 141, 1285-1286 (1963).
870 Markova, E.A., 1967. Occurrence of volynskite in a gold deposit of Central Asia. (in Russian). Zap. Vses. Mineralog. Obshchestva, 96, 324-326. Ref. in: Mineral. Abstr., 20, 25.
871 Martin, R., 1937. Courbes de dispersion des pouvoirs réflecteurs de quelques tellurures naturels. Compt. Rend. Acad. Sci. Paris, 204, 598-599.
872 Marumo, F. and Burri, G., 1965. Nowackiite, a new copper zinc arsenosulfosalt from Lengenbach (Binnatal, Switzerland). Chimia, 19, 500-501.
873 Marumo, F. and Nowacki, W., 1964. The crystal structure of lautite and of sinnerite, a new mineral from the Lengenbach quarry, Binnatal, Switzerland. Schweiz. Mineral. Petrog. Mitt., 44, 439-454.
874 Marumo, F. and Nowacki, W., 1965a. Die Kristallstruktur von Rathit-I. Phys. Verhandl., 16, 215. Ref. in: Zentr. Mineral., I, 1965, 35-36.
875 Marumo, F. and Nowacki, W., 1965b. The crystal structure of rathite-I. Z. Krist., 122, 433-456. Ref. in: Mineral. Abstr., 19, 16.
876 Marumo, F., Nowacki, W. and Engel, P., 1966. Kristallchemische Untersuchungen an Sulfosalzen. Schweiz. Mineral. Petrog. Mitt., 46, 694-695.
877 Maslenitzky, I.N., Faleev, P.V. and Iskyul, E.V., 1947. Tin-bearing minerals of the platinum group in sulfide copper-nickel ores. (in Russian). Dokl. Akad. Nauk SSSR, 58, 1137-1140. Ref. in: Am. Mineralogist, 35, 333.
878 Mason, B., 1942. Bixbyite from Långban, the identity of bixbyite and sitaparite. Geol. Fören. Stockholm Förhandl., 64, 117-125.
879 Mason, B., 1943a. Stockholm Högskola's collection of new or incompletely described minerals from Långban - a stocktaking. Geol. Fören. Stockholm Förhandl., 65, 80-82.
880 Mason, B., 1943b. Mineralogical aspects of the system $FeO-Fe_2O_3-MnO-Mn_2O_3$. Geol. Fören. Stockholm Förhandl., 65, 97-180.
881 Mason, B., 1943c. Alpha-vredenburgite. Geol. Fören. Stockholm Förhandl., 65, 263-270.
882 Mason, B., 1946. A zincian vredenburgite from Franklin, New Jersey. Geol. Fören. Stockholm Förhandl., 68, 51-55.
883 Mason, B., 1947. Mineralogical aspects of the system $Fe_3O_4-Mn_3O_4-ZnMn_2O_4-ZnFe_2O_4$. Am. Mineralogist, 32, 426-441.
884 Mather, W.B., 1937. Geology and paragenesis of the gold ores of the Howey mine, Red Lake, Ontario. Econ. Geol., 32, 131-153.
885 Mathieson, A.McL. and Wadsley, A.D., 1950. The crystal structure of cryptomelane. Am. Mineralogist, 35, 99-101.

886   Matias, V.V., 1963. Stannotantalite, a new variety of tantalite. New Data on Rare Element Mineralogy, Edited by A.I. Ginzburg, pp. 18-26. Consultants Bureau Enterprises, Inc., New York.
887   Matias, V.V., Rossovskiy, L.N., Shostatskiy, A.N. and Kumskova, N.M., 1963. Magnocolumbite, a new mineral. Dokl. Akad. Nauk SSSR, 148, 420-423. Translated in: Dokl. Acad. Sci. USSR, Earth Sci. Sect., 148, 97-100 (1964).
888   Maucher, A., 1938a. Ueber die Erzvorkommen von Keban-maden (Türkei). Z. Prakt. Geol., 46, 79-84.
889   Maucher, A., 1938b. Ueber Gudmundit aus der Antimonlagerstätte von Turkal (Türkei). Metallwirtschaft, 17, 617-619.
890   Maucher, A., 1939. Das Molybdänglanz- und Powellit-Vorkommen von Hüseyin beyobasi, Kasa Keskin, Vilayet Ankara, Türkei. Z. Angew. Mineral., I, 103-114.
891   Maucher, A., 1940. Ueber die Kieserzlagerstätte der Grube "Bayerland" bei Waldsassen in der Oberpfalz. Z. Angew. Mineral., II, 219-275.
892   Maucher, A. and Saupé, F., 1967. Sedimentärer Pyrit aus der Zinnober-Lagerstätte Almadén (Provinz Ciudad Réal, Spanien). Mineral. Deposita, 2, 312-317.
893   Mc Andrew, J., 1956a. Observations on hydrohetaerolite. Am. Mineralogist, 41, 268-275.
894   Mc Andrew, J., 1956b. Crystallography and composition of crednerite. Am. Mineralogist, 41, 276-287.
895   Mc Kie, D., 1963. The unit-cell of freudenbergite. Z. Krist., 119, 157-160. Ref. in: Mineral. Abstr., 17, 65.
896   Mc Kinstry, H:E., 1963. Mineral assemblages in sulfide ores: the system Cu-Fe-As-S. Econ. Geol., 58, 483-505.
897   Mc Kinstry, H.E. and Mikkola, A.K., 1954. The Elizabeth copper mine, Vermont. Econ. Geol., 49, 1-30.
898   McLellan, R.D., 1945. The occurrence and hardness of indium. Am. Mineralogist, 30, 635-638.
899   Mc Neil, R.D., 1966. Geology of the Orlando mine, Tennant Creek, Australia. Econ. Geol., 61, 221-242.
900   Mehnert, K.R., 1949. Die Kupfer-Wismut-Lagerstätte "Daniel" bei Wittichen (mittl. Schwarzwald). Neues Jahrb. Mineral. Geol. Paläontol., Monatsh., Abt. A, (1949) 217-241 and 243-260.
901   Meixner, H., 1935. Eine neue Manganparagenese vom Schwarzsee ("Kolsberger Alpe") bei Tweng in den Radstätter Tauern (Salzburg). Neues Jahrb. Mineral., Beil., 69 A, 500-514.
902   Meixner, H., 1950. Ueber Jordisit (amorphes Molybdänsulfid) von Bleiberg in Kärnten. Carinthia II, 139/140, 39-51.
903   Meixner, H., 1951. Zur erzmikroskopischen Unterscheidung der Tantalit-Tapiolit Phasen, unter besonderer Berücksichtigung eines neuen Vorkommen im Pegmatit von Spittal an der Drau. Bemerkungen zur Mineralisation des "Villacher Granits". Neues Jahrb. Mineral., Monatsh., 1951, 204-208.
904   Meixner, H., 1959. Kraubather Lagerstättestudien. V. Die Nickelmineralisation im Kraubather Serpentingebiet. Berg- und Hüttenmänn. Monatsh. Montan. Hochschule Leoben, 104, 83-87.
905   Mélon, J. and Toussaint, J., 1950. La thoreaulite de Kubitaka (Punia, Maniema, Congo belge) et la cristallographie de la thoreaulite. Ann. Soc. Géol. Belg., 74, Bull., 25-32.
906   Mendelssohn, E., 1932. Notes on a vein containing cobaltite, gold and apatite on the Far East Rand. Trans. Geol. Soc. S.Africa, 35, 191-192.
907   Merwin, H.E., Lombard, R.H. and Allen, E.T., 1923. Cubanite: identity with chalmersite: magnetic properties. Am. Mineralogist, 8, 135-138.
908   Michener, C.E. and Peacock, M.A., 1943. Parkerite from Sudbury, Ontario. Redefinition of the species. Am. Mineralogist, 28, 343-355.
909   Michener, C.E. and Yates, A.B., 1944. Oxidation of primary nickel sulphides. Econ. Geol., 39, 506-514.
910   Miers, H.A., 1893. Xanthoconite and rittingerite, with remarks on the red silvers. Mineral. Mag., 10, 185-216.
911   Mihálik, P. and Saager, R., 1968. Chromite grains showing altered borders from The Basal Reef, Witwatersrand System. Am. Mineralogist, 53, 1543-1550.
912   Mikheev, V.I., 1941. The structure of arsenosulvanite. (in Russian). Zap. Vses. Mineralog. Obshchestva, 70, 165-184. Ref. in: Am. Mineralogist, 40, 368-369.

913 Milton, C., Appleman, D., Chao, E.C.T., Cuttita, F., Dinnin, J.L., Dwornik, E.J., Hall, M., Ingram, B.L. and Rose, H.J., Jr., 1967. Mineralogy of merumite, a unique assemblage of chromium minerals from Guyana. Geol. Soc. Am., Program Ann. Meeting, 1967, 151-152.
914 Milton, C. and Chao, E.C.T., 1958. Eskolaite, Cr_2O_3, in "merumite" from British Guiana. Am. Mineralogist, 43, 1203.
915 Milton, C. and Ingram, B., 1959. Note on "revoredite" and related Pb-As-S glasses. Am. Mineralogist, 44, 1070-1076.
916 Milton, C. and Milton, D., 1958. Nickel-gold ore of the Mackinaw mine, Snohomish County, Washington. Econ. Geol., 53, 426-447.
917 Minčeva-Stefanova, J., 1958. Ueber die kristallochemischen Beziehungen zwischen Pyrit und Kobaltglanz. Chem. Erde, 19, 386-391.
918 Minčeva-Stefanova, J., 1960. Stromeyerit im Erzbezirk von Vraca. (in Russian). Compt. Rend. Acad. Bulgare Sci., 13, 579-582. Ref. in: Mineral. Abstr., 15, 486.
919 Minčeva-Stefanova, J. and Gorova, M., 1965. Die Wismutmineralisation in der Erzlagerstätte Gradiste, Bezirk Madan. (in Russian, German summary). Rev. Bulgar. Geol. Soc., 26, 193-198.
920 Minčeva-Stefanova, J., Gorova, M. and Pavlova, M., 1964. Zink-Tetraedrit aus den Blei-Zink-Lagerstätten im Erzberirk Madan. (in Russian, German summary). Rev. Bulgar. Geol. Soc., 25, 181-186.
921 Mintser, E.F., 1967. Benjaminite - $(Cu,Ag)_2Pb_2Bi_4S_9$. Dokl. Akad. Nauk SSSR, 174, 675-678. Translated in: Dokl. Acad. Sci. USSR, Earth Sci. Sect., 174, 127-131 (1967).
922 Mintser, E.F., Mymrin, V.A. and Isayeva, K.G., 1968. Joseite A from Central Asia. Dokl. Akad. Nauk SSSR, 178, 428-431. Translated in: Dokl. Acad. Sci. USSR, Earth Sci. Sect., 178, 114-117 (1968).
923 Mitcham, T.W., 1952. Indicator minerals, Coeur d'Alène silver belt. Econ. Geol., 47, 414-450.
924 Mitchell, R.S., 1967. Virginia metamict minerals: X-ray study of fergusonite. Southeastern Geology, 8, 145-153. Ref. in: Mineral. Abstr., 19, 106.
925 Mitchell, R.S. and Meintzer, R.E., 1967. Lithiophorite from Charlottesville, Virginia. Am. Mineralogist, 52, 1545-1549.
926 Mitreyeva, N.M., Yarenskaya, M.A., Koryak, E.A. and Muratova, D.N., 1968. Vanadium-arsenic germanite (in Russian). Zap. Vses. Mineralog. Obshchestva, 97, 325-331. Ref. in: Mineral. Abstr., 20, 59.
927 Miyahisa, M. and Noda, M., 1963. Herzenbergite and associated minerals from the Hoei mine, Kyushu, Japan (in Japanese). J. Mineral. Soc. Japan, 6, 349-360. Ref. in: Mineral. Abstr., 17, 688-689.
928 Mochnacka, K., 1966. Ore minerals of the polymetallic deposit at Kowary, Lower Silesia (in Polish). Polska Akad. Nauk, Práce Mineral., 4, 71 pp. Ref. in: Mineral. Abstr., 18, 91.
929 Moddle, D.A., 1957. Brannerite from Eastern Ontario. Can. Mineralogist, 6, 155-157.
930 Moench, R.H., 1962. Properties and paragenesis of coffinite from the Woodrow mine, New Mexico. Am. Mineralogist, 47, 26-33.
931 Mogensen, F., 1946. A ferro-ortho-titanate ore from Södra Ulvön. Geol. Fören. Stockholm Förhandl., 68, 578-588.
932 Moh, G.H., 1960. Experimentelle Untersuchungen an Zinnkiesen und analogen Germaniumverbindungen. Neues Jahrb. Mineral., Abhandl., 94, 1125-1146.
933 Moh, G.H., 1964. Experimentelle Untersuchungen bei niedrigen Temperaturen: "Blaubleibender Covellin", Bravoit und andere Sulfide des tiefhydrothermalen Bereiches. Fortschr. Mineral., 41, 165-166.
934 Moh, G.H., 1965. Das binäre System Zinn-Schwefel und seine Minerale. Fortschr. Mineral., 42, 211.
935 Moh, G.H., 1969. The tin-sulphur system and related minerals. Neues Jahrb. Mineral., Abhandl., 111, 227-263.
936 Moh, G.H. and Berndt, F., 1964. Two new natural tin sulphides, Sn_2S_3 and SnS_2. Neues Jahrb. Mineral., Monatsh., 1964, 94-95.
937 Moh, G.H. and Ottemann, J., 1962. Neue Untersuchungen an Zinnkiesen und Zinnkiesverwandten. Neues Jahrb. Mineral., Abhandl., 99, 1-28.
938 Montgomery, A., 1948. Mineralogy of the silver ores of Gowganda, Ontario. Univ. Toronto, Geol. Ser., 52, 23-38.
939 Moore, P.B., 1967a. Melanostibite - a novel composition in the ilmenite-pyrophanite group. Geol. Soc. Am., Program Ann. Meeting, 1967, 154-155.

940 Moore, P.B., 1967b. A classification of sulfosalt structures derived from the structure of aikinite. Am. Mineralogist, 52, 1874-1876.
941 Moore, P.B., 1968a. Contribution to Swedish mineralogy. II. Melanostibite and manganostibite, two unusual antimony minerals. The identity of ferrostibian with långbanite. Arkiv Mineral. Geol., 4, 449-458.
942 Moore, P.B., 1968b. Substitutions of the type $(Sb^{5+}_{0.5}Fe^{3+}_{0.5}) \rightleftharpoons (Ti^{4+})$: the crystal structure of melanostibite. Am. Mineralogist, 53, 1104-1109.
943 Morávek, P., 1956. The chemism and paragenesis of the tetrahedrite from the auriferous vein near Jílové, Central Bohemia. (in Czech, Engl. summary). Sb. Osmdesátinám Akad. Františka Slavíka, 287-304.
944 Moreau, J. and Tramasure, G., 1965. Contribution à l'étude des séries columbite-tantalite et tapiolite-mossite. Ann. Soc. Géol. Belg., 88, Bull., 301-392.
945 Morimoto, N., 1960. Symmetry and twinning of arsenopyrite. Ann. Rept. of the Dir. Geophys. Lab. Carnegie Inst. Wash., 1959-1960, 130. Zentr. Mineral, II, 1962, 121.
946 Morimoto, N., 1962. Djurleite, a new copper sulphide mineral. Mineral J. (Tokyo), 3, 338-344. Ref. in: Mineral. Abstr., 16, 180. Zentr. Mineral. I, (1966), 259.
947 Morimoto, N. and Clark, L.A., 1961. Arsenopyrite crystal-chemical relations. Am. Mineralogist, 46, 1448-1469.
948 Morimoto, N., Greig, J.W. and Tunell, G., 1960. Re-examination of a bornite from the Carn Brea mine, Cornwall. Carnegie Inst., Wash., Yearb., 59, 122-126.
948a Morimoto, N., Koto, K. and Shimazaki, Y., 1969. Anilite, Cu_7S_4, a new mineral. Am. Mineralogist, 54, 1256-1268.
949 Morimoto, N. and Kullerud, G., 1961. Polymorphism in bornite. Am. Mineralogist, 46, 1270-1282.
950 Morimoto, N. and Kullerud, G., 1963. Polymorphism in digenite. Am. Mineralogist, 48, 110-123.
951 Moritz, H., 1933. Die sulfidischen Erzen der Tsumeb-Mine vom Ausgehenden bis zum XVI. Sohle (-460 M). Neues Jahrb. Mineral., Beil., 67 A, 118-154.
952 Morris, D.F.C. and Short, E.L., 1966. Minerals of rhenium. Mineral. Mag., 35, 871-873.
953 Mörtsell, S., 1931. Gediget guld i Boliden-malmen. Geol. Fören. Stockholm Förhandl., 53, 394-414.
954 Moses, A.J., 1905. The crystallization of luzonite; and other crystallographic studies. Am. J. Sci., 20, 277-284.
955 Mouat, M.M., 1962. Manganese oxides from the Artillery Mountains area, Arizona. Am. Mineralogist, 47, 744-757.
956 Mozafari, Ch., 1968. Microdureté, susceptibilité magnétique et indice de réfraction des magnésiochromites. Bull. Soc. Belge Géol. Paléontol. Hydrol., 76, 44-59.
957 Mozgova, N.N., Borodayev, Yu.S., Nesterova, Yu.S. and Arapova, G.A., 1966a. Another find of zinckenite in Central Asia. Dokl. Akad. Nauk SSSR, 166, 1416-1419. Translated in: Dokl. Acad. Sci. USSR, Earth Sci. Sect., 166, 141-143 (1966).
958 Mozgova, N.N., Borodayev, Yu.S., Nesterova, Yu.S. and Arapova, G.A., 1966b. Another find of meneghinite in Eastern Transbaikal. Dokl. Akad. Nauk SSSR, 170, 1403-1406. Translated in: Dokl. Acad. Sci. USSR, Earth Sci. Sect., 170, 167-170 (1967).
959 Mozgova, N.N., Borodaev, Yu.S., Rakcheev, A.D. and Borishamskaya, S.S., 1969. On the diagnostics of lead sulphantimonides. Trans. Inst. Mining Met., 78, B 57-64.
960 Mrňa, F. and Pavlů, D., 1967. Ag-Bi-Co-Ni-As-Lagerstätten im Böhmischen Massiv. (in Czech, German summary). Sb. Geol. Věd (Praha), 9, 7-104.
961 Mukanov, K.M., Narkelyun, L.F. and Yakovlevskaya, T.A., 1960. On the occurrence of betekhtinite in the Dzhezkazgan ore deposit. Dokl. Akad. Nauk SSSR, 130, 404-407. Translated in: Dokl. Acad. Sci. USSR, Earth Sci. Sect., 130, 133-135 (1961).
962 Mukherjee, B., 1959. X-ray study of psilomelane and cryptomelane. Mineral. Mag., 32, 166-171.
963 Mukherjee, B., 1965. Crystallography of psilomelane. Mineral. Mag., 35, 643-655.
964 Mukherjee, B., 1966. Psilomelane from India. Mineral. Mag., 35, 971-974.
965 Müller, D. and Scherp, A., 1967. Die tertiäre Mineralisation auf der Blei-Zink-Erzlagerstätte Ramsbeck/Sauerland und ihre Genese. Neues Jahrb. Mineral., Abhandl., 106, 131-157.
966 Murdoch, J., 1941. Pyrostilpnite from Randsburg, California. Am. Mineralogist, 26, 130-132.

967 Murdoch, J., 1946. Microscopic determination of the opaque minerals. New York.
968 Murdoch, J., 1951. Perovskite. Am. Mineralogist, 36, 573-580.
969 Murdoch, J., 1953. X-Ray investigation of colusite, germanite, and renierite. Am. Mineralogist, 38, 794-801.
970 Murdoch, J. and Berry, L.G., 1954. X-Ray measurements on argentopyrite. Am. Mineralogist, 39, 475-485.
971 Murdoch, J. and Fahey, J.J., 1949. Geikielite, a new find from California. Am. Mineralogist, 34, 835-838.
972 Murdoch, J. and Gardner, D.L., 1942. Löllingite from the Philippine Islands. Econ. Geol., 37, 69-75.
973 Myer, G.H., 1962. Hydrothermal wurtzite at Thomaston Dam, Connecticut. Am. Mineralogist, 47, 977-979.
974 Naganna, C., 1962. Potassium-free cryptomelane from Sandur manganese ore deposits, Mysore State, India. Acta Univ. Carolinae, Geol., 55-59.
975 Naganna, C., 1964. Mineralogy of the manganese ores from Sandur ore deposits, Bellary District, Mysore State, India. Acta Univ. Carolinae, Geol., Monographia II.
976 Naganna, C. and Bouška, V., 1963. X-Ray study of woodruffite from Sandur ore dedeposits, Mysore State, India. Mineral. Mag., 33, 506-507.
977 Nagashima, K., Kato, A. and Chiba, M., 1965. Pegmatite minerals from Shimo-ono, Takahagi and its vicinity, Ibaragi Prefecture. J. Chem. Soc. Japan, Pure Chem. Sect., 86, 913-917. Ref. in: Mineral. Abstr., 19, 142.
978 Nakamura, T., 1951. High-temperature mineral associations in a certain quarz vein at the Ashio Mine. J. Fac. Sci. Univ. Tokyo, Sect. II, 8, 89-98.
979 Nakhla, F.M., 1956. The hardness of metallic minerals in polished sections. Econ. Geol., 51, 811-827.
980 Naldrett, A.J., 1965. Heazlewoodite in the Porcupine District (Ont.). Can. Mineralogist, 8, 383-385.
981 Naldrett, A.J., Craig, J.R. and Kullerud, G., 1967. The central portion of the Fe-Ni-S system and its bearing on pentlandite exsolution in iron-nickel sulfide ores. Econ. Geol., 62, 826-847.
982 Nambu, M. and Okada, K., 1963. Todorokite deposits of Iwasaki district, Aomori Prefecture (in Japanese). Bull. Res. Inst. Mineral Dressing Met. (Sendai, Japan), 19, 1-12. Ref. in: Mineral. Abstr., 19, 56.
983 Nambu, M. and Tanida, K., 1967. Manjiroite, a new manganese dioxide mineral, from Kohare Mine, Iwate Prefecture, Japan (in Japanese). J. Jap. Ass. Mineral., Petrol., Econ. Geol., 58, 39-54. Ref. in: Am. Mineralogist, 53, 2103.
984 Narita, E., 1963. Geology and ore deposits of the Onikobe-Hosokura District, northeastern Honshu, Japan. J. Fac. Sci., Hokkaidô Univ., Ser. IV, 16, 651-681.
985 Nash, J.T., 1968. Uranium deposits in the Jackpile sandstone, New Mexico. Econ. Geol., 63, 737-750.
986 Nayak, V.K., 1964. Origin of hollandite from Kajlidongri mine, Madhya Pradesh, India. Mineral. Mag., 33, 934-935.
987 Nayak, V.K., 1967. Bixbyite and manganophyllite from Kajlidongri, India. Mineral. Mag., 36, 294-296.
988 Nazirova, R., 1968. Galenobismutite and cosalite in the Chathal mountains (in Russian). Uzb. Geol. Zh., 12, 74-78. Ref. in: Chem. Abstr., 69, 9227.
989 Nechelyustov, G.N. and Lebedev, V.S., 1967. The first discovery of bonchevite in the USSR. Dokl. Akad. Nauk SSSR, 174, 679-682. Translated in: Dokl. Acad. Sci. USSR, Earth Sci. Sect., 174, 131-134 (1967).
990 Nechelyustov, G.M. and Mymrin, V.A., 1968. Kobellite found for the first time in the USSR. Dokl. Akad. Nauk SSSR, 181, 1223-1226. Translated in: Dokl. Acad. Sci. USSR, Earth Sci. Sect., 181, 128-131 (1969).
991 Nedashkovskii, P.G., Minaeva, N.A., Tolok, K.P. and Brovchuk, I.F., 1967. A new occurrence of magnocolumbite (in Russian). Zap. Vses. Mineralog. Obshchestva, 96, 720-723. Ref. in: Mineral. Abstr., 20, 56.
992 Nel, H.J., 1949. Hoegbomite from the corundum fields of the Eastern Transvaal. Geol. Surv. S. Africa, Dept. Mines, Mem. 43, 1-7.
993 Nelson, R., 1939. Colusite - its occurrence, paragenesis and genetic significance. Am. Mineralogist, 24, 369-376.
994 Němec, D., 1964. Lievrite from Županovice, S.W. Moravia. (in Czech). Časopis Mineral. Geol., 9, 333-335. Ref. in: Mineral. Abstr., 17, 339.
995 Neuhaus, A., 1936. Ueber Vorkommen von Kupfererz-führenden Spateisensteingängen im östlichen Bober-Katzbach-Gebirge (Schlesien). Chem. Erde, 10, 247-270.

996 Neuhaus, A., 1942. Ueber die Arsenführung der dichten Schwefelkiese (Melnikowit-Pyrite, Gelpyrite) von Wiesloch (Baden) und Deutsch Bleischarley (O.S.). Metall Erz. 39, 157-189.
997 Neumann, H., 1944. Silver deposits at Kongsberg. Norg. Geol. Undersøk., 162.
998 Neumann, H. and Bergstöl, S., 1964. Contribution to the mineralogy of Norway. No. 25: Pyrophanite in the southern part of the Oslo area. Norsk Geol. Tidsskr., 44, 39-42.
999 Neumann, H. and Sverdrup, T.L., 1960. Contribution to the mineralogy of Norway. No. 8. Davidite from Tuftan, Iveland. Norsk Geol. Tidsskr. 40, 277-288.
1000 Newhouse, W.H., 1931. A pyrrhotite-cubanite-chalcopyrite intergrowth from the Frood mine, Sudbury, Ontario. Am. Mineralogist, 16, 334-337.
1001 Newhouse, W.H., 1933a. Mercury in native silver. Am. Mineralogist, 18, 295-299.
1002 Newhouse, W.H., 1933b. The temperature of formation of the Mississippi Valley lead-zinc deposits. Econ. Geol., 28, 744-750.
1003 Newhouse, W.H. and Glass, J.P., 1936. Some physical properties of certain iron oxides. Econ. Geol., 31, 699-711.
1004 Nichol, I., 1962. Ph.D. Thesis, University of Durham.
1005 Nichol, I. and Phillips, R., 1965. Measurement of spectral reflectivity of manganese oxides. Mineral. Mag., 35, 200-213.
1006 Nickel, E.H., 1958. The composition and microtexture of an ulvöspinel-magnetite intergrowth. Can. Mineralogist, 6, 191-199.
1007 Nickel, E.H., 1959. The occurrence of native nickel-iron in the serpentine rocks of the Eastern Townships of Quebec Province. Can. Mineralogist, 6, 307-319.
1008 Nickel, E.H., 1964. Latrappite - a proposed new name for the perovskite-type calcium niobate mineral from the Oka area of Quebec. Can. Mineralogist, 8, 121-122.
1009 Nickel, E.H. and Mc Adam, R.C., 1963. Niobian perovskite from Oka, Quebec. A new classification for minerals of the perovskite group. Can. Mineralogist, 7, 683-697.
1010 Nickel, E.H., Rowland, J.F. and Mc Adam, R.C., 1963a. Ixiolite, a columbite substructure. Am. Mineralogist, 48, 961-979.
1011 Nickel, E.H., Rowland, J.F. and Mc Adam, R.C., 1963b. Wodginite, a new Sn-Mn tantalate from Wodgina, Australia and Bernic Lake, Manitoba. Can. Mineralogist, 7, 390-402.
1012 Noll, W., 1948. Zur Kristallchemie des Zinnsteins (Kassiterit). Heidelberger Beitr. Mineral. Petrog., 1, 593-625.
1013 Norrish, K., 1951. Priderite, a new mineral from the leucite-lamproites of the west Kimberley area, Western Australia. Mineral. Mag., 29, 496-501.
1014 Novák, F., Blüml, A. and Tacl, A., 1962. The origin of stannite as replacement of cassiterite in the Turkaňk zone of the Kutná Hora ore deposit. Mineral. Mag., 33, 339-342.
1015 Nowacki, W., 1964. Zur Kristallchemie der Sulfosalze, insbesondere aus dem Lengenbach (Binnatal, Kt. Wallis). Schweiz. Mineral. Petrog. Mitt., 44, 459-484.
1016 Nowacki, W., 1965. Ueber einige Mineralfunde aus dem Lengenbach (Binnatal, Kt. Wallis). Eclogae Geol. Helv., 58, 403-406.
1017 Nowacki, W., 1967a. Ueber die mögliche Identität von "Liveingit" mit Rathit-II. Neues Jahrb. Mineral., Monatsh., 1967, 353-354.
1018 Nowacki, W., 1967b. Ueberblick über einige Sulfid- und Arsenosulfosalz-Kristallstrukturen. Schweiz. Mineral. Petrog. Mitt., 47, 659-681.
1019 Nowacki, W., 1968. Ueber Hatchit, Lengenbachit und Vrbait. Neues Jahrb. Mineral., Monatsh., 1968, 69-75.
1020 Nowacki, W., 1969. Ueber Stephanit [$SbS_3/S/Ag_5^{III}$] und Pyrargyrit [SbS_3/Ag_3^{II}] aus dem Lengenbach (Binnatal, Schweiz). Schweiz. Mineral. Petrog. Mitt., 49, 381-384.
1021 Nowacki, W. and Bahezre, C., 1963. Die Bestimmung der chemischen Zusammensetzung einiger Sulfosalze aus dem Lengenbach (Binnatal, Kt. Wallis) mit Hilfe der elektronischeñ Mikrosonde. Schweiz. Mineral. Petrog. Mitt., 43, 407-411.
1022 Nowacki, W., Bahezre, C. and Marumo, F., 1963. Investigations on sulphosalts from Binnatal (Ct. Wallis, Switzerland), part 7. Acta Cryst., 16, A 11-12.
1023 Nowacki, W., Burri, G., Engel, P. and Marumo, F., 1967. Ueber einige Mineralstufen aus dem Lengenbach (Binnatal) II. Neues Jahrb. Mineral., Monatsh., 1967, 43-48.
1024 Nowacki, W., Engel, P., Matsumoto, T., Ohmasa, M., Ribar, B. and Takéuchi, Y., 1967. Die Kristallstruktur von Gratonit $Pb_9As_{14}S_{15}$, Xanthokon Ag_3AsS_3, Trechmannit $AgAsS_2$ und Wallisit $PbTlCuAs_2S_5$, dem Cu-Analogon von Hatchit $PbTlAgAs_2S_5$. Schweiz. Mineral. Petrog. Mitt., 47, 1138-1139.

1025 Nowacki, W., Iitaka, Y. and Bürki, H., 1960. Structural investigations on sulfosalts from the Lengenbach, Binn Valley, Switzerland. Acta Cryst., 13, 1006-1007.
1026 Nowacki, W. and Kunz, V., 1961. Strukturelle Untersuchungen an Sulfosalzen vom Lengenbach, Binnatal (Kt. Wallis). Neues Jahrb. Mineral., Monatsh., 1961, 94-95.
1027 Nowacki, W., Kunz, V. and Marumo, F., 1964. Ueber einige Mineralstufen aus dem Lengenbach (Binnatal). Schweiz. Mineral. Petrog. Mitt., 44, 129-132.
1028 Nowacki, W., Marumo, F. and Takéuchi, Y., 1964. Untersuchungen an Sulfiden aus dem Binnatal (Kt. Wallis, Schweiz). Schweiz. Mineral. Petrog. Mitt., 44, 5-9.
1029 Nowacki, W. and Stalder, H.A., 1969. Zwei Wismutsulfosalze von Sta. Maria, Val Medel, Kt. Graubünden, Schweiz. Mineral. Petrog. Mitt., 49, 97-101.
1030 Nowacki, W., Wuensch, B.J. and Kunz, V., 1964. Ueber eine stenglig-faserige Ausbildung von Jordanit aus dem Lengenbach. Schweiz. Mineral. Petrog. Mitt., 44, 455-458.
1031 Nuffield, E.W., 1944. Studies of mineral sulpho-salts: IX - Lengenbachite. Trans. Roy. Soc. Canada, Sect. IV, 38, 59-64.
1032 Nuffield, E.W., 1945a. Studies of mineral sulpho-salts: X - Andorite, ramdohrite, fizelyite. Trans. Roy. Soc. Canada, Sect. IV, 39, 41-50.
1033 Nuffield, E.W., 1945b. Mercurian silver from British Columbia. Univ. Toronto, Geol. Ser., 49, 71.
1034 Nuffield, E.W., 1946. Studies of mineral sulpho-salts: XII - Fülöppite and zinckenite. Univ. Toronto, Geol. Ser., 50, 49-62.
1035 Nuffield, E.W., 1947a. X-ray measurements on hutchinsonite. Univ. Toronto, Geol. Ser., 51, 79-81.
1036 Nuffield, E.W., 1947b. Studies of mineral sulpho-salts: XI - Wittichenite (klaprothite). Econ. Geol., 42, 147-160.
1037 Nuffield, E.W., 1948a. Observations on kobellite. Univ. Toronto, Geol. Ser., 52, 86-89.
1038 Nuffield, E.W., 1948b. Franckeite in relation to lengenbachite. Am. Mineralogist, 33, 203.
1039 Nuffield, E.W., 1952. Studies of mineral sulpho-salts: XVI: Cuprobismuthite. Am. Mineralogist, 37, 447-452.
1040 Nuffield, E.W., 1953. Benjaminite. Am. Mineralogist, 38, 550-552.
1041 Nuffield, E.W., 1954. Studies of mineral sulpho-salts: XVIII: Pavonite, a new mineral. Am. Mineralogist, 39, 409-415.
1042 Nuffield, E.W. and Harris, D.C., 1966. Studies of mineral sulpho-salts: XX: Berryite, a new species. Can. Mineralogist, 8, 407-413.
1043 Nuffield, E.W. and Peacock, M.A., 1945. Studies of mineral sulpho-salts: VIII - Plagionite and semseyite. Univ. Toronto, Geol. Ser., 49, 17-39.
1044 Ödman, O.H., 1932. Mineragraphic study of the opaque minerals in the lavas from Mt. Elgon, British East Africa. Geol. Fören. Stockholm Förhandl., 54, 285-304.
1045 Ödman, O.H., 1933. Erzmikroskopische Untersuchungen der Sulfiderze von Kaveltorp in Mittelschweden. Geol. Fören. Stockholm Förhandl., 55, 563-611.
1046 Ödman, O.H., 1939. The gold-copper-arsenic ore at Holmtjärn, Skelleftedistrict, N.Sweden. Geol. Fören. Stockholm Förhandl., 61, 91-111.
1047 Ödman, O.H., 1941a. Geology and ores of the Boliden deposit, Sweden. Sveriges Geol. Undersökn., Ser. C, 438, Årsbok 35 (1).
1048 Ödman, O.H., 1941b. Minerals of the Varuträsk pegmatite XXVIII. On "stibiomicrolite" and its decomposition products. Geol. Fören. Stockholm Förhandl., 63, 289-294.
1049 Ödman, O.H., 1942. Minerals of the Varuträsk pegmatite XXXIII. Native metals and sulphides. Geol. Fören. Stockholm Förhandl., 64, 277-282.
1050 Ödman, O.H., 1945. A nickel-cobalt-silver-mineralization in the Laver copper mine, N. Sweden. Sveriges Geol. Undersökn., Ser. C, 470, Årsbok 39 (3).
1051 Ödman, O.H., 1947. Manganese mineralization in the Ultevis district, Jokkmokk, N. Sweden. Part I: Geology. Sveriges Geol. Undersökn., Ser. C, 487, Årsbok 41 (6).
1052 Ödman, O.H., 1950. Manganese mineralization in the Ultevis District, Jokkmokk, North Sweden. Part II: Mineralogical notes. Sveriges Geol. Undersökn., Ser. C, 516, Årsbok 44 (2).
1053 Oelsner, O., 1956. Ueber einige neue Vorkommen von Jordisit und Ilsemannit. Geologie (Berlin), 5, 563-567.
1054 Oelsner, O., 1958. Ueber einige mexikanische Lagerstätten. Fortschr. Mineral., 36, 5-18.
1055 Oelsner, O., 1961a. Ueber ein neues Vorkommen von Coffinit im sächsischen Erzgebirge und Methoden zu seiner Erkennung. Geologie (Berlin), 10, 818-823.

1055a Oelsner, O., 1961b. Atlas der wichtigsten Mineralparagenesen im mikroskopischen Bild. (French and English editions have also been published). Bergakademie Freiberg, Fernstudium.
1056 Oelsner, O. and Tischendorf, G., 1957. Ueber einige Mineralvorkommen im Pyroxenitgranulit von Hartmannsdorf bei Karl-Marx-Stadt. Geologie (Berlin), 6, 278-288.
1057 Oen Ing Soen, 1959. On some sulphide minerals in the beryllium-lithium pegmatite of Mangualde, N. Portugal. Neues Jahrb. Mineral., Abhandl., 93, 192-208.
1058 Oen Ing Soen and Pauly, H., 1967. A sulphide paragenesis with pyrrhotite and marcasite in the siderite-cryolite ore of Ivigtut, South Greenland. Medd. Grønland, 175, Nr. 5, 55 pp.
1059 Oen Ing Soen and Sørensen, H., 1964. The occurrence of nickel-arsenides and nickel-antimonide at Igdlúnguaq, in the Ilímaussaq alkaline massif, South Greenland. Medd. Grønland, 172, nr. 1.
1060 Oftedal, I., 1959. Native bismuth in the molybdenite deposit at Skjoldevik, Haugesund peninsula, W. Norway. Norsk Geol. Tidsskr., 39, 81-82.
1061 Olsacher, J., 1939. Achavalita, seleniuro de hierro, nueva especie mineral. Bol. Fac. Cienc. Univ. Nat. Córdoba, Argentina, 2, 73-78. Ref. in: Min. Abstr., 1954, 12, 236.
1062 Onay, T.S., 1949. Ueber die Smirgelgesteine SW-Anatoliens. Schweiz. Mineral. Petrog. Mitt., 29, 357-491.
1063 Onorato, E., 1957. Röntgenographische Untersuchungen über den Kobaltglanz. Neues Jahrb. Mineral. Abhandl., 91, 41-51.
1064 Ontoev, D.O., 1959. Lillianite from the Bukuka deposit and conditions under which it formed. Dokl. Akad. Nauk SSSR, 126, 855-858. Translated in: Dokl. Acad. Sci. USSR, Earth Sci. Sect., 126, 591-593 (1960).
1065 Oosterbosch, R., 1951. Copper mineralization in the Fungurume region, Katanga. Econ. Geol., 46, 121-148.
1066 Oosterbosch, R., Picot, P. and Pierrot, R., 1964. La digénite sélénifère de Musonoï (Katanga). Bull. Soc. Franç. Minéral. Crist., 87, 613-617.
1067 Orcel, J., 1928. Notes sur les caractères microscopiques des minéraux opaques, principalement en lumière polarisée. Bull. Soc. Franç. Minéral. Crist., 51, 197-210.
1068 Orcel, J., 1932. Sur l'existence de la coronadite dans les minerais de manganèse de Bou Tazoult, région de l'Imini (Maroc). Compt. Rend. Acad. Sci. Paris, 194, 1956-1958.
1069 Orcel, J., 1935. La stannine de certains gisements d'étain de la France métropolitaine et coloniale. Compt. Rend. 68e Congr. Soc. Savantes, 11, 49-51.
1070 Orcel, J., 1937a. Notice sur les travaux scientifiques de M. Jean Orcel. Paris.
1071 Orcel, J., 1937b. Sur l'existence de l'érubescite orange et d'un nouveau sulfure double de cuivre et de fer dans l'érubescite de Vaulry (Haute-Vienne). Compt. Rend. 70e Congr. Soc. Savantes, 26, 135-139.
1072 Orcel, J., 1942. La coronadite et le minerai qui la renferme dans les gîtes de manganèse de l'Imini, Sud Marocain. Bull. Soc. Franç. Minéral. Crist., 65, 73-114.
1073 Orcel, J., 1943. Composition minéralogique et structure des zones cuprifères du minerai stannifère de Vaulry (Haute-Vienne); caractères du nouveau type de stannite qu'elles renferment. Bull. Soc. Franç. Minéral. Crist., 66, 435-451.
1074 Orcel, J. and Fastré, P., 1935. Courbes de dispersion de quelques étalons de pouvoirs réflecteurs utilisables dans l'étude microscopique des minerais métalliques. Compt. Rend. Acad. Sci. Paris, 200, 1485-1488.
1075 Orcel, J., Hénin, S. and Caillère, S., 1958. Sur la présence de stainiérite à Bou-Azzer et les propriétés de cet hydroxyde. Compt. Rend. Acad. Sci. Paris, 246, 792-795. Ref. in: Am. Mineralogist, 43, 1223-1224, Mineral Abstr., 15, 41.
1076 Orcel, J. and Jouravsky, G., 1935. Le minerai de cobalt de Bou Azzer (Maroc), sa composition minéralogique et sa structure. Congr. Intern. Mines Met. Geol. Appl., Sect. Geol. Appl., 207-216.
1077 Orcel, J. and Pavlovitch, S., 1931. Les caractères microscopiques des oxydes de manganèse et des manganites naturels. Bull. Soc. Franç. Minéral. Crist., 54, 108-179.
1078 Orlova, Z.V., 1956. Trudy Vses. Magadansk Nauch. Issled. Inst. Min. Tsvet. Met. SSSR, 2, 76. Ref. in: Am. Mineralogist, 43, 1222-1223.
1079 Ortlepp, R.J., 1964. Nsutite (battery grade manganese dioxide) from the Western Transvaal. Trans. Geol. Soc. S. Africa, 67, 149-161.
1080 Osipov, B.S., 1967. Certain optical and physical properties of synthetic uraninite. Dokl. Akad. Nauk SSSR, 176, 672-675. Translated in: Dokl. Acad. Sci. USSR, Earth Sci. Sect., 176 (1968), 126-129.

1081 Ottemann, J., 1963. Ueber den Chemismus eines Rutils mit optisch ungewöhnlichem Reflexionsverhalten. Neues Jahrb. Mineral., Abhandl., 100, 125-129.
1082 Ottemann, J. and Augustithis, S.S., 1967. Geochemistry and origin of "platinum-nuggets" in lateritic covers from ultrabasic rocks and birbirites of W.Ethiopia. Mineral. Deposita, 1, 269-277.
1083 Ottenburgs, R., 1964. Metallogenetische en geochemische studie van het blende-erts te Kipushi, Katanga. Ph.D. Thesis, Univ. Leuven (Belgium).
1084 Otto, H.H. and Strunz, H., 1968. Zur Kristallchemie synthetischer Blei-Wismut-Spiessglanze. Neues Jahrb. Mineral., Abhandl., 108, 1-19.
1085 Oztunali, O., 1959. Ueber die Struktur von Brannerit. Neues Jahrb. Mineral., Monatsh., 1959, 187-188.
1086 Pääkkönen, V., 1966. On the geology and mineralogy of the occurrence of native antimony at Seinäjoki, Finland. Bull. Comm. Géol. Finlande, 225, 70 pp.
1087 Pabst, A., 1946. Notes on the structure of delafossite. Am. Mineralogist, 31, 539-546.
1088 Pabst, A., 1951. X-ray examination of uranothorite. Am. Mineralogist, 36, 557-562.
1089 Pabst, A., 1954. Brannerite from California. Am. Mineralogist, 39, 109-117.
1090 Pabst, A., 1961. X-ray crystallography of davidite. Am. Mineralogist, 46, 700-718.
1091 Paděra, K., 1951. Tetradymite from Jílové. (in Czech, with Russian and Engl. summaries). Sb. Ustřed. Ustava Geol., oddíl geol., 18, 633-640.
1092 Paděra, K., 1954. Roentgenometric determination of the pararammelsbergite at Dobšiná. (in Czech, Russ. and Engl. summ.). Sb. Ústřed. Ústavu,Geol., 21, oddíl geol., 2. díl, 813-826.
1093 Paděra, K., 1956. Beitrag zur Revision der Mineralien aus der Gruppe von Wismutglanz und Aikinit. Chem. Erde, 18, 14-18.
1094 Paděra, K., 1962. Pararammelsbergite from Dobšiná and its distinction from rammelsbergite. (in Czech, Russ. and Engl. summary). Geol. Sborník (Bratislava), 13, 135-147.
1095 Paděra, K., Bouška, V. and Pelikán, J., 1955. Rezbanyit aus Dobšiná in der Ostslowakei, CSR. Chem. Erde, 17, 329-340.
1096 Page, N.J. and Jackson, E.D., 1967. Preliminary report on sulfide and platinum-group minerals in the chromitites of the Stillwater Complex, Montana. U.S., Geol. Surv., Profess. Papers, 575-D, 123-126.
1097 Palache, C., 1940. Cuprobismutite - a mixture. Am. Mineralogist, 25, 611-613.
1098 Palache, C., 1941. Contributions to the mineralogy of Sterling Hill, New Jersey: Morphology of graphite, arsenopyrite, pyrite and arsenic. Am. Mineralogist, 27, 709-717.
1099 Palache, C. and Berman, H., 1942. Boulangerite. Am. Mineralogist, 27, 552-562.
1100 Palache, C., Berman, H. and Frondel, C., 1944. The system of mineralogy of J.D. Dana and E.S. Dana, 7th Ed., Vol. I. Wiley and Sons, New York.
1101 Palache, C., Berman, H. and Frondel, C., 1951. The system of mineralogy of J.D. Dana and E.S. Dana, 7th Ed., Vol. II. Wiley and Sons, New York.
1102 Palache, C. and Fisher, J., 1939. Preliminary description of a new mineral from Cerro de Pasco, Peru. Am. Mineralogist, 24, 136.
1103 Palache, C. and Fisher, J., 1940. Gratonite - a new mineral from Cerro de Pasco, Peru. Am. Mineralogist, 25, 255-265.
1104 Palache, C., Richmond, W.E. and Winchell, H., 1938. Crystallographic studies of the sulpho-salts baumhauerite, meneghinite, jordanite, diaphorite, freieslebenite. Am. Mineralogist, 23, 821-836.
1105 Panagos, A. and Ottemann, J., 1966. Chemical differentiation of chromite grains in the nodular-chromite from Rodiani (Greece). Mineral. Deposita, 1, 72-75.
1106 Panagos, A. and Ramdohr, P., 1965. Ein bemerkenswertes Vorkommen von Valleriit, $CuFeS_2$ + n $Mg(OH)_2$, im Chromerz von Eretria, Griechenland. Neues Jahrb. Mineral., Monatsh., 1965, 149-151.
1107 Pantó, G. and Mikó, L., 1964. The Nagybörzsöny ore deposits. (in Hungarian, Engl. and Russ. summ.). Ann. Hung. Geol. Inst., 50, Fasc. 1, 153 pp.
1108 Papezik, V.S., 1955. Heazlewoodite from Miles Ridge, Yukon Territory. Am. Mineralogist, 40, 692-693.
1109 Papezik, V.S., 1966. Native arsenic in Newfoundland. Can. Mineralogist, 8, 670-671.
1110 Papezik, V.S., 1967. Native arsenic in Newfoundland. Can. Mineralogist, 9, 101-108.
1111 Parker, R.L., Adams, J.W. and Hildebrand, F.A., 1962. A rare sodium niobate mineral from Colorado. U.S., Geol. Surv., Profess. Papers, 450-C, 4-6.

1112   Parkin, L.W. and Glasson, K.R., 1954. The geology of the Radium Hill Uranium Mine, South Australia. Econ. Geol., 49, 815-825.
1113   Pärnamaa, E., 1963. On the use of Vickers microhardness (VH) in the microscopical identification of ore minerals, especially sulphides of Pb, Bi, Sb. Lunds Univ. Årsskr., N.F., Avd. 2, 59, Nr. 3, 51 pp.
1114   Partridge, F.C., 1946. Trevorite and a suggested new nickel-bearing silicate from Bon Accord, Sheba Siding, Barberton District. Trans. Geol. Soc. S. Africa, 46, 119-126.
1115   Patchett, J.E. and Nuffield, E.W., 1960. Studies of radioactive compounds: X: The synthesis and crystallography of brannerite. Can. Mineralogist, 6, 483-490.
1116   Pauling, L., 1930. The crystal structure of pseudobrookite. Z. Krist., A 73, 97-112.
1117   Pauling, L. and Hultgren, R., 1933. The crystal structure of sulvanite, Cu_3VS_4. Z. Krist., A 84, 204-212.
1118   Pauling, L. and Neumann, E.W., 1934. The crystal structure of binnite, $(Cu,Fe)_{12}As_4S_{13}$, and the chemical composition and structure of minerals of the tetrahedrite group. Z. Krist., A 88, 54-62.
1119   Pauling, L. and Weinbaum, S., 1934. The crystal structure of enargite, Cu_3AsS_4. Z. Krist., A 88, 48-53.
1120   Pauly, H., 1946. Mikroskopisk undersøgelse af nogle malmmineralier frå Grønland. Medd. Dansk Geol. Foren., 11, 30-46.
1121   Pavlov, N.V., 1957. On maghemite in the magnetite ores of the Kezhemsk deposit. (in Russian). Tr. Mineralog. Muzeya, Akad. Nauk SSSR, 8, 77-84.
1122   Pavlovitch, S., 1946. Etude microscopique de quelques minerais métalliques de Yougoslavie. Bull. Soc. Franç. Minéral. Crist., 55, 125-136.
1123   Peacock, M.A., 1940a. On maucherite (nickel-speiss, placodine, temiskamite). Mineral. Mag., 25, 557-572.
1124   Peacock, M.A., 1940b. On dyscrasite and antimonial silver. Univ. Toronto, Geol. Ser., 44, 31-46.
1125   Peacock, M.A., 1940c. Goldschmidtine, identical with stephanite. Am. Mineralogist, 25, 372-373.
1126   Peacock, M.A., 1941. On joseite, grünlingite, oruetite. Univ. Toronto, Geol. Ser., 46, 83-105.
1127   Peacock, M.A., 1942a. Studies of mineral sulpho-salts: VI - Aikinite. Univ. Toronto, Geol. Ser., 47, 63-69.
1128   Peacock, M.A., 1942b. On sternbergite and frieseite. Am. Mineralogist, 27, 229.
1129   Peacock, M.A., 1945. On potarite. Univ. Toronto, Geol. Ser., 49, 71-73.
1130   Peacock, M.A., 1947. On heazlewoodite and the artificial compound Ni_3S_2. Univ. Toronto, Geol. Ser., 51, 59-69.
1131   Peacock, M.A., 1950a. Studies of mineral sulpho-salts: XV - Xanthoconite and pyrostilpnite. Mineral. Mag., 29, 346-358.
1132   Peacock, M.A., 1950b. Hauchecornite. Am. Mineralogist, 35, 440-446.
1133   Peacock, M.A. and Berry, L.G., 1940. Röntgenographic observations on ore minerals. Univ. Toronto, Geol. Ser., 44, 47-69.
1134   Peacock, M.A. and Berry, L.G., 1947. Studies of mineral sulpho-salts: XIII - Polybasite and pearceite. Mineral. Mag., 28, 1-13.
1135   Peacock, M.A. and Dadson, A.S., 1940. On rammelsbergite and pararammelsbergite: distinct forms of nickel diarsenide. Am. Mineralogist, 25, 561-577.
1136   Peacock, M.A. and Henry, W.G., 1948. The crystal structure of cobaltite (CoAsS), gersdorffite (NiAsS) and ullmannite (NiSbS). Univ. Toronto, Geol. Ser., 52, 71-80.
1137   Peacock, M.A. and McAndrew, J., 1950. On parkerite and shandite and the crystal structure of $Ni_3Pb_2S_2$. Am. Mineralogist, 35, 425-439.
1138   Peacock, M.A. and Michener, C.E., 1939. On rammelsbergite from Ontario. Univ. Toronto, Geol. Ser., 42, 95-112.
1139   Peacock, M.A. and Thompson, R.M., 1946a. Melonite from Quebec and the crystal structure of $NiTe_2$. Univ. Toronto, Geol. Ser., 50, 63-73.
1140   Peacock, R.M. and Thompson, R.M., 1946b. Montbrayite, a new gold telluride. Am. Mineralogist, 31, 515-526.
1141   Peacor, D.R., 1967. New data on nigerite. Am. Mineralogist, 52, 864-866.
1142   Pearson, A.D. and Buerger, M.J., 1956. Confirmation of the crystal structure of pentlandite. Am. Mineralogist, 41, 804-805.
1143   Pellas, P., 1954. Sur une fergusonite anisotrope de Naegi (Japon). Bull. Soc. Franc. Minéral. Crist., 77, 461-473.

1144 Penfield, S.L. and Frenzel, A., 1897. On the identity of chalcostibite (wolfsbergite) and guejarite, and on chalcostibite from Huanchaca, Bolivia. Am. J. Sci., 4, 27-35.

1145 Penkov, I.P. and Safin, I.A., 1966. A study of stephanite (Ag_5SbS_4) by the method of nuclear quadropole resonance. Dokl. Akad. Nauk SSSR, 168, 1148-1150. Translated in: Dokl. Acad. Sci. USSR, Earth Sci. Sect., 168, 136-138 (1966).

1146 Perichaud, J.-J., Picot, P. and Pierrot, R., 1966. Sur l'existence d'une minéralisation stanno-argentifère exceptionnelle dans la région de Massiac (Cantal). Bull. Soc. Franç. Minéral. Crist., 89, 488-495.

1147 Permingeat, F., 1954. La stéphanite du gisement d'Azegour, Haut Atlas, Maroc. Bull. Soc. Franç. Minéral. Crist., 77, 1254-1259.

1148 Permingeat, F., 1955. Sur les niobo-tantalates de l'Anti-Atlas, Maroc: tapiolite et columbite. Bull. Soc. Franç. Minéral. Crist., 78, 123-156.

1149 Permingeat, F. and Weinryb, E., 1960. Sur les inclusions cobaltifères dans la chalcopyrite d'Azegour (Maroc). Exemple d'utilisation de la microsonde électronique. Bull. Soc. Franç. Minéral. Crist., 83, 65-66.

1150 Perseil, A., 1965. Existence d'un niveau à brèche du Dévonien supérieur de Brachy (Ariège) minéralisé en jacobsite. Bull. Soc. Franç. Minéral. Crist., 88, 349-350.

1151 Perseil, E.A., 1967. Nouvelles données sur la ranciéite du Rancié. Compt. Rend. Acad. Sci., Paris, 264, 1241-1244.

1152 Peters, T., 1963. Mineralogie und Petrographie des Totalserpentins bei Davos. Schweiz. Mineral. Petrog. Mitt., 43, 529-685.

1153 Peterson, V.E., 1942. A study of the geology and ore deposits of the Ashbrook silver mining district, Utah. Econ. Geol., 37, 466-502.

1154 Petruk, W., 1968. Mineralogy and origin of the Silverfields silver deposit in the Cobalt Area, Ontario. Econ. Geol., 63, 512-531.

1154a Petruk, W., Harris, D.C., Cabri, L.J. and Stewart, J.M., 1969. Native silver and silver-antimony minerals in the Cobalt-Gowganda ores. Can. Mineralogist, 10, 144.

1155 Petruk, W., Harris, D.C. and Stewart, J.M., 1969. Langisite, a new mineral, and the rare minerals cobalt, pentlandite, siegenite, parkerite and bravoite from the Langis mine, Cobalt-Gowganda area, Ontario. Can. Mineralogist, 9, 597-616.

1156 Picot, P., 1959. Sur la présence de minerais métalliques nickélifères dans les serpentines. Bull. Soc. Franç. Minéral. Crist., 82, 329-334.

1157 Picot, P. and Pierrot, R., 1963. La roquésite, premier minéral d'indium. Bull. Soc. Franç. Minéral. Crist., 86, 7-14.

1158 Picot, P., Sainfeld, P. and Vernet, J., 1963. Découverte de germanite en France. Bull. Soc. Franç. Minéral. Crist., 86, 299-300.

1159 Picot, P., Sainfeld, P. and Vernet, J., 1966. Présence d'arséniures de cuivre dans le dôme de Barrot (Alpes-Maritimes). Bull. Soc. Franç. Minéral. Crist., 89, 259-261.

1160 Picot, P., Scolari, G. and Troly, G., 1963. Nouvelles données sur la paragenèse du minerai de la mine de M'Passa (République du Congo) et comparaison avec d'autres gisements de l'Afrique centrale. Bull. Soc. Franc. Minéral. Crist., 86, 355-358.

1161 Picot, P., Troly, G. and Vincienne, H., 1963. Précisions nouvelles sur les minéralisations de Chizeuil (Saône-et-Loire). Bull. Soc. Franç. Minéral. Crist., 86, 373-375.

1162 Picot, P. and Vernet, J., 1963. Sur la présence de quelques minéraux rares de bismuth dans une minéralisation plombifère de la région de Tende (Alpes-Maritimes). Bull. Soc. Franç. Minéral. Crist., 86, 87-88.

1163 Picot, P. and Vernet, J., 1967. Un nouveau gisement de koutekite: le dôme du Barrot (Alpes-Maritimes). Bull. Soc. Franç. Minéral. Crist., 90, 82-89.

1164 Piepoli, P., 1933. Etude microscopique de quelques minerais des filons argentifères du Sarrabus (Sardaigne). Bull. Soc. Franç. Minéral. Crist., 56, 277-302.

1165 Piepoli, P., 1934. Etude microscopique de quelques minerais du filon cobalto-nickélifère de Riu Planu is Castangias (Gonnosfanadiga, Sardaigne). Bull. Soc. Franç. Minéral. Crist., 57, 270-282.

1165a Piller, H. and Gehlen, K. von 1964. On errors of reflectivity measurements and of calculations of refractive index n and absorption coefficient k. Am. Mineralogist, 49, 867-882.

1166 Piotrovskii, G.L., 1958. Zur Genese von Zinnober und Metacinnabarite. (in Russian). Mineral. Sb. L'vovsk. Geo. Obshchestvo, 12, 225-232. Ref. in: Zentr. Mineral., II, 1962, 149.

1167 Pokrovskaya, I.V., 1966. Collomorphic molybdenite in ores of the Rdder-Sokolnyi deposit (in Russian). Zap. Vses. Mineralog. Obshchestva, 95, 602-606. Ref. in: Mineral. Abstr., 18, 202.
1168 Poliakova, O.P., 1957a. Geocronite from the Smirnovsk ore deposit (West Transbaikal). (in Russian). Tr. Mineralog. Muzeya, Akad. Nauk SSSR, 8, 99-102.
1169 Poliakova, O.P., 1957b. On franckeite in the lead-polymetallic ores of the Smirnovsk deposit (West-Transbaikal). (in Russian). Tr. Mineralog. Muzeya, Akad., Nauk SSSR, 8, 103-107.
1170 Pollock, D.W. 1959. Sulfide paragenesis in the Eastern Metals Depont, Montgomery County, Quebec. Econ. Geol., 54, 234-247.
1171 Polushkina, A.P. and Sidorenko, G.A., 1963. Structural variant of cobaltite. Dokl. Akad. Nauk SSSR, 153, 1420-1423. Translated in: Dokl. Acad. Sci. USSR, Earth Sci. Sect., 153, 167-170. (1965).
1172 Poplavko, E.M., Marchukova, I.D. and Zak, S.Sh., 1962. A rhenium mineral in the ores of the Dzhezkazgan deposits. Dokl. Akad. Nauk SSSR, 146, 433-436. Translated in: Dokl. Acad. Sci. USSR, Earth Sci. Sect., 146, 110-112 (1964).
1173 Portnov, A.M., Nikolaeva, L.E. and Stolyarova, T.I., 1966. The new titanium mineral landauite. Dokl. Akad. Nauk SSSR, 166, 1420-1421. Translated in: Dokl. Acad. Sci. USSR, Earth Sci. Sect., 166, 143-145 (1966).
1174 Pöschl, A., 1968. Die Pb-Ag-Erzlagerstätten des Rio Ribeira de Iguapé Bezirkes (Südbrasilien). Neues Jahrb. Mineral., Monatsh., 1968, 33-41.
1175 Posnjak, E. and Merwin, H.E., 1922. The system Fe_2O_3 - SO_3 - H_2O. J. Am. Chem. Soc., 44, 1965-1994.
1176 Pouba, Z. and Vejnar, Z., 1955. The polymetallic ore veins at Jasenic in the Nízké Tatry. (in Czech, Engl. summary). Sb. Ústřed. Ustavu Geol., oddíl geol., 22, 485-555.
1177 Prévot, M., Rémond, G. and Caye, R., 1968. Etude de la transformation d'une titanomagnétite en titanomaghémite dans une roche volcanique. Bull. Soc. Franç. Minéral. Crist., 91, 65-74.
1178 Prinz, W.C., 1961. Manganese oxide minerals at Philipsburg, Montana. U.S., Geol. Surv., Profess. Papers, 424-B, 296-297.
1179 Prouvost, J., 1960. Transformations expérimentales des sulfures métalliques naturels, étude de leur mécanisme. Bull. Soc. Franç. Minéral. Crist., 83, 265-294.
1180 Pyatenko, Yu.A. and Pudovkina, Z.V., 1961. The crystal structure of calzirtite - a new derivative of the structural type of CaF_2 - CeO_2. Kristallografiya (USSR), 6, 196-199. Ref. in: Am. Mineralogist, 46, 1515.
1181 Pijpekamp, B. van de, Burke, E.A.J. and Maaskant, P., 1969. Magnesioferrite, a mineral new for Långban, Sweden. Arkiv Mineral. Geol., Band 5, Nr. 1.
1182 Quensel, P.D., 1940. Minerals of the Varuträsk pegmatite XIX. The uraninite minerals (ulrichite and pitchblende). Geol. Fören. Stockholm Förhandl., 62, 391-396.
1183 Quensel, P.D., Ahlborg, K. and Westgren, A., 1937. Minerals of the Varuträsk pegmatite II. Allemontite. With an X-ray analysis of the mineral and of other arsenic-antimony alloys. Geol. Fören. Stockholm Förhandl., 59, 135-144.
1184 Quervain, F. de., 1936. Chalkographische Beobachtungen am Lherzolithserpentin von Selva (Poschiavo). Schweiz. Mineral. Petrog. Mitt., 16, 404.
1185 Quervain, F. de., 1945. Awaruit und pentlandit im Serpentin von Selva bei Poschiavo. Schweiz. Mineral. Petrog. Mitt., 25, 305-310.
1186 Quervain, F. de., 1963. Die Erzmineralien des Serpentins von Selva-Quadrada (Puschlav). Schweiz. Mineral. Petrog. Mitt., 43, 295-310.
1187 Quervain, F., 1967. Das Nickelerzvorkommen Val Boschetto im Centovalli (Tessin). Schweiz. Mineral. Petrog. Mitt., 47, 633-641.
1188 Radcliffe, D., 1967. Some properties of rammelsbergite and pararammelsbergite. Can. Mineralogist, 9, 128-131.
1189 Radcliffe, D., 1968a. Structural formula and composition of skutterudite. Can. Mineralogist, 9, 559-563.
1190 Radcliffe, D., 1968b. Chemistry of safflorite-loellingite. Can. Mineralogist, 9, 579.
1191 Radcliffe, D. and Berry, L.G., 1968. The safflorite-loellingite solid solution series. Am. Mineralogist, 53, 1856-1881.
1192 Radkewitsch, E.A. and Poljakowa, O.P., 1960. Umwandlung von Sulfostannaten innerhalb Zinnstein-Sulfid-Lagerstätten. Neues Jahrb. Mineral., Abhandl., 94, 191-207.

1193 Radtke, A.S., 1962. Coulsonite, FeV_2O_4, a spinell-type mineral from Lovelock, Nevada. Am. Mineralogist, 47, 1284-1291.
1194 Radtke, A.S., 1963. Data on cuprian coloradoite from Kalgoorlie, Western Australia. Econ. Geol., 58, 593-598.
1195 Radtke, A.S., 1964. Geology and mineralogy of the Buena Vista iron ores, Churchill County, Nevada. Econ. Geol., 59, 279-290.
1196 Radtke, A.S. and Jones, L.M., 1966. Strontium-bearing todorokite from Soganliyürük, Turkey. U.S., Geol. Surv., Profess. Papers, 550-C, 158-161.
1197 Radtke, A.S., Taylor, C.M. and Frost, J.E., 1967. Bismuth and tin minerals in gold- and silver-bearing sulfide ores, Ohio Mining District, Marysvale, Utah. U.S., Geol. Surv., Profess. Papers, 575-D, 127-130.
1198 Radtke, A.S., Taylor, C.M. and Hewett, D.F., 1967. Aurorite, argentian todorokite, and hydrous silver-bearing lead manganese oxide. Econ. Geol., 62, 186-206.
1199 Radusinović, D.R., 1966. Greigite from Lojane chromium deposit, Macedonia. Am. Mineralogist, 51, 209-215.
1200 Rakić, S.D., 1952. Kubanit iz Suplje Stijene (Crna Gora). (in Serbian, German summary). Bull. Muséum Hist. Nat. Pays Serbe, Sér. A, 5, 171-176.
1201 Rakić, S.D., 1960. Pseudo-Entmischungen von Zinkblende in Magnetkies. Ein Beispiel aus der Pb-Zn-Erzlagerstätte Rudnik (Serbien). Neues Jahrb. Mineral., Abhandl., 94, 584-590.
1202 Rakić, S.D., 1963. Liquidmagmatische Co-Ni-Cu Erzvorkommen bei Petković im Peridotitmassiv von Orahovac (Kosmet-Serbien). Bull. Muséum Hist. Nat. Pays Serbe, Sér. A, 18, 17-34.
1202a Ramberg, I.B., 1969. Contributions to the mineralogy of Norway, No. 41. Lepidocrocite at Rössvatn, North Norway. An example of pseudomorphism after pyrite cubes. Norsk Geol. Tidsskr., 49, 251-256.
1203 Ramdohr, P., 1928. Neue mikroskopische Beobachtungen am Cubanit (Chalmersit) und Ueberlegungen über seine lagerstättenkundliche Stellung. Z. Prakt. Geol., 36, 169-178.
1204 Ramdohr, P., 1931. Neue Beobachtungen über die Verwendbarkeit opaker Erze als geologische Thermometer. Z. Prakt. Geol., 39, 70-89.
1205 Ramdohr, P., 1932. Die Goldlagerstätte des Eisenbergs bei Corbach. Abhandl. Prakt. Geol. Bergwirtschaftslehre, 21, 1-39.
1206 Ramdohr, P., 1935. Vorkommen und Eigenschaften des Herzenbergits. Z. Krist., A 92, 186-189.
1207 Ramdohr, P., 1936a. Ein Zinnerzvorkommen im Marmor bei Arandis, Südwestafrika. Neues Jahrb. Mineral., Beil., 70 A, 1-48.
1208 Ramdohr, P., 1936b. Bleiglanz, Schapbachit, Matildit. Fortschr. Mineral., 20, 56-57.
1209 Ramdohr, P., 1937a. Erzmikroskopische Untersuchungen an einigen seltenen oder bisher wenig beachteten Erzmineralien, Teil I. Zentr. Mineral., A, 193-211.
1210 Ramdohr, P., 1937b. Erzmikroskopische Untersuchungen an einigen seltenen oder bisher wenig beachteten Erzmineralien, Teil 2. Zentr. Mineral., A, 289-303.
1211 Ramdohr, P., 1938a. Erzmikroskopische Untersuchungen an einigen seltenen oder bisher wenig beachteten Erzmineralien, Teil 3. Zentr. Mineral., A, 129-136.
1212 Ramdohr, P., 1938b. Antimonreiche Paragenesen von Jakobsbakken bei Sulitelma. Norsk Geol. Tidsskr., 18, 275-289.
1213 Ramdohr, P., 1938c. Ueber Schapbachit, Matildit und den Silber- und Wismutgehalt mancher Bleiglänze. Sitzber. Preuss. Akad. Wiss. Kl. Phys. Math., VI, 71-91.
1214 Ramdohr, P., 1938d. Die Kahn-Grube bei Arandis, Südwestafrika. Z. Prakt. Geol., 46, 41-50.
1215 Ramdohr, P., 1939. Wichtige neue Beobachtungen an Magnetit, Hämatit, Ilmenit und Rutil. Abhandl. Preuss. Akad. Wiss., Berlin, Math. Naturwiss. Kl., 14, 1-14.
1216 Ramdohr, P., 1940. Die Erzmineralien in gewöhnlichen magmatischen Gesteinen. Abhandl. Preus. Akad. Wiss., Berlin, Math. Naturwiss. Kl., 2, 1-43.
1217 Ramdohr, P., 1941. Erzmikroskopie und Röntgenaufnahme bei der Erkennung der Selbständigkeit von Mineralien. Tschermaks Mineral. Petrog. Mitt., 52, 269-271.
1218 Ramdohr, P., 1942. Gratonit aus den oberschlesischen Bleizinkerzgruben. Zentr. Mineral., A, 17-32.
1219 Ramdohr, P., 1943. Die Mineralien im System Cu_2S - CuS. Z. Prakt. Geol., 51, 1-12.
1220 Ramdohr, P., 1944. Zum Zinnkiesproblem. Abhandl. Preuss. Akad. Wiss., Berlin, Math. Naturwiss. Kl., 4, 1-30.

1221 Ramdohr, P., 1945a. Myrmekitische Verwachsungen von Erzen. Neues Jahrb. Mineral., 79 A, 161-191.
1222 Ramdohr, P., 1945b. Beobachtungen an durchbewegten Erzen von Routivare, Nordschweden. Geol. Fören. Stockholm Förhandl., 67, 367-388.
1223 Ramdohr, P., 1947. Orientierte Aufwachsung von Scheelit auf Wolframit. Heidelberger Beitr. Mineral. Petrog., 1, 105-109.
1224 Ramdohr, P., 1948. Pararammelsbergite: old and new observations. Univ. Toronto, Geol. Ser., 52, 9-22.
1225 Ramdohr, P., 1950a. Ueber das Vorkommen von Heazlewoodit Ni_3S_2 und über ein neues ihn begleitendes Mineral: Shandit, $Ni_3Pb_2S_2$. Sitzber. Deut. Akad. Wiss. Berlin, Math. Naturwiss. Kl., 1949, Nr. VI, 3-29.
1226 Ramdohr, P., 1950b. Ueber Josephinit, Awaruit, Souesit, ihre Eigenschaften, Entstehung und Paragenesis. Mineral. Mag., 29, 374-394.
1227 Ramdohr, P., 1950c. Die Lagerstätte von Broken Hill in New South Wales. Heidelberger Beitr. Min. Petr., 2, 291-333.
1228 Ramdohr, P., 1952a. Ueber den Mineralbestand der Zink- und Bleilagerstätte von Wiesloch in Baden. Fortschr. Mineral., 31, 13-14.
1229 Ramdohr, P., 1952b. Maldonit. Sitzber. Deut. Akad. Wiss. Berlin, Math. Naturwiss. Kl., 1952, No. V, 1-8.
1230 Ramdohr, P., 1952c. Neue Beobachtungen am Bühl-Eisen. Sitzber. Deut. Akad. Wiss. Berlin, Math. Naturwiss. Kl., 1952, No. V, 9-24.
1231 Ramdohr, P., 1953a. Ulvöspinel and its significance in titaniferous iron ores. Econ. Geol., 48, 677-688.
1232 Ramdohr, P., 1953b. Mineralbestand, Strukturen und Genesis der Rammelsberg-Lagerstätte. Geol. Jahrb., 67, 367-494.
1233 Ramdohr, P., 1956a. Ueberraschende Verwachsungsstrukturen. Fortschr. Mineral. 34, 23-24.
1234 Ramdohr, P., 1956b. Stilleit, ein neues Mineral, natürliches Zinkselenid, von Shinkolobwe. Geotekton. Symp. Ehren Hans Stille, Stuttgart 1956, 481-483.
1235 Ramdohr, P., 1957. Eisenalabandin, ein merkwürdiger natürlicher Hochtemperatur-Mischkristall. Neues Jahrb. Mineral., Abhandl., 91, 89-93.
1236 Ramdohr, P., 1958. Die Uranvorkommen vom Witwatersrand, Blind River und Dominion Reef. Fortschr. Mineral., 36, 52.
1237 Ramdohr, P., 1960. Die Erzmineralien und ihre Verwachsungen. 3. Auflage. Akademie-Verlag, Berlin, 1089 pp.
1238 Ramdohr, P., 1961a. Das Vorkommen von Coffinit in hydrothermalen Uranerzgängen, besonders vom Co-Ni-Bi-Typ. Neues Jahrb. Mineral., Abhandl., 95, 313-324.
1239 Ramdohr, P., 1961b. Magnetische Cassiterite. Bull. Comm. Géol. Finlande, 196, 473-481.
1240 Ramdohr, P., 1967. A widespread mineral association, connected with serpentinization, with notes on some new or insufficiently defined minerals. Neues Jahrb. Mineral., Abhandl., 107, 241-265.
1241 Ramdohr, P. and Frenzel, G., 1956. Die Manganerze. Intern. Geol. Congr., 20th, Mexico, Symp. Yacim. Manganeso, Tomo I, 19-73.
1242 Ramdohr, P. and Lawrence, L.J., 1958. Radioactive haloes in a davidite-ilmenite ore from Cloncurry, Queensland. J. Geol. Soc. Australia, 5, 33-35.
1243 Ramdohr, P. and Ödman, O.H., 1932. Valleriit. Geol. Fören. Stockholm Förhandl., 54, 89-97.
1244 Ramdohr, P. and Ödman, O.H., 1939. Falkmannit, ein neues Bleispiessglanzerz und sein Vorkommen, besonders in Boliden und Grube Bayerland. Mit einem Beitrag zur Kenntnis des Geokronits. Neues Jahrb. Mineral., 75 A, 315-350.
1245 Ramdohr, P. and Schidlowski, M., 1965. Ein radioaktiver Hof in Chromit. Neues Jahrb. Mineral., Monatsh., 1965, 225-227.
1246 Ramdohr, P. and Schmitt, M., 1955. Vier neue natürliche Kobaltselenide vom Steinbruch Trogtal bei Lautenthal im Harz. Neues Jahrb. Mineral., Monatsh., 1955, 133-142.
1247 Ramdohr, P. and Schmitt, M., 1959. Oregonit, ein neues Ni-Fe-arsenid mit metallartigen Eigenschaften. Neues Jahrb. Mineral., Monatsh., 1959, 239-247.
1248 Ramdohr, P. and Strunz, H., 1967. Klockmanns Lehrbuch der Mineralogie, 15^e umgearbeitete Auflage. Ferdinand Enke, Stuttgart, 820 pp.
1249 Rammelsberg, C., 1859. Ueber den sogenanten octaëdrischen Eisenglanz vom Vesuv, und über die Bildung von Magneteisen durch Sublimation. Poggendorff's Ann. Phys. Chem., 107, 451-454.

1250 Ramsdell, L.S., 1929. An X-ray study of the domeykite group. Am. Mineralogist, 14, 188-196.
1251 Ramsdell, L.S., 1932. An X-ray study of psilomelane and wad. Am. Mineralogist, 17, 143-149.
1252 Ramsdell, L.S., 1942. The unit cell of cryptomelane. Am. Mineralogist, 27, 611-613.
1253 Ramsdell, L.S., 1943. The crystallography of acanthite, Ag_2S. Am. Mineralogist, 28, 401-425.
1254 Rao, N.K. and Rao, G.V.U., 1968. Ore microscopic study of copper ore from Kolihan, Rajasthan - India. Econ. Geol., 63, 277-287.
1255 Rasor, C.A., 1939. Manganese mineralization at Tombstone, Arizona. Econ. Geol., 34, 790-804.
1256 Rasor, C.A., 1943. Bravoite from a new locality. Econ. Geol., 38, 399-407.
1257 Rasul, S.H., 1965. The manganese ores of Shivrajpur, Districht Panch Mahals, Gujarat State, India. Econ. Geol., 60, 149-162.
1258 Razenkova, N.I. and Poplavko, E.M., 1963. The occurrence of rhenium in the ores of the Dzhezkazgan deposit. Geochemistry (USSR; English Transl.), 1963, 808-815.
1259 Rensburg, W.C.J. van, and Liebenberg, L., 1967. Mackinawite from South Africa. Am. Mineralogist, 52, 1027-1035.
1260 Rentzsch, J., 1963. Zur Entstehung der Blei-Zink-Kupfer-Lagerstätten in triassischen Karbonatgesteinen des Nortwestbalkans. Freiberger Forschungsh., C,166, 1-102.
1261 Rentzsch, J. and Starke, R., 1961. Silberkiese in Galenit des fluobarytischen Bleierzformation im Brander Revier (Freiberger Erzbezirk). Bergakademie, 13, 349-350. Ref. in: Zentr. Mineral., I, 1961, 446.
1262 Ribeiro Filho, E., 1966. Jacobsita de Licínio de Almeida, Bahia. Bol. Soc. Brasil. Geol., 15, 43-48. Ref. in: Mineral. Abstr., 20, 55.
1263 Richards, S.M., 1966. Mineragraphy of fault-zone sulphides, Broken Hill, N.S.W. Commonwealth Sci. Ind. Res. Org., Mineragr. Invest. Tech. Papers, 1966, No. 5.
1264 Richmond, W.E., 1936. Crystallography of livingstonite. Am. Mineralogist, 21, 719-720.
1265 Richmond, W.E. and Fleischer, M., 1942. Cryptomelane, a new name for the commonest of the psilomelane minerals. Am. Mineralogist, 27, 607-610.
1266 Richmond, W.E., Fleischer, M. and Mrose, M.E., 1969. Studies on manganese oxide minerals. IX. Rancieite. Bull. Soc. Franç. Minéral. Crist., 92, 191-195.
1267 Ridland, G.C., 1941. Mineralogy of the Negus and Con mines, Yellowknife, Nw. Territories, Canada. Econ. Geol., 36, 45-70.
1268 Riley, J.F., 1965. An intermediate member of the binary system FeS_2 (pyrite) - CoS_2 (cattierite). Am. Mineralogist, 50, 1083-1086.
1269 Rimskaya-Korsakova, O.M. and Troyanov, M.D., 1956. New data relating to tungstenite. (in Russian). Zap. Vses. Mineralog. Obshchestva, 85, 277-285. Ref. in: Mineral. Abstr., 13, 651.
1270 Roberts, W.M.B., 1960. Mineralogy, and genesis of White's orebody, Rum Jungle uranium field, Australia. Neues Jahrb. Mineral., Abhandl., 94, 868-889.
1271 Robinson, S.C., 1948a. The identity of falkmannite and yenerite with boulangerite. Am. Mineralogist, 33, 716-723.
1272 Robinson, S.C., 1948b. Synthesis of lead sulphantimonites. Econ. Geol., 43, 293-312.
1273 Robinson, S.C., 1948c. Studies of mineral sulpho-salts: XIV - Artificial sulphantimonites of lead. Univ. Toronto, Geol. Ser., 52, 54-70.
1274 Robinson, S.C., 1949. Owyheeite. Am. Mineralogist, 34, 398-402.
1275 Robinson, S.C., 1955. Mineralogy of uranium deposits, Goldfields, Saskatchewan. Geol. Surv. Can., Dept. Mines Tech. Surv., Bull. 31, 128 pp.
1276 Robinson, S.C. and Brooker, E.J., 1952. A Co-Ni-Cu selenide from the Goldfields District, Saskatchewan. Am. Mineralogist, 37, 542-544.
1277 Robinson, S.C., Evans, H.T., Jr., Schaller, W.T. and Fahey, J.J., 1957. Nolanite, a new iron-vanadium mineral from Beaverlodge, Saskatchewan. Am. Mineralogist, 42, 619-628.
1278 Robinson, S.C. and Sabina, A.P., 1955. Uraninite and thorianite from Quebec and Ontario. Am. Mineralogist, 40, 624-633.
1279 Roedder, E., 1968. The noncolloidal origin of "colloform" textures in sphalerite ores. Econ. Geol., 63, 451-471.
1280 Roedder, E. and Dwornik, E.J., 1968. Sphalerite color banding: lack of correlation with iron content, Pine Point, Northwest Territories, Canada. Am. Mineralogist, 53, 1523-1529.

1281   Roger, G., 1969. Sur la minéralogie et le mode de gisement des filons à antimoine du district de Brioude-Massiac (Haute-Loire, Cantal), Massif Central français. Bull. Soc. Franç. Minéral. Crist., 92, 76-85.
1282   Rogers, A.F., 1947. Uraninite and pitchblende. Am. Mineralogist, 32, 90-91.
1283   Roland, G.W., 1968. The system Pb-As-S. Composition and stability of jordanite. Mineral. Deposita, 3, 249-260.
1284   Roland, G.W., 1968b. Synthetic trechmannite, Am. Mineralogist, 53, 1208-1214.
1285   Rösch, H., 1963. Zur Kristallstruktur des Gratonits. Neues Jahrb. Mineral., Abhandl., 99, 307-337.
1286   Roseboom, E.H., Jr., 1962a. Skutterudites $(Co,Ni,Fe)As_{3-x}$: composition and cell-dimensions. Am. Mineralogist, 47, 310-327.
1287   Roseboom, E.H., Jr., 1962b. Djurleite, $Cu_{1.96}S$, a new mineral. Am. Mineralogist, 47, 1181-1184.
1288   Roseboom, E.H., Jr., 1963. Co-Ni-Fe diarsenides: composition and cell dimensions. Am. Mineralogist, 48, 271-299.
1289   Roseboom, E.H., Jr., 1966. An investigation of the system Cu-S and some natural copper sulphides between 25° and 700°C. Econ. Geol., 61, 641-672.
1290   Ross, C.S., 1937. Sphalerite from a pegmatite near Spruce Pine, North Carolina. Am. Mineralogist, 22, 643-650.
1291   Rouse, R.C. and Peacor, D.R., 1968. The relationship between senaite, magnetoplumbite, and davidite. Am. Mineralogist, 53, 869-879.
1292   Rowland, J.F. and Berry, L.G., 1951. The structural lattice of hessite. Am. Mineralogist, 36, 471-479.
1293   Roy, S., 1959. Mineralogy and texture of the manganese ore bodies of Dongari Buzurg, Bhandara District, Bombay State, India, with a note on their genesis. Econ. Geol., 54, 1556-1574.
1294   Roy, S., 1959b. Variation in etch behaviour of jacobsite with different cell dimension. Nature, 183, 1256-1257.
1295   Roy, S., 1960. Mineralogy and texture of the manganese ores of Kodur, Srikakulam District, Andhra Pradesh, India. Can. Mineralogist, 6, 491-503.
1296   Roy, S., 1962. Study of the metamorphic manganese ores of Bharweli Mine-area, Madhya Pradesh, India, and their genesis. Econ. Geol., 57, 195-208.
1297   Roy, S., 1964. Manganese ores, Andhra Pradesh, India. Econ. Geol., 59, 154-157.
1298   Roy, S., 1966. Syngenetic manganese formations of India. Jadavpur University Edition, 219 pp.
1299   Roy, S., 1968. Mineralogy of different genetic types of manganese deposits. Econ. Geol., 63, 760-786.
1299a Rozhkov, I.S., Kitsul, V.I., Razin, L.V. and Borishanskaya, S.S., 1962. Platinum of the Aldan Shield (in Russian). Akad. Nauk SSSR, Sibirsk. Otdel., Yakutsk, Filial, Moskva. Ref. in: Leonard et al., 1969.
1300   Rucklidge, J.C., 1967. Electron probe studies of some Canadian telluride minerals. Can. Mineralogist, 9, 305.
1301   Rucklidge, J., 1968. Electron probe data for some Canadian platinum metal minerals. Can. Mineralogist, 9, 580.
1302   Rucklidge, J., 1969a. Electron microprobe investigations of platinum metal minerals from Ontario. Can. Mineralogist, 9, 617-628.
1303   Rucklidge, J., 1969b. Frohbergite, montbrayite, and a new Pb-Bi telluride. Can. Mineralogist, 9, 709-726.
1304   Rucklidge, J.C. and Stumpfl, E.F., 1968. Changes in the composition of petzite (Ag_3AuTe_2) during analysis by electron probe. Neues Jahrb. Mineral., Monatsh., 1968, 61-68.
1305   Ruiz Elizondo, J., Avila Ibarra, G., Cano Corona, O. and Ayala Rojas, G., 1965. Nota preliminar sobre la identificación por rayos X, de óxido tálico Tl_2O_3, como mineral en minerales de Vizarrón, Municipia de Cadereyta, Querétaro. Univ. Nac. Autónoma Méx., Inst. Geol., Bol. 76, 71-80.
1306   Rumyantsev, G.S., 1965. Composition and properties of minerals of the magnetite-jacobsite series redetected in the Magnetitovoye deposit (Buryat ASSR). Dokl. Akad. Nauk SSSR, 164, 1143-1146. Translated in: Dokl. Acad. Sci. USSR, Earth Sci. Sect., 164, 135-138 (1966).
1307   Runnells, D.D., 1969. The mineralogy and sulfur isotopes of the Ruby Creek Copper Prospect, Bornite, Alaska. Econ. Geol., 64, 75-90.
1308   Russel, H.D., Hiemstra, S.A. and Groeneveld, D., 1954. The mineralogy and petrology of the carbonatite at Loolekop, Eastern Transvaal. Trans. Geol. Soc. S. Africa, 57, 197-208.

1309 Rust, G.W., 1940. Geologic occurrence of gratonite at Cerro de Pasco, Peru. Am. Mineralogist, 25, 266-270.
1310 Saager, R., 1968. Newly observed ore-minerals from the Basal Reef in the Orange Free State goldfield in South Africa. Econ. Geol., 63, 116-123.
1311 Saager, R. and Mihálik, P., 1967. Two varieties of pyrite from the Basal Reef of the Witwatersrand System. Econ. Geol., 62, 719-731.
1312 Saari, E., Knorring, O. von, and Sahama, T.G., 1968. Niobian wolframite from the Nuaparra Pegmatite, Zambezia, Mozambique. Lithos, 1, 164-168.
1313 Safiannikoff, A., 1959. Un nouveau mineral de niobium. Acad. Roy. Sci. Outre-Mer (Brussels), Bull. Séance, 5, 1251-1255. Ref. in: Am. Mineralogist, 46, 1004.
1314 Safiannikoff, A. and Wambeke, L. van, 1961. Sur un terme plombifère du groupe pyrochlore-microlite. Bull. Soc. Franç. Minéral. Crist., 84, 382-384.
1315 Sağiroğlu, G.N., 1956. Sur quelques gisements de scheelite en Turquie. Berg- und Hüttenmänn. Monatsh. Montan. Hochschule Leoben, 101, 32-34. Ref. in: Zentr. Mineral., II, 1959, 670.
1316 Sakharova, M.S., 1955. Bismuth sulphosalts of the Ustarasaisk deposit. (in Russian). Tr. Mineralog. Muzeya, Akad. Nauk SSSR, 7, 112-116. Ref. in: Am. Mineralogist, 41, 814.
1317 Saksela, M., 1947. Ueber eine antimonreiche Paragenese in Ylöjärvi, SW-Finland. Bull. Comm. Géol. Finlande, 140, 199-222.
1318 Saksela, M., 1952. Das Fahlerzvorkommen von Seinäjoki, Südostbothnien, Finnland. Bull. Comm. Géol. Finlande, 157, 41-52.
1319 Saksela, M., 1960. Beiträge zur Kenntnis der sog. chloritischen Kupferformationen im fennoskandischen Grundgebirge. Neues Jahrb. Mineral., Abhandl., 94, 319-351.
1320 Samoyloff, V., 1934. The Llallagua - Uncia tin deposits. Econ. Geol., 29, 481-499.
1321 Sampson, E., 1941. Notes on the occurrence of gudmundite. Econ. Geol., 36, 175-184.
1322 Sampson, E. and Hriskevich, M.E., 1957. Co-As minerals associated with aplites at Cobalt, Ontario. Econ. Geol., 52, 60-75.
1323 Sandefur, B.T., 1942. The geology and paragenesis of the nickel ores of the Guniptau mine, Goward Nipissing District, Ontario. Econ. Geol., 37, 173-187.
1324 Santos, E.S., 1968. Reflectivity and microidentation hardness of ferroselite from Colorado and New Mexico. Am. Mineralogist, 53, 2075-2077.
1324a Sarkar, S.C. and Deb, M., 1969. Tetradymite and wehrlite from Singhbhum Copper-Belt, India. Mineral. Mag., 37, 423-425.
1325 Satpayev, K.I., 1965. Rhenium in the Dzhezkazgan ores. Geochemistry Intern., 2, 728 (1965).
1326 Satpayeva, T.A., 1959. Betechtinit in Erzen der Lagerstätte Dzhezkazgan. (in Russian). Izv. Akad. Nauk Kaz. SSR, Ser. Geol., 1959, 95-103. Ref. in: Zentr. Mineral., I, 1961, 270.
1327 Satpayeva, T.A., Safargaliev, G.S., Polyakova, T.P., Satpaeyva, M.K., Marzuvanov, V.L. and Eursova, M.Z., 1964. A new complex sulphide in the ores of the Dzhezkazgan deposits. (in Russian). Izv. Akad. Nauk Kaz. SSR, Ser. Geol., 1964, 29-41. Ref. in: Am. Mineralogist, 50, 809-810, Mineral. Abstr., 1965, 17, 302-303.
1328 Sattran, V., 1958. Polymetallische Vererzung im östlichen Erzgebirge. (in Czech, German summary). Sb. Ustřed. Ustavu Geol., 25, oddfl geol., 135-185.
1329 Schachner-Korn, D., 1960. Bravoitführende Blei-Zinkvererzungen im Devon und Buntsandstein der Nordeifel. Neues Jahrb. Mineral., Abhandl., 94, 469-478.
1330 Schachner-Korn, D. and Springer, G., 1967. Kobalt- und nickelhaltiger Markasit. Neues Jahrb. Mineral., Monatsh., 1967, 152-154.
1331 Schaller, W.T. and Vlisidis, A.C., 1962. Ludwigite from the Read magnetite deposit, Stevens County, Washington. Econ. Geol., 57, 950-953.
1332 Schempp, C.A. and Schaller, W.T., 1931. Sulvanite from Utah. Am. Mineralogist, 16, 557-562.
1333 Scherbina, A.S., 1941. Die goldreiche Kupferglanz-Bornit-Lagerstätte von Glava in Värmland, Schweden, und ihre geologische Stellung. Neues Jahrb. Mineral., Beil., 76 A, 377-458.
1334 Schermerhorn, L.J.G., 1956. Igneous, metamorphic and ore geology of the Castro Dairo - São Pedro do Sul - Sátão region. Comm. Serv. Geol. Port., 37, 617 pp.
1335 Schidlowski, M., 1965. Ein bemerkenswerte Verwachsung von Uranpecherz und Gold aus dem Witwatersrand-Konglomeraten. Naturwissenschaften, 52, 11-12.
1336 Schidlowski, M., 1966. Beiträge zur Kenntnis der radioaktiven Bestandteile der Witwatersrand-Konglomerate. II. Brannerit und "Uranpecherzgeister". Neues Jahrb. Mineral., Abhandl., 105, 310-324.

1337 Schidlowski, M., 1967. Note on graphite in the Witwatersrand conglomerates. Trans. Geol. Soc. S. Africa, 70, 65-66.
1338 Schidlowski, M. and Ottemann, J., 1966. Mackinawite from the Witwatersrand conglomerates. Am. Mineralogist, 51, 1535-1541.
1339 Schneiderhöhn, H., 1929. The mineragraphy, spectrography and genesis of the platinumbearing nickel-pyrrhotite ores of the Bushveld igneous complex. (in: P.A. Wagner, The platinum deposits and mines of South Africa, p. 206-246).
1340 Schneiderhöhn, H. and Ramdohr, P., 1931. Lehrbuch der Erzmikroskopie, Bd. II. Berlin.
1341 Schoeller, W.R. and Powell, A.R., 1920. Villamaninite, a new mineral. Mineral. Mag., 19, 14-18.
1342 Scholtz, D.L., 1936. The magmatic nickeliferous ore deposits of East Griqualand and Pondoland. Trans. Geol. Soc. S.Africa, 39, 81-210.
1342a Schot, E.H. and Ottemann, J., 1969. Elektrum und Kobellit im Meliert-Erz von Rammelsberg. Neues Jahrb. Mineral., Abhandl., 112, 101-115.
1343 Schouten, C., 1928. Mineragrafisch onderzoek van goudertsen van Lebong Bahroe en Tandaiberg. Verhandel. Koninkl. Geol. Mijnbouwk. Genoot. Ned. Kol., Mijnbouwk. Ser., II, 161-233.
1344 Schouten, C., 1934. Structures and textures of synthetic replacements in "open space". Econ. Geol., 29, 611-658.
1345 Schouten, C., 1937. Metasomatische Probleme. Diss. Amsterdam.
1346 Schouten, C., 1946a. Some notes on micro-pseudomorphism. Econ. Geol., 41, 348-382.
1347 Schouten, C., 1946b. Synthetic replacements as an aid to ore-genetic studies. Econ. Geol., 41, 659-667.
1348 Schouten, C., 1962. Determination tables for ore microscopy. Elseviers Publishing Company, Amsterdam.
1349 Schrader, R. and Petzold, D., 1967. Beitrag zum natürlichen Vorkommen der Phase CoOOH. Neues Jahrb. Mineral., Monatsh., 1967, 215-223.
1350 Schreiter, P., 1967. Ueber Erzgehalt und magnetische Suszeptibilität basaltischer Gesteine am Beispiel des Nephelinbasanites vom Steinberg bei Ostritz/Lausitz. Chem. Erde, 26, 151-187.
1351 Schröder, A., 1952. Der Elementarkörper und die Dichte des Ramsdellite, MnO_2. Fortschr. Mineral., 31, 11.
1352 Schüller, A., 1960. Zur Kenntnis des Betechtinit. Neues Jahrb. Mineral., Monatsh., 1960, 121-131.
1353 Schüller, A. and Ottemann, J., 1963. Castaingit, ein neues mit Hilfe der Elektronen-Mikrosonde bestimmtes Mineral aus den Mansfelder "Rücken". Neues Jahrb. Mineral., Abhandl., 100, 317-321.
1354 Schüller, A. and Wohlmann, E., 1955. Betechtinit, ein neues Blei-Kupfer-Sulfid aus den Mansfelder Rücken. Geologie (Berlin), 4, 535-555.
1355 Schulz, H., 1964. Beobachtungen an westerzgebirgischen Quecksilbervorkommen. Z. Angew. Geologie, 10, 588-592.
1356 Schürenberg, H., 1963. Ueber iranische Kupfererzvorkommen mit komplexen Co-Ni-Erzen. Neues Jahrb. Mineral., Abhandl., 99, 200-230.
1357 Schwartz, G.M., 1931a. Intergrowths of bornite and chalcopyrite. Econ. Geol., 26, 186-201.
1358 Schwartz, G.M., 1931b. Textures due to unmixing of solid solutions. Econ. Geol., 26, 739-763.
1359 Schwartz, G.M., 1932. Microscopic criteria of hypogene and supergene origin of ore minerals. Econ. Geol., 27, 533-553.
1360 Schwartz, G.M., 1934. Paragenesis of the oxidized ores of copper. Econ. Geol., 29, 55-75.
1361 Schwartz, G.M., 1935. Relations of chalcocite-stromeyerite-argentite. Econ. Geol., 30, 128-146.
1362 Schwartz, G.M., 1937a. The paragenesis of pyrrhotite. Econ. Geol., 32, 31-55.
1363 Schwartz, G.M., 1937b. Paragenesis of iron sulphides in a Black Hill deposit. Econ. Geol., 32, 810-825.
1364 Schwartz, G.M., 1939. Significance of bornite-chalcocite micro-textures. Econ. Geol., 34, 399-418.
1365 Schwartz, G.M., 1942. Progress in the study of exsolution in ore minerals. Econ. Geol., 37, 345-364.
1366 Schwartz, G.M., 1944. The host minerals of native gold. Econ. Geol., 39, 371-411.

1366a Schwartz, G.M., 1951. Classification and definitions of textures and mineral structures in ores. Econ. Geol., 46, 578-591.
1367   Schwartz, G.M. and Park, C.F., Jr., 1932. A microscopic study of ores from the Campbell mine, Bisbee, Arizona. Econ. Geol., 27, 39-51.
1368   Schwartz, G.M. and Ronbeck, A.C., 1940. Magnetite in sulphide ores. Econ. Geol., 35, 585-610.
1369   Schwartz, M. von, 1937. Gefügeausbildung der terrestrischen Eiseneinschlüsse im Basalt vom Bühl bei Kassel. Zentr. Mineral., A, 74-91.
1370   Sclar, C.B. and Drovenik, M., 1960. Lazarevićite, a new cubic copper-arsenic sulphide from Bor, Jugoslavia. Bull. Geol. Soc. Am., 71, 1970.
1371   Sclar, C.B. and Geier, B.H., 1957. The paragenetic relationship of germanite and renierite from Tsumeb, South West Africa. Econ. Geol., 52, 612-631.
1372   Seeliger, E., 1952. Die Mineralparagenesen einiger Bohrungen im Muschelkalk südlich und südwestlich von Wiesloch in Baden. Fortschr. Mineral., 31, 14-17.
1373   Seeliger, E., 1954. Ein neues vorkommen von Hutchinsonit in Wiesloch in Baden. Neues Jahrb. Mineral., Abhandl., 86, 163-178.
1374   Seeliger, E., 1956. Untersuchungen an Quecksilbererzen der Pfalz. Fortschr. Mineral., 34, 26-27.
1375   Seeliger, E. and Mücke, A., 1969. Donathit, ein tetragonaler, Zn-reicher Mischkristall von Magnetit und Chromit. Neues Jahrb. Mineral., Monatsh., 1969, 49-57.
1376   Seeliger, E. and Strunz, H., 1962. Reaktionen an Einschlüssen im Basalt vom Parkstein (Weiden, Oberpfalz). Erzmineralien vom Typus β-Korund (Freudenbergit, Högbomit). Chem. Erde, 22, 681-708.
1377   Seeliger, E. and Strunz, H., 1965. Erzpetrographie der Uran-Mineralien von Wölsendorf. II. Mitteilung: Brannerit, Lermontovit (?), Selen und Selenide, Niund Bi-Begleitmineralien etc. Neues Jahrb. Mineral., Abhandl., 103, 163-178.
1378   Segeler, C.G., 1959. Notes on a second occurrence of groutite. Am. Mineralogist, 44, 877-878.
1379   Sekanina, M.J., 1937. Etude métallographique des minerais de Pribram (Tchécoslovaquie). Bull. Soc. Franç. Minéral. Crist., 60, 152-223.
1380   Semenov, E.I., Sørensen, H., Bessmertnaya, M.S. and Novorossova, L.E., 1967. Chalcothallite - a new sulphide of copper and thallium from the Ilímaussaq alkaline intrusion, South. Greenland. Medd. Grønland, 181. Nr. 5, II, 26 pp.
1381   Shadlun, T.N., 1961. Tellurobismuthite from the Uchalinski pyrite ore deposit, S. Urals. (in Russian). Zap. Vses. Mineralog. Obshchestva, 90, 294-296. Ref. in: Mineral. Abstr., 16, 618.
1382   Shadlun, T.N., 1965. Ueber kombinierte röntgenometrische und mikrospektralanalytische Bestimmung kleinster Körner von Erzmineralien. Neues Jahrb. Mineral., Abhandl., 102, 115-122.
1382a Vyalsov, L.N., 1970. Private communication.
1383   Shannon, E.V., 1926. The identity of carrollite with linnaeite. Am. J. Sci., 11, 489-493.
1384   Shenon, P.J., 1932a. A massive sulphide deposit of hydrothermal origin in serpentine. Econ. Geol., 27, 596-613.
1385   Shenon, P.J., 1932b. Chalcopyrite and pyrrhotite inclusions in sphalerite. Am. Mineralogist, 17, 514-518.
1386   Shibuya, G., 1959. On the single crystals of jacobsite from Sannotaké, Fukuoka Pref., and their oxydation by heating. J. Mineral. Soc. Japan, 4, 157-176. Ref. in: Zentr. Mineral., I, 1960, 466.
1387   Shishkin, N.N., 1957. On a nickel variety of cobaltite. (in Russian). Dokl. Akad. Nauk SSSR, 114, 414-415. Ref. in: Mineral. Abstr., 14, 53.
1388   Shishkin, N.N., 1965. Not an orthorhombic variety of cobaltite but a high-cobalt glaucodot. (in Russian). Zap. Vses. Mineralog. Obshchestva, 94, 477-481. Ref. in: Mineral. Abstr., 17, 689.
1389   Shnaĭder, M.S. and Shnaider, A.A., 1966. The occurrence of realgar in sulphide ores of Novo-Zolotushiskii and Kamyshinskii (Rudny Altai) deposits. (in Russian). Zap. Vses. Mineralog. Obshchestva, 95, 108-109. Ref. in: Mineral. Abstr., 18, 249.
1390   Shnyukov, Ye.F., 1963. Realgar in Kerch iron ores. Dokl. Akad. Nauk SSSR, 150, 1347-1348. Translated in: Dokl. Acad. Sci. USSR, Earth Sci. Sect., 150, 129-131 (1965).
1391   Short, M.N., 1937. Etch tests on calaverite, krennerite and sylvanite. Am. Mineralogist, 22, 667-674.

1392   Short, M.N., 1948. Microscopic determination of the ore minerals. 2nd.Ed. U.S., Geol. Surv., Bull., 914, repr. 1948 with corrections and revisions.
1393   Shulhof, W.P. and Wright, H.D., 1959. Unusual galena from the Boulder batholith, Montana. Am. Mineralogist, 44, 1096-1098.
1394   Silberman, Ya.R., Baljter, B.L. and Erenburg, B.G., 1958. Der Kobaltglanz aus der Lagerstätte Wladimirowskoe im Gornyj Altai. (in Russian). Vest. Zapadno Sibirsk Novosib. Geol. Upr., 1958, 52-56. Ref. in: Zentr. Mineral., I, 1961, 95, 96.
1394a   Sillitoe, R.H., 1969. Delafossite from Mina Remolinos Nuevo, Atacama province, northern Chile. Mineral. Mag., 37, 425-427.
1395   Silman, J.F.B., 1954. Native tin associated with pitchblende at Nesbitt La Bine uranium mine, Beaverlodge, Saskatchewan. Am. Mineralogist, 39, 529-531.
1396   Sims, P.K., Young, E.Y. and Sharp, W.N., 1961. Coffinite in uranium vein deposits of the Front Range, Colorado. U.S. Geol. Surv., Profess. Papers, 424-B, 3-5.
1397   Sindeeva, N.D., 1964. Mineralogy and types of deposits of selenium and tellurium. New York.
1398   Singh, D. Santokh, 1967a. Tables for the microscopic identification of tin minerals. International Tin Council, London, 20 pp.
1399   Singh, D. Santokh, 1967b. Some general aspects of tin minerals in Malaya. International Tin Council, London, 22 pp.
1400   Sinha Roy, S., 1968. Vredenburgite from the peroxide manganese ores of Dongri Buzurg, Maharashtra, India. Mineral. Mag., 36, 1034-1036.
1401   Sinyakov, V.I., 1966. Microhardness of magnetite from deposits of diverse origin. Dokl. Akad. Nauk SSSR, 169, 929-932. Translated in: Dokl. Acad. Sci. USSR, Earth Sci. Sect., 169, 145-147 (1967).
1402   Sitnin, A.A. and Bykova, A.V., 1962. The first specimen of microlite from granite. Dokl. Akad. Nauk SSSR, 147, 203-206. Translated in: Dokl. Acad. Sci. USSR, Earth Sci. Sect., 147, 141-144 (1964).
1403   Sitnin, A.A. and Leonova, T.I., 1961a. An occurrence of strüverite in exo-contact greisens of one of the granite massifs of Eastern Siberia. Dokl. Akad. Nauk SSSR, 137, 685-687. Translated in: Dokl. Acad. Sci. USSR, Earth Sci. Sect., 137, 447-448 (1962).
1404   Sitnin, A.A. and Leonova, T.N., 1961b. Loparite, a new accessory mineral of albitized and greisenized granites. Dokl. Akad. Nauk SSSR, 140, 1407-1410. Translated in: Dokl. Acad. Sci. USSR, Earth Sci. Sect., 140, 1090-1093 (1963).
1405   Skinner, B.J., 1966. The system Cu-Ag-S. Econ. Geol., 61, 1-26.
1406   Skinner, B.J., Erd, R.C. and Grimaldi, F.S., 1964. Greigite, the thio-spinel of iron, a new mineral. Am. Mineralogist, 49, 543-555.
1407   Skinner, B.J., Jambor, J.L. and Ross, M., 1966. McKinstryite, a new Cu-Ag sulfide. Econ. Geol., 61, 1383-1389.
1408   Skinner, B.J. and McBriar, E.M., 1958. Minium from Broken Hill, N.S.W. Mineral. Mag., 31, 947-950.
1409   Skvortsova, K.V., Sidorenko, G.A., Darà, A.D., Silanteva, N.I. and Mendoeva, M.M., 1964. Femolite, a new Mo-sulfide. (in Russian). Zap. Vses. Mineralog. Obshchestva, 93, 436-443. Ref. in: Am. Mineralogist, 50, 261, and Mineral. Abstr., 17, 181-182.
1410   Smirnov, F.L., 1959. Seltene Mineralien in den Kupferkieserzen einer Kupferkieslagerstätte im Nordkaukasus. (in Russian). Tr. Mineralog. Muzeya Akad. Nauk SSSR, 10, 171-179. Ref. in: Zentr. Mineral., I, 1962, 236-237.
1411   Smirnov, F.L. and Yakovlev, L.I., 1959. On germanite from the ores from one of the pyrite deposits of Central Kazakhstan. (in Russian). Tr. Mineralog. Muzeya Akad. Nauk SSSR, 10, 180-184. Ref. in: Mineral. Abstr., 14, 480, and Zentr. Mineral., I, 1962, 235.
1411a   Smith, D.G.W., 1969. Contributions to the mineralogy of Norway, No. 42. A reinvestigation of the pseudobrookite from Havredal (Bamble), Norway. Norsk. Geol. Tidsskr., 49, 284-288.
1412   Smith, F.G., Dasgupta, S.K. and Hill, V.G., 1957. Manganoan, ferroan wurtzite from Llallagua, Bolivia (I). Can. Mineralogist, 6, 128-135.
1413   Snetsinger, K.G., 1969. Manganoan ilmenite from a Sierran adamellite. Am. Mineralogist, 54, 431-436.
1414   Sobotka, J., 1954. Aurostibite, $AuSb_2$, in the ore veins of Krásná Hora and Milešov, first macroscopical occurrence. (in Czech). Rozpravy Česk. Akad. Věd., 64, 43-60. Ref. in: Mineral. Abstr., 12, 574.

1415 Sobotka, J., 1956. Chalcostibite, a new mineral for Czechoslovakia. (in Czech). Časopis Mineral. Geol., 1, 269. Ref. in: Mineral. Abstr., 14, 224.
1416 Sobotka, J., 1961. Beiträge zur topographischen Mineralogie des Erzgebietes von Jílové. Acta Univ. Carolinae, Geol., 193-227.
1417 Socolescu, M. and Superceanu, C., 1960. Neue Kupfervererzungen mit Domeykit-Algodonit-Whitneyit, bei Intregalde, im Siebenbürgischen Erzgebirge, Rumänien. (in Rumanian). Rev. Minelor. (Bucharest), 1960, 431-448. Ref. in: Zentr. Mineral., II, 1961, 538-540.
1418 Socolescu, M., Superceanu, C. and Vlaicu, Z., 1958. Die Kieslagerstätten von Kobåsel-Rodna Veche in Nordsiebenbürgen. (in Rumanian). Rev. Minelor., (Bucharest), 1958, 542-548. Ref. in: Zentr. Mineral., II, 1960, 333.
1419 Somanchi, S. and Clark, L.A., 1966. The occurrence of an Ag_6Sb phase at Cobalt, Ontario. Can. Mineralogist, 8, 610-619.
1420 Somina, M.Ya., 1966. Trigonal molybdenite from east Siberian carbonatites. Dokl. Akad. Nauk SSSR, 167, 898-901. Translated in: Dokl. Acad. Sci. USSR, Earth Sci. Sect., 167, 96-99 (1966).
1421 Sorem, R.K. and Cameron, E.N., 1960. Manganese oxides and associated minerals of the Nsuta manganese deposits, Ghana, West Africa. Econ. Geol., 55, 278-310.
1422 Sorem, R.K. and Gunn, D.W., 1967. Mineralogy of manganese deposits, Olympic Peninsula, Washington. Econ. Geol., 62, 22-56.
1423 Sorokin, Yu.P. and Melkozerov, Yu.N., 1965. Aikinite from the "Nevyanskaya-Seredovina" deposit. (in Russian). Zap. Vses. Mineralog. Obshchestva, 94, 693-697. Ref. in: Mineral. Abstr., 17, 689.
1424 Sosedko, A.F., 1958. Stibiotantalit aus einem Pegmatitgang im Norden der Kola-Halbinsel. (in Russian). Dokl. Akad. Nauk SSSR, 118, 1025-1026. Ref. in: Zentr. Mineral., I, 1959, 60.
1425 Spencer, L.J., 1928. Potarite, a new mineral discovered by the late Sir John Harrison in British Guiana. Mineral. Mag., 21, 397-406.
1426 Springer, G., 1968. Electronprobe analyses of mackinawite and valleriite. Neues Jahrb. Mineral., Monatsh., 1968, 252-258.
1427 Springer, G., 1968b. Electronprobe analyses of stannite and related tin minerals. Mineral. Mag., 36, 1045-1051.
1428 Springer, G., 1969a. Electronprobe analyses of tetrahedrite. Neues Jahrb. Mineral., Monatsh., 1969, 24-32.
1429 Springer, G., 1969b. Compositional variations in enargite and luzonite. Mineral. Deposita, 4, 72-74.
1429a Springer, G., 1969c. Naturally occurring compositions in the solid-solution series Bi_2S_3-Sb_2S_3. Mineral. Mag., 37, 294-296.
1429b Springer, G., 1969d. Microanalytical investigations into germanite, renierite, briartite and gallite. Neues Jahrb. Mineral., Monatsh., 1969, 435-441.
1430 Springer, G. and Demirsoy, S., 1969. Beitrag zur Klärung der Existenz von Klaprothit. Neues Jahrb. Mineral., Monatsh., 1969, 32-37.
1431 Springer, G. and Schachner-Korn, D., 1967. Electron-probe analyses of mackinawite and valleriite from Palabora and other localities. Trans. Inst. Min. Met., 76, B 230.
1432 Springer, G., Schachner-Korn, D. and Long, J.V.P., 1964. Metastable solid solution relations in the system FeS_2 - CoS_2 - NiS_2. Econ. Geol., 59, 475-491.
1433 Stankeev, E.A. and Aristov, V.V., 1958. Boulangerit in polymetallischen Erzen des Erzfeldes Algatschinskoe, Osttransbajkalien. (in Russian). Izv. V. Ysshikh Uchebn-Zavedenii, Geol. Razvedka, Nr. 8, 66-74. Ref. in: Zentr. Mineral., I, 1961, 273.
1434 Stanton, R.L., 1957. Studies of polished sections of pyrite, and some implications. Can. Mineralogist, 6, 87-118.
1435 Staples, A.B. and Warren, H.V., 1946. Minerals from the Highland-Bell silver mine, Beaverdell, British Columbia. Univ. Toronto, Geol. Ser., 50, 27-33.
1436 Staples, L.W., 1951. Ilsemannite and jordisite. Am. Mineralogist, 36, 609-614.
1437 Staritskii, Yu.G., 1965. Native iron and copper from the Kureika river. (in Russian). Zap. Vses. Mineralog. Obshchestva, 94, 580-582. Ref. in: Mineral. Abstr., 17, 723.
1438 Stearn, N.H., 1936. The cinnabar deposits in Southwestern Arkansas. Econ. Geol., 31, 1-28.
1439 Steclaci, L., 1961/2. Ueber das Germanit-Vorkommen in der Erzlagerstätte von Toroiaga-Baia Borsa. Rev. Geol. Geograph. (Bucharest), V, 313-316. Ref. in: Zentr. Mineral., II, 1963, 777.

1440 Steinike, K., 1959. Ueber optische Anomalien eines Magnetits von der Roteisenlagerstätte Pörnitz bei Schleiz. Neues Jahrb. Mineral., Monatsh., 1959, 169-174.
1440a Štemprok, M., 1953a. Die Methode des schnellen Nachweises von Schwefel in einigen natürlichen Sulfiden und Sulfosalzen. Z. Geol., 1, 453-461.
1440b Štemprok, M., 1953b. Elektrographische Methode in der Erzmikroskopie unter Anwendung des Cellophans. Chem. Erde, 16, 232-238.
1441 Stephens, M.M., 1931. Effect of light on polished surfaces of silver minerals. Am. Mineralogist, 16, 532-549.
1442 Stephens, M.M., 1935. The identification of types of chalcocite by use of the carbon arc. Econ. Geol., 30, 604-629.
1443 Stern, T.W., Stieff, L.R., Evans, H.T., Jr. and Sherwood, A.M., 1957. Doloresite, a new vanadium oxide mineral from the Colorado Plateau. Am. Mineralogist, 42, 587-593.
1444 Stevenson, J.S., 1933. Vein-like masses of pyrrhotite in chalcopyrite from the Waite-Ackerman-Montgomery mine, Quebec. Am. Mineralogist, 18, 445-449.
1445 Stevenson, J.S., 1943. Hypogene native arsenic from Criss Creek, British Columbia. Univ. Toronto, Geol. Ser., 48, 83-91.
1446 Stevenson, J.S., 1947. Geology of the Red Rose tungsten mine, Hazelton, British Columbia. Econ. Geol., 42, 433-464.
1447 Stevenson, J.S., 1951. Uranium mineralization in British Columbia. Econ. Geol., 46, 353-366.
1448 Stevenson, J.S. and Jeffery, W.G., 1964. Colloform magnetite in a contact metasomatic iron deposit, Vancouver Island, British Columbia. Econ. Geol., 59, 1298-1305.
1449 Stieff, L.R., Stern, T.W. and Sherwood, A.M., 1956. Coffinite, a uranous silicate with hydroxyl substitution, a new mineral. Am. Mineralogist, 41, 675-688.
1450 Stillwell, F.L., 1926. Observations on the mineral constitution of the Broken Hill lode. Australasian Inst. Mining Met. Proc., 64, 1-76.
1451 Stillwell, F.L., 1931a. Stannite ore from Oonah mine, Zeehan, Tasmania. Australasian Inst. Mining Met. Proc., 81, 1-7.
1452 Stillwell, F.L., 1931b. The occurrence of telluride minerals at Kalgoorlie. Australasian Inst. Mining Met. Proc., 84, 115-190.
1453 Stillwell, F.L., 1933. The occurrence of gold in King Cassilis ore. Australasian Inst. Mining Met. Proc., 90, 227-236.
1454 Stillwell, F.L., 1934. Observations on the zinc-lead lode at Rosebery, Tasmania. Australasian Inst. Mining Met. Proc., 94, 43-67.
1455 Stillwell, F.L., 1935. An occurrence of gersdorffite in North-East Dundas, Tasmania. Australasian Inst. Mining Met. Proc., 100, 465-476.
1456 Stillwell, F.L., 1940. The occurrence of gold in the Broken Hill lode. Australasian Inst. Mining Met. Proc., 117, 23-28.
1457 Stillwell, F.L., 1944. The occurrence of tellurides at Norseman, Western Australia. Australasian Inst. Mining Met. Proc., 135, 1-8.
1458 Stillwell, F.L. and Edwards, A.B., 1939. Note on loellingite and the occurrence of cobalt and nickel in the Broken Hill lode. Australasian Inst. Mining Met. Proc., 114, 111-124.
1459 Stillwell, F.L. and Edwards, A.B., 1941. Coated gold from Cobar, N.S.W. Australasian Inst. Mining Met. Proc., 121, 1-10.
1460 Stillwell, F.L. and Edwards, A.B., 1942a. The mineragraphic investigation of mill products of lead-zinc ores. J. Council Sci. Industr. Research, 15, 161-174.
1461 Stillwell, F.L. and Edwards, A.B., 1942b. The mineral association of Tennant Creek gold ores. Australasian Inst. Mining Met. Proc., 126, 139-144.
1462 Stillwell, F.L. and Edwards, A.B., 1943. Mineral composition of the tin ores of Renison Bell, Tasmania. Australasian Inst. Mining Met. Proc., 131/132, 173-186.
1463 Stillwell, F.L. and Edwards, A.B., 1944. Cobaltite in the Broken Hill lode. Australasian Inst. Mining Met. Proc., 133, 21-27.
1464 Stillwell, F.L. and Edwards, A.B., 1945. The mineral composition of the Black Star copper ore body, Mount Isa, Queensland. Australasian Inst. Mining Met. Proc., 139, 149-159.
1465 Stillwell, F.L. and Edwards, A.B., 1946. An occurrence of submicroscopic gold in the Dolphin East lode, Fiji. Australasian Inst. Mining Met. Proc., 141, 31-46.
1466 Stillwell, F.L. and Edwards, A.B., 1951. Jacobsite from the Tamworth District of New South Wales. Mineral. Mag., 29, 538-541.
1467 Stoiber, R.E., 1940. Minor elements in sphalerite. Econ. Geol., 35, 501-519.

1468 Straczek, J.A., Horen, A., Ross, M. and Warshaw, C.M., 1960. Studies of the manganese oxides. IV. Todorokite. Am. Mineralogist, 45, 1174-1184.
1469 Strunz, H., 1943. Beitrag zum Pyrolusitproblem. Naturwissenschaften, 31, 89-91.
1470 Strunz, H., 1959. Tsumeb, seine Erze und Sekundärmineralien, insbesondere der zweiten Oxydationszone. Fortschr. Mineral., 37, 87-90.
1471 Strunz, H., 1961a. Epitaxis von Uraninit auf Columbit. Aufschluss, 12, 81-84. Ref. in: Zentr. Mineral., I, 1961, 282.
1472 Strunz, H., 1961b. "Chromrutil" von der Red Ledge Mine ist kein Rutil. Redledgeite. Neues Jahrb. Mineral., Monatsh., 1961, 107-111.
1473 Strunz, H., 1961c. Zur Kristallchemie des Uranminerals Davidit. Naturwissenschaften, 48, 597.
1474 Strunz, H., 1963. Redledgeit, eine TiO_2 - Einlagerungsstruktur analog Kryptomelan. Neues Jahrb. Mineral., Monatsh., 1963, 116-119.
1475 Strunz, H., 1966. Mineralogische Tabellen. 4. Auflage. Akademische Verlagsgesellschaft Geest und Portig K.-G., Leipzig, 560 pp.
1476 Strunz, H., Geier, B.H. and Seeliger, E., 1958. Gallit, $CuGaS_2$, das erste selbständige Galliummineral, und seine Verbreitung in den Erzen der Tsumeb- und Kipushi-Mine. Neues Jahrb. Mineral., Monatsh., 1958, 241-264.
1477 Strunz, H. and Seeliger, E., 1960. Erzpetrographie der primären Uranmineralien von Wölsendorf. Erste Feststellung von Coffinit auf einer Uranlagerstätte Mitteleuropas. Neues Jahrb. Mineral., Abhandl., 94, 681-719.
1478 Strunz, H. and Seeliger, E., 1961. Ueber das Uran-Erz Coffinit von Wölsendorf. Aufschluss, 12, 353-359.
1479 Strunz, H., Seeliger, E. and Tennyson, Ch., 1961. Mineralien aus der Fürstenzeche im Bayerischen Wald. Aufschluss, 12, 145-151.
1480 Strunz, H. and Tennyson, Ch., 1961. Ueber den Columbit vom Hühnerkobel im Bayr. Wald und seine Uran-Paragenese. Aufschluss, 12, 313-324.
1481 Stumpfl, E.F., 1961. Some new platinoid-rich minerals, identified with the electron microanalyser. Mineral. Mag., 32, 833-847.
1482 Stumpfl, E.F., 1966. On the occurrence of native platinum with copper sulphides at Congo Dam, Sierra Leone. Overseas Geol., Mineral Resources (Gt. Brit.), 10, 1-10.
1483 Stumpfl, E.F. and Clark, A.M., 1964. A natural occurrence of Co_9S_8, identified by X-ray microanalysis. Neues Jahrb. Mineral., Monatsh., 1964, 240-245.
1484 Stumpfl, E.F. and Clark, A.M., 1965a. Hollingworthite, a new rhodium mineral, identified by electron probe microanalysis. Am. Mineralogist, 50, 1068-1074.
1485 Stumpfl, E.F. and Clark, A.M., 1965b. Electron-probe analysis of gold-platinoid concentrates from southeast Borneo. Trans. Inst. Mining Met., 74, 933-946.
1486 Stumpfl, E.F. and Rucklidge, J., 1968. New data on natural phases in the system Ag-Te. Am. Mineralogist, 53, 1513-1522.
1487 Sun, M.S. and Weege, R.J., 1959. Native selenium from Grants, New Mexico. Am. Mineralogist, 44, 1309-1311.
1488 Superceanu, C., 1957a. Ueber das neue Kobalterzvorkommen von Eibenthal im südlichen Banat. Neues Jahrb. Mineral., Monatsh., 1957, 149-165.
1489 Superceanu, C., 1957b. Neue Scheelitvorkommen in der Erzlagerstätte von Baia Sprie (Felsöbanya), Nordsiebenbürgen. Neues Jahrb. Mineral., Monatsh., 1957, 165-173.
1489a Sutherland, J.K. and Boorman, R.S., 1969. A new occurrence of roquesite at Mount Pleasant, New Brunswick. Am. Mineralogist, 54, 1202-1203.
1490 Sveshnikova, O.L. and Rakcheyev, A.D., 1965. Owyheeite from the Darasun ore fields (Eastern Transbaikal). Dokl. Akad. Nauk SSSR, 165, 1164-1167. Translated in: Dokl. Acad. Sci. USSR, Earth Sci. Sect., 165, 144-146 (1966).
1490a Swjaginzeff, O.E., see also Zvyagintsev, O.E.
1491 Swjaginzeff, O.E., 1932. Ueber das Osmiridium. Abhandlung I. Z. Krist., A 83, 172-186.
1492 Swjaginzeff, O.E. and Brunowski, B.K., 1932. Ueber das Osmiridium. Abhandlung II: Röntgenographische Untersuchung. Z. Krist., A 83, 187-192.
1493 Syritso, L.F. and Senderova, V.M., 1964. The problem of the existence of lillianite. (in Russian). Zap. Vses. Mineralog. Obshchestva, 93, 468-471. Ref. in: Am. Mineralogist, 50, 811 and Mineral Abstr., 17, 174-175.
1494 Sztrókay, K. von, 1944. Erzmikroskopische Beobachtungen an Erzen von Recsk (Mátra Bánya) in Ungarn. Neues Jahrb. Mineral., Abhandl., A 79, 104-128.

1495 Sztrókay, K., 1946. Ueber den Wehrlit (Pilsenit). Ann. Hist.-Naturales Musei Nationalis Hung., 39, 75-103.
1496 Sztrókay, K.I., 1952. Neue Beobachtungen an Ungarischen Erzmineralien (in Hungarian, German summary). Földt. Közl., 82, 1-3 sz.
1497 Sztrókay, K.I. and Balyi, K., 1953. Reflexionsmessung und theoretische Wertbestimmung an opaken Erzmineralien. Acta Geol., Acad. Sci. Hung., 2, 169-184.
1498 Taber, S., 1948. Gold crystals from the Southern Appalachians. Am. Mineralogist, 33, 482-488.
1499 Takasu, S., 1965. Silver tellurides from the Kawazu mine, Shizuoka Prefecture. J. Mineral. Soc. Japan, 7, 350-355. Min. Abstr., 18, 231.
1500 Takeuchi, T. and Nambu, M., 1957. On neodigenite in Japan. (Studies on the minerals of Cu-Fe-S series in Japan, first report). Sci, Rept. Tôhoku Univ., Third Ser., 5, 305-315.
1501 Takeuchi, T. and Nambu, M., 1958. On cubanite in Japan. (Studies on the minerals of Cu-Fe-S series in Japan, second report). Sci. Rept. Tôhoku Univ., Third Ser., 6, 1-10.
1502 Takeuchi, T. and Nambu, M., 1959. On valleriite in Northeast Japan. (Studies on the minerals of Cu-Fe-S series in Japan, fourth report.) Sci. Rept. Tôhoku Univ., Third Ser., 6, 323-329.
1503 Takeuchi, T. and Nambu, M., 1961. On idaite in Japan. Sci. Rept. Tôhoku Univ., Third Ser., 7, 189-198.
1504 Takéuchi, Y., 1956. The crystal structure of vonsenite. Mineral. J. (Tokyo), 2, 19-26. Ref. in: Zentr. Mineral., I, 1959, 136.
1505 Takéuchi, Y., Ghose, S. and Nowacki, W., 1965. The crystal structure of hutchinsonite. Z. Krist., 121, 321-348. Ref. in: Mineral. Abstr., 17, 459.
1506 Takéuchi, Y. and Nowacki, W., 1964. Detailed crystal structure of rhombohedral MoS_2 and systematic deduction of possible polytypes of molybdenite. Schweiz. Mineral. Petrog. Mitt., 44, 105-120.
1507 Talapatra, A.K., 1968. Sulfide mineralization associated with migmatization in the southeastern part of the Singhbhum shear zone, Bihar, India. Econ. Geol., 63, 156-165.
1508 Talmage, S.B., 1925. Quantitative standards for hardness of the ore minerals. Econ. Geol., 20, 531-553.
1509 Tavora, E., 1949. Simetria de kermesita. Anais Acad. Brasil. Cienc., 21, 75-100. Ref. in: Mineral. Abstr., 11, 421.
1510 Taylor, C.M. and Radtke, A.S., 1967. New occurrence and data of nolanite. Am. Mineralogist, 52, 734-743.
1511 Taylor, C.M. and Radtke, A.S., 1969. Micromineralogy of silver-bearing sphalerite from Flat River, Missouri. Econ. Geol., 64, 306-318.
1512 Taylor, K., Bowie, S.H.U. and Horne, J.E.T., 1962. Radioactive minerals in the Dominion Reef. Mining Mag., 107, 329-332.
1513 Taylor, L.A., 1969. The significance of twinning in Ag_2S. Am. Mineralogist, 54, 961-963.
1514 Temple, A.K., 1954. Rammelsbergite from the southern uplands of Scotland. Mineral. Mag., 30, 541-543.
1515 Temple, A.K., 1966. Alteration of ilmenite. Econ. Geol., 61, 695-714.
1516 Terada, K. and Cagle, F.W., Jr., 1960. The crystal structure of potarite (PdHg) with some comments on allopalladium. Am. Mineralogist, 45, 1093-1097.
1517 Terziev, G., 1966. Kostovite, an Au-Cu telluride from Bulgaria. Am. Mineralogist, 51, 29-36.
1518 Terziyev, G.I., 1966. The conditions of germanium accumulation in hydrothermal mineralization (as illustrated by a Bulgarian copper-pyrite deposit). Geochemistry Intern., 3, 341-346 (1966).
1519 Tex, E. den, 1955. Secondary alteration of chromite. Am. Mineralogist, 40, 353-355.
1520 Thayer, T.P., Milton, C., Dinnin, J. and Rose, H., Jr., 1964. Zincian chromite from Outokumpu, Finland. Am. Mineralogist, 49, 1178-1183.
1521 Thompson, M.E., Roach, C. and Braddock, W., 1956. New occurrences of native selenium. Am. Mineralogist, 41, 156-157.
1522 Thompson, R.M., 1946a. Goldfieldite - tellurian tetrahedrite. Univ. Toronto, Geol. Ser., 50, 77-79.
1523 Thompson, R.M., 1946b. Antamokite discredited. Univ. Toronto, Geol. Ser., 50, 79.
1524 Thompson, R.M., 1947a. Berthonite identical with bournonite. Univ. Toronto, Geol. Ser., 51, 81-83.

1525 Thompson, R.M., 1947b. Frohbergite, $FeTe_2$, a new member of the marcasite group. Am. Mineralogist, 32, 210.
1526 Thompson, R.M., 1948. Pyrosynthesis of telluride minerals. Am. Mineralogist, 33, 209-210.
1527 Thompson, R.M., 1949a. The telluride minerals and their occurrence in Canada. Am. Mineralogist, 34, 342-382.
1528 Thompson, R.M., 1949b. Goongarrite and warthaite discredited. Am. Mineralogist, 34, 459-460.
1529 Thompson, R.M., 1950a. Mineral occurrences in western Canada. Am. Mineralogist, 35, 451-455.
1530 Thompson, R.M., 1950b. The probable non-existence of alaskaite. Am. Mineralogist, 35, 456-457.
1531 Thompson, R.M., 1951. Mineral occurrences in western Canada. Am. Mineralogist, 36, 504-509.
1532 Thompson, R.M., 1953. Mineral occurrences in western Canada. Am. Mineralogist, 38, 545-549.
1533 Thompson, R.M., 1954a. Naumannite from Republic, Washington. Am. Mineralogist, 39, 525.
1534 Thompson, R.M., 1954b. Mineral occurrences in western Canada. Am. Mineralogist, 39, 525-528.
1535 Thompson, R.M., Peacock, M.A., Rowland, J.F. and Berry, L.G., 1951. Empressite and "stuetzite". Am. Mineralogist, 36, 458-470.
1536 Thomson, J.E., 1930. A mineragraphic study of the marcasite group. Univ. Toronto, Geol. Ser. 29, 75-83.
1537 Thomson, J.E., 1932. Mineralogy of the Eldorado mine, Great Bear Lake, N.W.T. Univ. Toronto, Geol. Ser., 32, 43-51.
1538 Thomson, J.E., 1934a. The mineralogy of the silver-uraninite deposits of Great Bear Lake, N.W.T. Univ. Toronto, Geol. Ser., 36, 25-31.
1539 Thomson, J.E., 1934b. Telluride ores at Straw Lake, Ontario, and Eureka mine, Quebec. Univ. Toronto, Geol. Ser., 36, 33-36.
1540 Thomson, J.E., 1935. Mineralization of the Little Long Lac and Sturgeon River areas. Univ. Toronto, Geol. Ser., 38, 37-45.
1541 Thomson, J.E., 1937. A review of the occurrence of tellurides in Canada. Univ. Toronto, Geol. Ser., 40, 95-101.
1542 Thomson, J.E., 1943. Boulangerite and columbite-tantalite from Manitoba. Univ. Toronto, Geol. Ser., 48, 103.
1543 Thoreau, J. and Du Trieu de Terdonck, R., 1933. Le gîte d'uranium de Shinkolobwe-Kasolo (Katanga). Mém. Inst. Colonial. Belge, I (8), 1-46.
1544 Thoreau, J., Gastellier, S. and Herman, P., 1950. Sur une thoreaulite du Ruanda Occidental. Ann. Soc. Géol. Belg., 73, Bull., 213-220.
1545 Threadgold, I.M., 1958a. Mineralization at the Morning Star gold mine, Wood's Point, Victoria. Australasian Inst. Mining Met. Proc., 185, 1-27.
1546 Threadgold, I.M., 1958b. Antimony-gold mineralization at Steel's Creek, near Yarra Glen, Victoria. Australasian Inst. Mining Met. Proc., Stillwell Anniv. Volume, 241-248.
1547 Threadgold, I.M., 1960. Mineral composition of some uranium ores from the South Alligator river area, Northern Territory. Commonwealth Sci. Ind. Res. Org., Mineragr. Invest. Tech. Papers, 1960, No. 2, 53 pp.
1548 Tikhonenkov, I.P. and Kazakova, M.E., 1957. Nioboloparite - a new mineral of the perovskite group. (in Russian). Zap. Vses. Mineralog. Obshchestva, 86, 641-644. Ref. in: Am. Mineralogist, 32, 792.
1549 Tikhonenkov, I.P. and Semenov, Ye.I., 1963. Arsenides of cobalt, nickel and iron in alkalic pegmatites. Dokl. Akad. Nauk SSSR, 150, 888-889. Translated in: Dokl. Acad. Sci. USSR, Earth Sci. Sect., 150, 121-122 (1965).
1550 Timofeyevskiy, D.A., 1967a. Sulphantimonites of the lead-silver owyheeite deposits of east Transbaikal. Zap. Vses. Mineralog. Obshchestva, 96, 30-44. Ref. in: Geologie (Berlin), 17, 479; and Mineral. Abstr., 20, 148-149.
1551 Timofeyevskiy, D.A., 1967b. First find of silver-rich freibergite in the USSR. Dokl. Akad. Nauk.SSSR, 176, 1388-1391. Translated in: Dokl. Acad. Nauk SSSR, Earth Sci. Sect., 176 (1968), 148-151.
1552 Tischendorf, G., 1959. Zur Genesis einiger Selenidvorkommen, insbesondere von Tilkerode im Harz. Freiberger Forschungsh., C 69, 168 pp.
1553 Tischendorf, G., 1960. Ueber Eskebornite von Tilkerode im Harz. Neues Jahrb. Mineral., Abhandl., 94, 1169-1182.

1554 Tokody, L., 1932. Ueber Hessit. Z. Krist., A 82, 154-157.
1555 Tokody, L., 1934. Berichtigung zu meiner Mitteilung "Ueber Hessit". Z. Krist., A 89, 416.
1556 Tokody, L., 1953. Proustit und Xanthokon von Baia Laposului (Láposbánya), Rumänien. (in German, Russ. summ.). Acta Geol., Acad. Sci. Hung., 2, 185-190.
1557 Tokody, L. and Vavrinecs, G., 1935. A vasköi ankerit és cosalit (Hungarian with German summary). Földt. Közl., 65, 301-305.
1558 Tolman, C.F. and Ambrose, J.W., 1934. The rich ores of Goldfield, Nevada. Econ. Geol., 29, 255-279.
1559 Tolun, R., 1954-'55. A study on the concentration tests and benefication of the Uludağ tungsten ore. Bull. Mineral Res. Exploration Inst. Turkey, Foreign Ed., 46-47, 106-127.
1560 Toubeau, G., 1962. Mesure de la microdureté Vickers des minéraux opaques et son intérêt en minéralogie. Bull. Soc. Belge Géol. Paléontol. Hydrol., 71, 242-262.
1561 Toulmin, P. the 3D., 1963. Proustite, pyrargyrite solid solutions. Am. Mineralogist, 48, 725-736.
1562 Traill, R.J., 1963. A rhombohedral polytype of molybdenite. Can. Mineralogist, 7, 524-526.
1563 Traill, R.J. and Boyle, R.W., 1955. Hawleyite, isometric cadmium sulphide, a new mineral. Am. Mineralogist, 40, 555-559.
1564 Trdlička, Z. and Kupka, F., 1957. Kobellite and native bismuth from the locality of Fichtenhübel, Slovakia. (in Czech, Engl. summary). Sb. Osmdesátinám Akad. Frentiška Slavika, 453-466.
1565 Trdlička, Z., Kvaček, M. and Kupka, F., 1963. Mineralogisch-chemische Erforschung des Kobellits aus den Sideritgängen des Fichtenhübel-Erzgebietes, CSSR. Sb. Národ. Musea Praze (Acta Musei Nationalis Pragae), XIX B, H. 3, 116-134. Ref. in: Zentr. Mineral., I, 1963, 238.
1566 Trdlička, Z. and Hoffman, V., 1967. Vorkommen von Kobellit in der Lagerstätte von Smolník in der Slowakei. (in Czech, German summary). Časopis Národního Musea (Prague), 136, 13-16. Ref. in: Zentr. Mineral. I, 1966, 262.
1567 Trojer, F.J., 1966. Refinement of the structure of sulvanite. Am. Mineralogist, 51, 890-894.
1568 Tufar, W., 1966. Bemerkenswerte Myrmekite aus Erzvorkommen vom Alpen-Ostrand. Neues Jahrb. Mineral., Monatsh., 1966, 246-252.
1569 Tufar, W., 1967. Der Bornit von Trattenbach (Niederösterreich). Neues Jahrb. Mineral., Abhandl., 106, 334-351.
1570 Tunell, G. and Ksanda, C.J., 1935. The crystal structure of calaverite. J. Wash. Acad. Sci., 25, 32-33. Ref. in: Mineral. Abstr., 6, 170.
1571 Tunell, G. and Ksanda, C.J., 1936a. The crystal structure of krennerite. J. Wash. Acad. Sci., 26, 507-509. Ref. in: Mineral. Abstr., 7, 81.
1572 Tunell, G. and Ksanda, C.J., 1936b. The strange morphology of calaverite in relation to its internal properties. J. Wash. Acad. Sci., 26, 509-528. Ref. in: Mineral. Abstr., 7, 81-82.
1573 Tunell, G. and Ksanda, C.J., 1937. The space group and unit cell of sylvanite. Am. Mineralogist, 22, 728-730.
1574 Tunell, G. and Murata, K.J., 1950. The atomic arrangement and chemical composition of krennerite. Am. Mineralogist, 35, 959-984.
1575 Tunell, G., Posnjak, E. and Ksanda, C.J., 1935. Geometrical and optical properties and crystal structure of tenorite. Z. Krist., A 90, 120-142.
1576 Turneaure, F.S., 1935a. The tin deposits of Llallagua, Bolivia, Part I. Econ. Geol., 30, 14-60.
1577 Turneaure, F.S., 1935b. The tin deposits of Llallagua, Bolivia, Part II. Econ. Geol., 30, 170-190.
1578 Turneaure, F.S., 1945. Tin deposits of Carguaicollo, Bolivia. Am. J. Sci., 243-A (Daly Volume), 523-541.
1579 Turnock, A.C., 1966. Synthetic wodginite, tapiolite and tantalite. Can. Mineralogist, 8, 461-470.
1580 Tweto, O., 1947. Scheelite in the Boulder District, Colorado. Econ. Geol., 42, 47-56.
1581 Ulrich, G.H.F., 1869. Observations on the "Nuggety Reef", Mount-Tarrangower gold-field. Quart. J. Geol. Soc. London, 25, 326-335.
1582 Uytenbogaardt, W., 1949. Names of ore minerals, disqualified by röntgenographic and microscopic research. Geol. Fören. Stockholm Förhandl., 71, 285-292.

1583   Uytenbogaardt, W., 1951. Tables for microscopic identification of ore minerals. Princeton, University Press.
1584   Uytenbogaardt, W., 1960. Uranium mineralization in the Västervik area. Intern. Geol. Congr., 21st, Copenhagen, Part XV, 114-122.
1585   Uytenbogaardt, W., 1967a. Results of Vickers Hardness measurements on ore minerals, achieved in different laboratories since 1956 (Second summary). Publication nr. 46 of the Department of Mineralogy and Petrology, Institute of Earth Science, Free University, 1085 De Boelelaan, Amsterdam-11, Netherlands.
1585a Uytenbogaardt, W., 1967b. Principles of micro-indentation hardness measurement. Publication nr. 45 of the Department of Mineralogy and Petrology, Institute of Earth Sciences, Free University, 1085 De Boelelaan, Amsterdam-11, Netherlands.
1586   Vaasjoki, O. and Kaitaro, S., 1951. "Lillianite" from Iilijärvi, Orijärvi Region. Bull. Comm. Géol. Finlande, 154, 123-126.
1587   Vaes, J.F., 1947. Quelques sulfures de Shinkolobwe. Ann. Soc. Géol. Belg. 70, Bull., 227-232.
1588   Vaes, J.F., 1948. La reniérite (anciennement appelée "bornite orange"). Un sulfure germanifère provenant de la Mine Prince-Léopold, Kipushi (Congo belge). Ann. Soc. Géol. Belg., 72, Bull., 19-32.
1589   Vähätalo, V.O., 1953. On the geology of the Outokumpu ore deposit in Finland. Bull. Comm. Géol. Finlande, 164, 98 pp.
1590   Van Tassel, R., 1959. Strengite, phosphosidérite, cacoxénite et apatite fibroradiée de Richelle. Bull. Soc. Belge Géol. Paléontol. Hydrol., 68, 360-368.
1591   Varček, C., 1965. Geokronit aus Nizná Slaná im Zips-Gömörer Erzgebirge. Geol. Sbornîk (Bratislava), 16, 175-184.
1592   Varček, C., Háber, M., Streško, V. and Samajová, E., 1968. Beziehungen zwischen Chemismus, Thermalität und physikalischen Eigenschaften von Sphaleriten. (in Czech, German summary). Acta Geol. Geograph. Univ. Comenianae, Geol., 13, 13-54.
1593   Vasilyev, V.I., 1963. A zinc-bearing variety of metacinnabarite, guadalcazarite, in the mercury ores of Gorno-Altai. Dokl. Akad. Nauk SSSR, 153, 676-678. Translated in: Dokl. Acad. Sci. USSR, Earth Sci. Sect., 153, 132-134 (1965).
1594   Vasilyev, V.I., 1965. Tetrahedrite as a source of secondary cinnabar in Gornyy Altai. Dokl. Akad. Nauk SSSR, 162, 901-904. Translated in: Dokl. Acad. Sci. USSR, Earth Sci. Sect., 162, 169-173 (1965).
1595   Vasilyev, V.I., 1966. Saukovite, a new zinc- and cadmium-bearing sulfide of mercury. Dokl. Akad. Nauk SSSR, 168, 182-185. Translated in: Dokl. Acad. Sci. USSR, Earth Sci. Sect., 168, 123-127 (1966).
1596   Vasilyev, V.I. and Berdichevskiy, G.V., 1967. Discovery of chalcostibite in mercury ores of Gornyy Altai. Dokl. Akad. Nauk SSSR, 175, 429-431. Translated in: Dokl. Acad. Sci. USSR, Earth Sci. Sect., 175, 124-127 (1968).
1596a Vasilyev, V.I. and Lavrentyev, Yu.G., 1968. A find of onofrite, a seleniferous species of black mercury sulfide, in mercury ore of the Altai-Sayan region. Dokl. Akad. Nauk SSSR, 182, 430-433. Translated in: Dokl. Acad. Sci. USSR, Earth Sci. Sect., 182, 119-122 (1969).
1597   Vasilyev, V.I. and Obolenskii, A.A., 1966. Geocronite in lead-zinc ores of the Tyute deposit, Gornyi Altai (in Russian). Zap. Vses Mineralog. Obshchestva, 95, 479-483. Ref. in: Mineral. Abstr., 18, 202.
1597a Vaughan, D.J., 1969a. Zonal variation in bravoite. Am. Mineralogist, 54, 1075-1083.
1597b Vaughan, D.J., 1969b. Nickelian mackinawite from Vlakfontein, Transvaal. Am. Mineralogist, 54, 1190-1193.
1598   Vaux, G., 1937. X-ray studies on pyrolusite (including polianite) and psilomelane. Mineral. Mag., 24, 521-526.
1599   Vaux, G. and Bannister, F.A., 1938. The identity of zinckenite and keeleyite. Mineral. Mag., 25, 221-227.
1600   Väyfynen, H., 1938. Petrologie des Nickelerzfeldes Kaulatunturi-Kammikivitunturi in Petsamo. Bull. Comm. Géol. Finlande, 116.
1601   Veen, A.H. van der, 1963. A study of pyrochlore. Verhandel. Koninkl. Ned. Geol. Mijnbouwk. Genoot., Geol. Ser., 22, 188 pp.
1602   Veen, A.H. van der, 1965. Calzirtite and associated minerals from Tapira, Brazil. Mineral. Mag., 35, 544-546.
1603   Veen, R.W. van der, 1925. Mineragraphy and ore-deposition, Vol. I. The Hague.
1604   Velde, D., 1968. A new occurrence of priderite. Mineral. Mag., 36, 867-870.

1605 Venerandi, I., 1965. Nuove osservazioni nel giacimento ferrifero della Nurra. Rend. Soc. Mineral. Ital., 21, 317-333.
1606 Vershkovskaya, O.V. and Lebedeva, S.I., 1968. Gudmundite from the Gorev deposit. Dokl. Akad. Nauk SSSR, 178, 424-427. Translated in: Dokl. Acad. Sci. USSR, Earth Sci. Sect., 178, 111-114 (1968).
1607 Viaene, W. and Moreau, J., 1968. Contribution à l'étude de la germanite, de la reniérite et de la briartite. Ann. Soc. Géol. Belg., 91, 127-143.
1608 Victor, I., 1957. Brunt Hill wolframite deposit, New Brunswick, Canada. Econ. Geol., 52, 149-168.
1609 Villiers, J.E. De, 1943. A preliminary description of the new mineral partridgeite. Am. Mineralogist, 28, 336-338.
1610 Villiers, P.R. De, and Herbstein, F.H., 1967. Distinction between two members of the braunite group. Am. Mineralogist, 52, 20-30.
1611 Villiers, P.R. De, and Herbstein, F.H., 1968. Second occurrence of marokite. Am. Mineralogist, 53, 495-496.
1612 Villwock, R., 1961. Ueber eine Paragenese von Cattierit, Melnikowit-Pyrit und Baryt in der Tonlagerstätte Kärlich (Neuwieder Becken). Neues Jahrb. Mineral., Abhandl., 95, 325-336.
1613 Vincent, E.A., 1960. Ulvöspinel in the Skaergaard Intrusion, Greenland. Neues Jahrb. Mineral., Abhandl., 94, 993-1016.
1614 Vincent, E.A. and Phillips, R., 1954. Iron-titanium oxide minerals in layered gabbros of the Skaergaard intrusion, East Greenland. Part I. Chemistry and ore-microscopy. Geochim. Cosmochim. Acta, 6, 1-26.
1615 Vlasov, K.A. (Editor), 1966. Geochemistry and mineralogy of rare elements and genetic types of their deposits. Volume II: Mineralogy of rare elements. (Translated from Russian). Israel Program for Scientific Translations, Jerusalem, 945 pp.
1616 Vokes, F.M., 1957a. Some copper sulphide parageneses from the Raipas formation of Northern Norway. Norg. Geol. Undersøk., 200, 74-111.
1617 Vokes, F.M., 1957b. On the presence of minerals of the linnaeite series in some copper ores from the Raipas formation of Northern Norway. Norg. Geol. Undersøk., 200, 112-120.
1618 Vokes, F.M., 1960. Contributions to the mineralogy of Norway. No. 7. Cassiterite in the Bleikvassli ore. Norsk Geol. Tidsskr., 40, 193-201.
1619 Vokes, F.M., 1962. Contributions to the mineralogy of Norway. No. 15. Gahnite in the Bleikvassli ore. Norsk Geol. Tidsskr., 42, 317-329.
1620 Vokes, F.M., 1967. Linnaeite from the Precambrian Raipas Group of Finnmark, Norway. An investigation with the electron microprobe. Mineral. Deposita, 2, 11-25.
1621 Vokes, F.M., Bergstøl, S. and Vralstad, T., 1969. Studies on some Norwegian ilmenites and pyrophanites. Third Annual Regional Conference, Copenhagen. Medd. Dansk Geol. Foren., 19, 330.
1622 Volborth, A., 1960. Gediegen Wismutantimon und andere Erzmineralien im Li-Be-Pegmatit von Viitaniemi, Eräjärvi, Zentralfinnland. Neues Jahrb. Mineral., Abhandl., 94, 140-149.
1623 Vorma, A., 1960. Laitakarite, a new Bi-Se-mineral. Bull. Comm. Géol. Finlande, 188, 1-10.
1624 Vorma, A., Kallio, P. and Meriläinen, K., 1966. Molybdenite 3 R from Inari, Finnish Lapland. Bull. Comm. Géol. Finlande, 222, 67-68.
1625 Vorma, A. and Siivola, J., 1967. Sukulaite - $Ta_2Sn_2O_7$ - and wodginite as inclusions in cassiterite in the granite pegmatite in Sukula, Tammela in SW Finland. Bull. Comm. Géol. Finlande, 229, 173-187.
1626 Vrabka, F. and Paděra, K., 1957. Geocronite from Bohutín near Příbram, Central Bohemia. (in Czech, Engl. summary). Sb. Osmdesátinám Akad. Františka Slavika, 497-502.
1627 Vujanović, V., 1954. Chalkophanit aus Janjevo bei Pristina (Serbien). Neues Jahrb. Mineral., Monatsh., 1954, 40-47.
1628 Vujanović, V., 1961. Sind Entmischungen von Sphalerit im Pyrrhotin der Grube Rudnik vorhanden? Neues Jahrb. Mineral., Monatsh., 1961, 209-215.
1629 Vultée, J. von, 1951. Die orientierte Verwachsungen der Mineralien. Fortschr. Mineral., 30, 297-378.
1630 Vuorelainen, Y. and Häkli, A., 1964. Eräistä uusista Nikkelimineraaleista. (in Finnish, Engl. summary). Eripainos Geol., lehdestä n:o 5/1964.

1631 Vuorelainen, Y., Huhma, A. and Häkli, A., 1964. Sederholmite, wilkmanite, kullerudite, mäkinenite and trüstedtite, five new nickel selenide minerals. Bull. Comm. Géol. Finlande, 215.
1632 Waal, S.A. de 1969a. Ramsdellite from the Hozatel mine, northern Cape Province, South Africa. Econ. Geol., 64, 221-223.
1632a Waal, S.A. de, 1969b. Nickel minerals from Barberton, South African: I. Ferroan trevorite. Am. Mineralogist, 54, 1204-1208.
1633 Wadsley, A.D., 1952. The crystal structure of lithiophorite. Acta Cryst., 5, 676-680.
1634 Wadsley, A.D., 1953. The crystal structure of psilomelane. Acta Cryst., 6, 433-438.
1635 Wadsley, A.D., 1955a. The crystal structure of chalcophanite. Acta Cryst., 8, 165-172.
1636 Wadsley, A.D., 1955b. Hydrohausmannite and hydrohetaerolite. Am. Mineralogist, 40, 349-353.
1637 Wagner, P.A., 1929. The platinum deposits and mines of South Africa. Oliver and Boyd, Edinburgh, 326 pp.
1638 Wahlstrom, E.E., 1950. Melonite in Boulder County, Colorado. Am. Mineralogist, 35, 948-953.
1639 Walenta, K., 1960. Natürliches Eisen (II)-oxyd (Wüstit) aus der vulkanischen Tuffbreccie von Scharnhausen bei Stuttgart. Neues Jahrb. Mineral., Monatsh., 1960, 150-159.
1640 Walker, T.L., 1923. Trevorite, a distinct mineral species. Univ. Toronto, Geol. Ser., 16, 53-54.
1641 Wambeke, L. van, 1954. Sur la présence de tétradymite (Bi_2Te_2S) dans les filons hydrothermaux de la tonalite de la Helle et sur une nouvelle texture de ce minerai. Bull. Soc. Belge Géol. Paléontol. Hydrol., 63, 260-276.
1642 Wambeke, L. van, 1968. A second occurrence of non-metamict davidite. Mineral. Deposita, 3, 178-181.
1643 Wang (Gi-Gyn) and Ye (Shu-kin), 1963. Biogenic pyrite and siegenite from Szechuan. Sci. Sinica (Peking), 12, 617-620. Ref. in: Mineral. Abstr., 16, 642.
1644 Warren, H.V., 1936. A gold-bismuth occurrence in British Columbia. Econ. Geol., 31, 205-211.
1645 Warren, H.V., 1939. An occurrence of cosalite in British Columbia. Univ. Toronto, Geol. Ser., 42, 151-155.
1646 Warren, H.V., 1946. Bismuth tellurides from the White Elephant Claim, British Columbia. Univ. Toronto, Geol. Ser., 50, 75-77.
1647 Warren, H.V., 1947. New occurrence of antimony and tellurium minerals in Western Canada. Univ. Toronto, Geol. Ser., 51, 71-78.
1648 Warren, H.V., 1948. Ore minerals from Western Canada. Univ. Toronto, Geol. Ser., 52, 82-86.
1649 Warren, H.V. and Davis, P., 1940. Some bismuth minerals from British Columbia. Univ. Toronto, Geol. Ser., 44, 107-111.
1650 Warren, H.V. and Lord, C.S., 1935. An occurrence of schwatzite in British Columbia. Econ. Geol., 30, 67-71.
1651 Warren, H.V. and Peacock, M.A., 1945. Hedleyite, a new bismuth telluride from British Columbia, with notes on wehrlite and some bismuth-tellurium alloys. Univ. Toronto, Geol. Ser., 49, 55-69.
1652 Warren, H.V. and Thompson, R.M., 1944. Minor elements in gold. Econ. Geol., 39, 457-471.
1653 Warren, H.V. and Thompson, R.M., 1945a. Antimony minerals from British Columbia and Yukon Territory: boulangerite, chalcostibite, jamesonite, zinckenite. Univ. Toronto, Geol. Ser., 49, 78-84.
1654 Warren, H.V. and Thompson, R.M., 1945b. Sphalerites from Western Canada. Econ. Geol., 40, 309-335.
1655 Warren, H.V. and Watson, K. DeP., 1937. A pyrrhotite ruby silver occurrence in British Columbia. Econ. Geol., 32, 826-831.
1656 Watanabe, T., 1936. Crystals of native tellurium from Japan. J. Fac. Sci., Hokkaidô Imp. Univ., Ser. IV, Vol. III, 101-111.
1657 Watanabe, T. and Kato, A., 1966. Ore microscopy and electronprobe microanalysis of some manganese minerals with vredenburgite-type intergrowth. Min. Soc. India, IMA Volume, International Mineralogical Association, Papers, 4th General Meeting, 197-202.

1658 Watanabe, T., Kato, A. and Ito, J., 1960. The minerals of the Noda-Tamagawa mine, Iwate Prefecture, Japan. II: Pyrochroite ore and its origin. Mineral J. (Tokyo), 3, 30-41.
1659 Watson, K. DeP., 1943. Colloform sulphide veins of Port au Port peninsula, Newfoundland. Econ. Geol., 38, 621-647.
1660 Watson, K.D., 1954. Paragenesis of the Pb-Zn-Cu deposits at the Mindamar mine, Nova Scotia. Econ. Geol., 49, 389-412.
1661 Watson, T.L., 1925. Hoegbomite from Virginia. Am. Mineralogist, 10, 1-9.
1662 Wayland, R.G., 1960. The Alaska Juneau gold ore body. Neues Jahrb. Mineral., Abhandl., 94, 267-279.
1663 Webb, R.W., 1935. Tetradymite from Inyo Mountains, California. Am. Mineralogist, 20, 399-400.
1664 Weeks, A.D., Cisney, E.A. and Sherwood, A.M., 1953. Montroseite, a new vanadium oxide from the Colorado Plateau. Am. Mineralogist, 38, 1235-1241.
1665 Weeks, A.D. and Truesdell, A.H., 1958. Mineralogy and geochemistry of the uranium deposits of the Grants District, New Mexico. Bull. Geol. Soc. Am., 69, 1658-1659.
1666 Weibel, M. and Köppel, V., 1962. Boulangerit und Jamesonit als alpine Kluftmineralien. Schweiz. Mineral. Petrog. Mitt., 42, 579-581.
1667 Weibel, M. and Köppel, V., 1963. Bleiglanz von der Grimsel und Zinkblende von Sedrun. Schweiz. Mineral. Petrog. Mitt., 43, 339-343.
1668 Weiser, T., 1967. Zink- und Vanadium-führende Chromite von Outokumpu/Finnland. Neues Jahrb. Mineral., Monatsh., 1967, 234-243.
1669 Weissberg, B.G., 1965. Getchellite, $AsSbS_3$, a new mineral from Humboldt County, Nevada. Am. Mineralogist, 50, 1817-1826.
1670 Welin, E., 1959. Till kännedomen om sulfidmalmerna och de zonerade zinkbländerna i Boda och Solleröns Kommuner, Kopparbergs Län. Geol. Fören. Stockholm. Förhandl., 81, 495-513.
1671 Welin, E., 1961. Uranium Mineralization in a skarn iron ore at Håkantorp, County of Örebro, Sweden. Geol. Fören Stockholm Förhandl., 83, 129-143.
1672 Welin, E., 1965. Two occurrences of uranium in Sweden. The Los cobalt deposit and the iron ores of the Västervik area. Geol. Fören. Stockholm Förhandl., 87, 492-508.
1673 Welin, E., 1966. Notes on the mineralogy of Sweden. 5. Bismuth-bearing sulphosalts from Gladhammar, a revision. Arkiv Mineral. Geol., 4, 377-386.
1674 Welin, E. and Uytenbogaardt, W., 1963. Notes on the mineralogy of Sweden. 3. A davidite-thorite paragenesis on the island of Björkö, south of Västervik, Sweden. Arkiv Mineral. Geol., 3, 277-292.
1675 Wells, R.C., 1949. Antimonial silver from Cobalt, Ontario. Am. Mineralogist, 34, 456-457.
1676 Wernick, J.H., 1960. Constitution of the $AgSbS_2$ - PbS, $AgBiS_2$ - PbS, and $AgBiS_2$-$AgBiSe_2$ systems. Am. Mineralogist, 45, 591-598.
1677 Wernicke, F., 1933. Magnetkiesparagenesen von sächsischen Lagerstätten. Tschermaks Mineral. Petrog. Mitt., 44, 463-469.
1678 Westerveld, J., 1941. Mineralisatie op de tineilanden (Billiton, Banka, Riouw-Lingga archipel). Jaarboek Mijnbouwk. Stud. Delft, 1938-1941, 187-233.
1679 Westerveld, J. and Uytenbogaardt, W., 1948. Mineragrafische notities betreffende het erts van de mijn Salida, S.W.K. Verhandel. Koninkl. Ned. Geol. Mijnbouwk. Genoot., Mijnbouwk. Ser., IV, 59-70.
1680 Westland, A.D. and Beamish, F.E., 1958. The chemical analysis of iridosmines and other platinum-metal minerals. Am. Mineralogist, 43, 503-516.
1681 Westland, A.D. and Beamish, F.E., 1961. Native palladium from Colombia. Can. Mineralogist, 6, 689-691.
1682 White, W.S. and Wright, J.C., 1966. Sulfide-mineral zoning in the basal Nonesuch Shale, Northern Michigan. Econ. Geol., 61, 1171-1190.
1683 Whittle, A.W.G., 1959. The nature of davidite. Econ. Geol., 54, 64-81.
1684 Wickman, F.E., 1951. The crystal structure of galenobismutite, $PbBi_2S_4$. Arkiv Mineral. Geol., 1, 219-225.
1685 Wickman, F.E., 1953. The crystal structure of aikinite, $CuPbBiS_3$. Arkiv Mineral. Geol., 1, 501-507.
1686 Wiedersich, H., Savage, J.W., Muir, A.H., Jr. and Swarthout, D.G., 1968. On the composition of delafossite. Mineral. Mag., 36, 643-650.

1687 Wilke, A., 1952. Die Erzgänge von St. Andreasberg im Rahmen des Mittelharz-Ganggebietes. Geol. Jahrb., Beih., 7, 228 pp.
1688 Wilkerson, A.S., 1939. Telluride-tungsten mineralization of the Magnolia mining district, Colorado. Econ. Geol., 34, 437-450.
1689 Wilkinson, J.F.G., 1957. Titanomagnetites from a differentiated teschenite sill. Mineral. Mag., 31, 443-454.
1690 Willard, M.E., 1941. Mineralization at the Polaris mine, Idaho. Econ. Geol., 36, 539-550.
1691 Williams, G., 1934. The genetic significance of some tin-tungsten lodes in Stewart Island, New Zealand. Econ. Geol., 29, 411-434.
1692 Williams, K.L., 1958a. Nickel mineralization in Western Tasmania. Australasian Inst. Mining Met. Proc., Stillwell Anniv. Volume, 263-302.
1693 Williams, K.L., 1958b. Tin-tungsten mineralization at Moina, Tasmania. Australasian Inst. Mining Met. Proc., 185, 29-50.
1694 Williams, K.L., 1960a. An association of awaruite with heazlewoodite. Am. Mineralogist, 45, 450-453.
1695 Williams, K.L., 1960b. Some less common minerals in the Roseberry and Hercules zinc-lead ores. Australasian Inst. Mining Met. Proc., 196, 51-59.
1696 Williams, K.L., 1967. Electron probe microanalysis of sphalerite. Am. Mineralogist, 52, 475-492.
1697 Williams, S.A., 1962. Paramelaconite and associated minerals from the Algomah mine, Ontonagon County, Michigan. Am. Mineralogist, 47, 778-779.
1698 Williams, S.A., 1965. A new occurrence of andorite. Am. Mineralogist, 50, 1498-1500.
1699 Williams, S.A., 1968a. Complex silver ores from Morey, Nevada. Can. Mineralogist, 9, 478-484.
1700 Williams, S.A., 1968b. More data on greigite. Am. Mineralogist, 53, 2087-2088.
1701 Wise, W.S., 1959. An occurrence of geikielite. Am. Mineralogist, 44, 879-882.
1702 Wittich, E. and Neumann, B., 1901. Ein neues Cadmium-Mineral. Zentr. Mineral. Geol. Palaeontol., 1901, 549-551.
1703 Woodhouse, C.D. and Norris, R.M., 1957. A new occurrence of millerite. Am. Mineralogist, 42, 113-115.
1704 Wretblad, P.E., 1941. Die Allemontite. Geol. Fören. Stockholm Förhandl., 63, 19-48.
1705 Wright, H.D., 1954. Mineralogy of a uraninite deposit at Caribou, Colorado. Econ. Geol., 49, 129-174.
1706 Wuensch, B.J. and Nowacki, W., 1963. Zur Kristallchemie des Sulfosalzes Marrit. Chimia (Aarau), 17, 381-382. Ref. in: Am. Mineralogist, 50, 812.
1707 Wuensch, B.J. and Nowacki, W., 1966. The substructure of the sulfosalt jordanite. Schweiz. Mineral. Petrog. Mitt., 46, 89-96.
1708 Wuensch, B.J., Takéuchi, Y. and Nowacki, W., 1965. Die Verfeinerung der Kristallstruktur von Binnit. Phys. Verhandl., 16, 215. Ref. in: Zentr. Mineral., Tl. I, 1965, 36.
1709 Wijkerslooth, P. de., 1955. Morphological and optical properties of bursaite ($Pb_5Bi_4S_{11}$). Unesco Symp. Appl. Geol. Near East, Ankara, 1955.
1710 Wijkerslooth, P. de, 1956. Erzmikroskopische Beobachtungen an einigen selteneren Erzmineralien der Türkei. Bull. Mineral Res. Exploration Inst. Turkey, Foreign Ed., 48, 112-117.
1711 Wijkerslooth, P. de, 1957. Ueber die primären Erzmineralien der Kupferlagerstätte von Ergani-Maden (Vilayet-Elazig), Türkei. Geol. Fören. Stockholm Förhandl., 79, 257-273.
1712 Yagoda, H., 1945. The localization of copper and silver sulphide minerals in polished section by the potassium cyanide etch pattern. Am. Mineralogist, 30, 51-64.
1713 Yagoda, H., 1946. The localization of uranium and thorium minerals in polished section. Part I: the alpha-ray emission pattern. Am. Mineralogist, 31, 87-124.
1714 Yoshimura, T., 1934. "Todorokite", a new manganese mineral from the Todoroki mine, Hokkaidô, Japan. J. Fac. Sci., Hokkaidô Univ., Ser. IV, 2, 289-297.
1715 Young, B.B. and Millman, A.P., 1964. Microhardness and deformation characteristics of ore minerals. Trans. Inst. Mining Met., 73, 437-466.
1716 Youngberg, E.A. and Wilson, T.L., 1952. The geology of the Holden mine. Econ. Geol., 47, 1-12.
1717 Ypma, P.J.M., 1968. Pyrite group: an unusual member: $Cu_{0.60}Ni_{0.14}Co_{0.03}Fe_{0.23}S_2$. Science, 159, p. 194.

1718 Ypma, P.J.M., Evers, H.J. and Woensdregt, C.F., 1968. Mineralogy and petrology of the Providencia mine (León, Spain), type-locality of villamaninite. Neues Jahrb. Mineral., Monatsh., 1968, 174-191.

1719 Yund, R.A., 1962. The system Ni-As-S: phase relations and mineralogical significance. Am. J. Sci., 260, 761-782.

1720 Yund, R.A., 1963. Crystal data for synthetic $Cu_{5,5x}Fe_xS_{6,5x}$ (idaite). Am. Mineralogist, 48, 672-676.

1721 Yund, R.A. and Kullerud, G., 1964. Stable mineral assemblages of anhydrous copper and iron oxides. Am. Mineralogist, 49, 689-696.

1722 Yund, R.A. and Kullerud, G., 1966. Thermal stability of assemblages in the Cu-Fe-S system. J. Petrol., 7, 454-488.

1723 Yushko, S.A., 1961. Sulvanit in Pb-Zn-Erzen des Karatau-Gebirges Südkasachstan (in Russian). Tr. Mineral. Muzea Akad. Nauk SSSR, 11, 215-219. Ref. in: Zentr. Mineral., I, 1962, 236.

1724 Yushko-Zakharova, O.Ye., 1964. Nickel telluride, a new mineral. Dokl. Akad. Nauk SSSR, 154, 613-614. Translated in: Dokl. Acad. Sci. USSR, Earth Sci. Sect., 154, 103-105 (1964).

1725 Yushko-Zakharova, O.E. and Chernayev, L.A., 1966a. Palladium bismuthide from Monchegorsk ores. Dokl. Akad. Nauk SSSR, 170, 183-185. Translated in: Dokl. Acad. Sci. USSR, Earth Sci. Sect., 170, 140-142 (1967).

1726 Yushko-Zakharova, O.E. and Chernayev, L.A., 1966b. The composition and properties of niggliite from copper-nickel ores of the Monchegorsk deposit. Dokl. Akad. Nauk SSSR, 170, 1164-1166. Translated in: Dokl. Acad. Sci. USSR, Earth Sci. Sect., 170, 148-150 (1967).

1727 Zábranský, F. and Radzo, V., 1966. Occurrence of kobellite and paragenetic conditions in the Oriešková-Vein north or Nižný Medzev. (in Czech, German and Engl. summary). Sb. Východoslovenského Músea Košiciach, Sér. A, VII A, 25-37.

1728 Zachariasen, W.H., 1933. X-ray examination of colusite, $(Cu,Fe,Mo,Sn)_4(S,As,Te)_{3-4}$. Am. Mineralogist, 18, 534-537.

1729 Zdorik, T.B., Sidorenko, G.A. and Bykova, A.V., 1961. A new titanozirconate of calcium - calzirtite. Dokl. Akad. Nauk SSSR, 137, 681-684. Translated in: Dokl. Acad. Sci. USSR, Earth Sci. Sect., 137, 443-446 (1962).

1730 Zemann, A. and Zemann, J., 1959. Zur Kenntnis der Kristallstruktur von Lorandit. Acta Cryst., 12, 1002-1006.

1731 Zhabin, A.G., Pudovkina, Z.V. and Bykova, A.V., 1962. Calzirtite from carbonatites of the Guly massif of hyperbasites and alkaline rocks in Northern Siberia. Dokl. Akad. Nauk SSSR, 146, 1339-1400. Translated in: Dokl. Acad. Sci. USSR, Earth Sci. Sect., 146, 140-141 (1964).

1732 Zimmer, E., 1935. Beiträge zur Kenntnis der Edlen Quarzformation Freibergs. Tschermaks Mineral. Petrog. Mitt., 47, 328-370.

1733 Zsivny, V. and Náray-Szabó, I. von, 1947. Parajamesonit, ein neues Mineral von Kisbánya. Schweiz. Mineral. Petrog. Mitt., 27, 183-189.

1734 Zsivny, V. and Zombory, L., 1934. Berthierite from Kisbánya, Carpathians. Mineral. Mag., 23, 566-568.

1735 Zubkov, L.B. and Galetskiy, L.S., 1966. Niobium- and tantalum-bearing cassiterite from bedrock in the northwestern Ukrainian shield. Dokl. Akad. Nauk SSSR, 169, 660-663. Translated in: Dokl. Acad. Sci. USSR, Earth Sci. Sect., 169, 135-138 (1967).

1736 Zuyev, V.N., 1959. On an unusual zonal structure in scheelite and wolframite. (in Russian). Tr. Mineral. Muzea Akad. Nauk SSSR, 9, 185-189. Ref. in: Mineral. Abstr., 14, 478.

1737 Zuyev, V.N., Zubkov, L.B., Zubynina, K.B., Utkina, T.F. and Chistov, L.B., 1966. New data on the modes of occurrence of tantalum and niobium in wolframite. Dokl. Akad. Nauk SSSR, 166, 194-197. Translated in: Dokl. Acad. Sci. USSR, Earth Sci. Sect., 166, 105-107 (1966).

1738 Zvyagintzev, O.E., 1934. Ein neues, Metalle der Platingruppe enthaltendes, Mineral. Dokl. Akad. Nauk SSSR, 4, 178-179. Ref. in: Mineral. Abstr., 6, 51.

1739 Zwicker, W.K., Groeneveld Meijer, W.O.J. and Jaffe, H.W., 1962. Nsutite, a widespread manganese oxide mineral. Am. Mineralogist, 47, 246-266.

INDEX[1]

Acanthite, 34
Achavalite, 216
Aguilarite, 218
Aikinite, 292
Akaganeite, 161
Alabandite, 124
-, Fe-, 125
Alaskaite,* see Pavonite, 282
Algodonite, 114
Allargentum, 82
Allemontite I*, 77, 85
- III*, 85
Allopalladium, 322
Altaite, 240
Anatase, 178
Andorite, 260
Anilite, 61
Annivite, 109
Antimonial silver, 80
Antimonite*, see Stibnite, 42
Antimonpearceite, 262
Antimony, 76
Aramayoite, 264
Argentite, 34
Argentopyrite, 268
Argyrodite, 256
Argyropyrite, 268
Arsenargentite, 85
Arsenic, 84
Arsenolamprite, 86
Arsenopalladinite, 325
Arsenopyrite, 190
Arsenpolybasite, 262
Arsensulvanite, 100
Auricuprid, 71
Aurorite, 344
Aurosmiridium, 329
Aurostibite, 72
Avicennite, 166
Awaruite, 128

Baumhauerite, 300
Benjaminite, 284
Berndtite, 308
Berryite, 284
Berthierite, 48
Berzelianite, 216
Betafite, 189
Betekhtinite, 94
Billingsleyite, 260
Birnessite, 352
Bismuth, 36
Bismuthinite, 66

Bismuth-sulphosalts, 28, 281
Bismutotantalite, 200
Bi_2Te_5, 251
Bixbyite, 352
Blaubleibender covellite, 54
Bohdanowiczyte*, 283
Bonchevite, 288
Bornhardtite, 220
Bornite, 88
-, X-*, 89
-, Orange*, 89
Boulangerite, 272
Bournonite, 68
Bracewellite, 213
Braggite, 330
Brannerite, 196
Braunite, 350
Bravoite, 132
Breithauptite, 148
Briartite, 90
Bursaite, 288

Cadmoselite, 216
Cafetite, 181
Calaverite, 240
Calzirtite, 188
Canfieldite, 256
Cannizzarite, 288
Carrollite, 146
Cassiterite, 208
Castaingite, 102
Cattierite, 134
Cementite*, see Cohenite, 142
Cesarolite, 354
Chalcocite, 58
-, lamellar*, 59
-, pinkish grey*, 59
-, sooty*, 59
-, steely*, 59
Chalcophanite, 344
Chalcopyrite, 92
-, cubic*, see Talnakhite, 94
Chalcostibite, 86
Chalcothallite, 46
Chaoite, 104
Chilenite, 75
Chloanthite*, see Skutterudite, 152
Chromite, 174
Chromrutile*, see Redledgeite, 181
Cinnabar, 78
Clausthalite, 220
Cobaltite, 192
Cobaltpentlandite, 136

---

[1] Ore mineral names marked with an asterisk are considered to be superfluous, being disapproved by the IMA-CNMMN; applied to insufficiently described material; or being synonyms.

Coffinite, 196
Cohenite, 142
Coloradoite, 234
Columbite, 202
Colusite, 108
Cooperite, 322
Copper, 116
Coronadite, 356
Corynite, 159
Cosalite, 286
Coulsonite, 172
Covellite, 54
-, blaubleibender, 54
Crednerite, 356
Crichtonite, 175
Crookesite, 218
Cryptomelane, 354
Csiklovaite, 244
Cubanite, 96
-, cubic, 96
Cuprite, 118
Cuprobismutite, 294
Custerite*, see Kësterite, 310
Cylindrite, 316

Dadsonite, 272
Danaite*, see Arsenopyrite, 190
Davidite, 174
Delafossite, 116
Derbylite, 179
Descloizite, 144
Diaphorite, 264
Dienerite, 151
Digenite, 60
Dimorphite, 41
Djalindite*, see Dzhalindite, 136
Djanlindite*, see Dzhalindite, 136
Djurleite, 60
Doloresite, 100
Domeykite, 114
Donathite, 172
Dufrenoysite, 302
Dysanalyte*, see Perovskite, 182
Dyscrasite, 84
Dzhalindite, 136
Dzhezkazganite*, 86

Electrum, 72
Emplectite, 294
Empressite, 244
Enargite, 110
-, mottled*, 111
-, grüner*, 111
Eskebornite, 222
Eskolaite, 212
Eucairite, 224

Famatinite*, see Stibioluzonite, 110
Feitknechtite, 341
Femolite*, see Molybdenite, 102
Ferberite, 186
Fergusonite, 202

Ferrite*, see Iron, 128
Ferroplatinum, 327
Ferrovonsenite*, see Vonsenite, 164
Ferroselite, 228
Fizelyite*, see Andorite, 260
Formanite, 202
Franckeite, 316
Franklinite, 342
Freboldite, 226
Freibergite, 106
Freieslebenite, 264
Freudenbergite, 182
Frieseite, 269
Frohbergite, 244
Froodite, 330
Fülöppite, 272

Gahnite, 210
Galaxite, 210
Galena, 64
Galenobismutite, 286
Galenobornite*, see Bornite, 88
Gallite, 112
Gaudefroyite, 338
Geikielite, 176
Geocronite, 274
Germanite, 106
Gersdorffite, 158
Getchellite, 42
Geversite, 326
Giessenite, 290
Gladite*, see Aikinite, 292
Glaucodot, 190
Goethite, 160
Gold, 70
Goldfieldite, 106
Graphite, 104
Gratonite, 298
Greenockite, 124
Greigite, 140
Grimaldiite, 213
Groutite, 340
Gruenlingite*, see Joseite, 236
Guadalcazarite, 78
Guanajuatite, 230
Gudmundite, 130
Guettardite, 274
Guyanaite, 213

Häggite, 101
Hammarite*, see Aikinite, 292
Hastite, 226
Hatchite, 299
Hauchecornite, 122
Hauerite, 162
Hausmannite, 346
Hawleyite, 125
Heazlewoodite, 122
Hedleyite, 246
Hematite, 198
Hercynite, 210
Herzenbergite, 318

Hessite, 242
Hetaerolite, 338
Heterogenite, 162
Heteromorphite, 274
Hexastannite*, 311
Hocartite, 308
Högbomite, 182
Hollandite, 356
Hollingworthite, 322
Horobetsuite, 52
Horsfordite, 115
Huebnerite, 186
Huntilite, 75
Hutchinsonite, 298
Hydrohausmannite*, see Feitknechtite, 341
Hydrohetaerolite, 340
Hydroxidepearlite*, 129

Idaite, 56
Ikunolite, 230
Ilmenite, 176
-, white*, 199
Ilmenorutile*, see Nb-rutile, 180
Ilvaite, 198
Imgreite*, see Melonite, 252
Imhofite, 44
Indite, 136
Indium, 104
Irarsite, 322
Iridium, 328
Iridosmium, 334
Irinite*, see Perovskite, 182
Iron, 128
Isostannite*, see Kësterite, 310
Ixiolite, 200

Jacobsite, 348
Jaipurite, 139
Jalindite*, see Dzhalindite, 136
Jalpaite, 38
Jamesonite, 48
Jordanite, 302
Jordisite, 103
Joseite A, 236
- B, 236

Kallilite, 159
Karelianite, 212
Kennedyite, 179
Kermesite, 42
Kesterite*, see Kësterite, 310
Kësterite, 310
Kitkaite, 230, 248
Klaprothite*, 294
Klockmannite, 222
Knopite*, see Perovskite, 182
Kobellite, 290
Kongsbergite, 75
Koppite, 189
Kösterite*, see Kësterite, 310
Kostovite, 250
Kotulskite, 332

Koutekite, 112
Krennerite, 252
Kullerudite, 226
Kusterite*, see Kësterite, 310

Laitakarite, 228
Landauite, 179
Langisite, 148
Latrappite, 183
Launayite, 274
Laurite, 322
Lautite, 68
Lazarevicite*, see Arsensulvanite, 100
Lead, 36
Lead-antimony-sulphosalts, 27, 271
Lead-arsenic-sulphosalts, 29, 297
Lengenbachite, 300
Lepidocrocite, 160
Lillianite, 286
Lindströmite*, see Aikinite, 292
Linnaeite, 146
Lithargite, 50
Lithiophorite, 344
Liveingite, 300
Livingstonite, 80
Loellingite, 156
Loparite, 183
Lorandite, 44
Ludwigite, 164
Lueshite, 183
Luzonite, 110

Mackinawite, 140
Madocite, 274
Maghemite, 170
Magnesiochromite, 175
Magnesiocolumbite, 203
Magnesioferrite, 170
Magnesioludwigite*, see Ludwigite, 164
Magnesioniobite*, see Magnesiocolumbite, 203
Magnetite, 168
Magnetoplumbite, 184
Mäkinenite, 226
Maldonite, 72
Manganite, 342
Manganocolumbite, 202
Manganoniobite*, see Manganocolumbite, 202
Manganosite, 338
Manganotantalite, 202
Manjiroite, 355
Marcasite, 204
Marokite, 340
Marrite, 298
Massicotite, 50
Matildite*, see Schapbachite, 282
Maucherite, 148
Mawsonite, 90
Mcconnellite, 213
Mckinstryite, 44
Melaconite*, see Tenorite, 118
Melanostibite, 177
Melnicovite-pyrite*, 207

Melonite, 252
Meneghinite, 50
Merenskyite, 332
Merumite*, see Eskolaite, 212
Metacinnabarite, 78
Metastibnite, 44
Miargyrite, 262
Michenerite, 324
Microlite, 189
-, plumbian, 189
Millerite, 122
Minium, 50
Modderite, 151
Mohsite, 175
Molybdenite, 102
Moncheite, 332
Montbrayite, 238
Monteponite, 125
Montesite*, 319
Montroseite, 100
Moschellandsbergite, 75
Mossite*, 202
Murdochite, 166

Nagyagite, 236
Nakaseite*, see Andorite, 260
Naumannite, 224
Nb-rutile, 180
Neodigenite*, see Digenite, 60
Nevyanskite*, see Iridosmium, 334
Neyite, 284
Niccolite, 150
Nickel, 122
Nigerite, 210
Niggliite, 330
Niobite*, see Columbite, 202
Nioboloparite*, see Perovskite, 182
Nolanite, 192
Novakite, 112
Nowackiite, 69
Nsutite, 358
Nuffieldite, 292

Obruchevite, 189
Olovotantalite*, see Tin-tantalite, 200
Onofrite, 76
Orcelite, 148
Oregonite, 142
Orpiment, 40
Oruetite*, see Joseite, 236
Osmiridium, 328
Osmium, 334
Ottemannite, 316
Owyheeite, 266
Oxidepearlite*, 129
Oxidic manganese minerals, 31, 337

Palladium, 326
Palladium bismuthide, 330
Pandaite, 189
Paraguanajuatite, 228
Parajamesonite, 49

Paramelaconite*, see Paratenorite, 118
Paramontroseite, 101
Pararammelsbergite, 154
Paratenorite, 118
Parkerite, 62
Partridgeite, 353
Patronite, 38
Pavonite, 282
Paxite, 112
Pearceite, 262
Pearlite*, 129
Penroseite, 218
Pentlandite, 130
Perovskite, 182
Petzite, 234
Pitchblende, 194
Plagionite, 276
Platiniridium, 327
Platinoid minerals, 30, 321
Platinum, 326
Plattnerite, 162
Platynite, 228
Playfairite, 276
Plumboferrite, 184
Polybasite, 262
Polydymite, 146
Polyxen, 327
Porpezite, 71
Potarite, 324
Priderite, 181
Proustite, 258
Pseudobrookite, 178
Psilomelane, 354
Pyrargyrite, 258
Pyrite, 206
-, gel-*, 207
-, melnicovite-*, 207
Pyrochlore, 188
Pyrochroite, 340
Pyrolusite, 358
Pyrophanite, 177
Pyrostilpnite, 258
Pyrrhotite, 138

Quenselite, 348

Raguinite, 46
Ramdohrite*, see Andorite, 260
Rammelsbergite, 154
Ramsdellite, 352
Rancieite, 338
Rathite-I, 300
-II*, see Liveingite, 300
Realgar, 40
Redledgeite, 181
Renierite, 90
Rezbanyite*, see Aikinite, 292
Rhodite, 71
Rhodostannite, 314
Rhombomagnojacobsite*, see Hausmannite, 346
Rickardite, 242
Robinsonite, 276

Roquesite, 98
Roseite*, see Iridosmium, 334
Rutheniridosmium, 335
Rutile, 180
-, chromrutile*, see Redledgeite, 181
-, Nb-, 180
-, Ta-, 180
Rijkeboerite, 189

Safflorite, 156
Sakuraiite, 308
Samsonite, 260
Sanmartinite, 187
Sartorite, 302
Saukovite*, see Metacinnabarite, 78
Schapbachite, 282
Scheelite, 184
Schirmerite, 282
Schreibersite, 142
Schwazite, 106
Sederholmite, 226
Selenides, 22, 215
Selenidspinell*, see Tyrrellite, 220
Selenium, 224
Selenjoseite*, see Laitakarite, 228
Selentellurium, 247
Seligmannite, 304
Semseyite, 276
Senaite, 175
Shandite, 62
Siegenite, 146
Silver, 74
-, antimonial, 80
Silver-sulphosalts and Ag-Fe-sulphides, 26, 255
Sinnerite, 68
Skutterudite, 152
Smaltite*, see Skutterudite, 152
Smithite, 266
Smythite, 140
Sorbyite, 276
Sperrylite, 324
Sphalerite, 126
Spinel, 210
Stainierite*, see Heterogenite, 162
Stannite, 312
- (?) I*, 311
- (?) II*, 311
- (?) III*, 314
- (?) IV*, 314
- jaune, 31'
Stannoenargite, 111
Stannoidite, 310
Stannoluzonite*, see Luzonite, 110
Stannopalladinite, 325
Stannotantalite*, see Tin-tantalite, 200
Staringite, 208
Stephanite, 256
Sternbergite, 268
Sterryite, 278
Stibarsenic, 84
Stibiocolumbite, 201
Stibiodufrenoysite*, see Veenite, 278

Stibioenargite, 110
Stibioluzonite, 110
Stibioniobite*, see Stibiocolumbite, 201
Stibiopalladinite, 324
Stibiotantalite, 200
Stibnite, 42
Stilleite, 216
Stromeyerite, 52
Strueverite*, see Ta-rutile, 180
Stuetzite, 242
Sukulaite, 208
Sulphur, 40
Sulvanite, 100
Sylvanite, 248
Sysertskite*, see Iridosmium, 334

Talnakhite, 94
Tantalite, 202
Tantalobetafite, 189
Tantalo-obruchevite, 189
Tapiolite, 202
Ta-rutile, 180
Teallite, 318
Tellurides, 24, 233
Tellurium, 246
Tellurobismuthite, 250
Tennantite, 108
Tenorite, 118
Teremkovite*, see Owyheeite, 266
Tetradymite, 248
Tetrahedrite, 108
Thoreaulite, 200
Thorianite, 196
Tiemannite, 216
Tin, 34
Tin-sulphosalts and tin-sulphides, 29, 307
Tin-tantalite, 200
Tintinaite, 290
Titanhematite*, see Hematite, 198
Titanobetafite, 189
Titanomaghemite, 171
Titanomagnetite, 168
Titano-obruchevite, 189
Titanopyrochlore, 189
Todorokite, 350
Trechmannite, 266
Trevorite, 172
Trogtalite, 218
Troilite, 138
Trüstedtite, 220
Tungstenite, 102
Twinnite, 278
Tyrrellite, 220

Uhligite*, see Perovskite, 182
Ullmannite, 158
Ulvite, 170
Ulvöspinel, 170
Umangite, 222
Uraninite, 194
Uranothorite, 197
Ustarasite, 290

Vaesite, 134
Valleriite, 98
Veenite, 278
Villamaninite, 144
Violarite, 132
Voltzite*, see Wurtzite, 126
Volynskite, 238
Vonsenite, 164
Vrbaite, 46
Vredenburgite
-, α-, 346
-, β-, 347, 349
-, zincian*, 347
Vulcanite, 244
Vysotskite, 330

Wairauite, 128
Wallisite, 299
Wehrlite, 238
Weissite, 234
Westgrenite, 189

Whitneyite, 114
Wilkmanite, 224
Willyamite, 159
Wittichenite, 292
Wodginite, 200
Wolframite, 186
Woodruffite, 352
Wuestite, 166
Wulfenite, 144
Wurtzite, 126

Xanthoconite, 258
X-bornite*, see Bornite, 88

Zinc, 74
Zincite, 162
Zinckenite*, see Zinkenite, 278
Zinkenite, 278
Zirconolite, 189
Zvyagintsevite, 326

**A CATALOGUE OF SELECTED DOVER BOOKS
IN ALL FIELDS OF INTEREST**

A CATALOG OF SELECTED DOVER BOOKS IN ALL FIELDS OF INTEREST

LASERS AND HOLOGRAPHY, Winston E. Kock. Sound introduction to burgeoning field, expanded (1981) for second edition. 84 illustrations. 160pp. 5⅜ × 8¼. (EUK) 24041-X Pa. $3.50

FLORAL STAINED GLASS PATTERN BOOK, Ed Sibbett, Jr. 96 exquisite floral patterns—irises, poppie, lilies, tulips, geometrics, abstracts, etc.—adaptable to innumerable stained glass projects. 64pp. 8¼ × 11. 24259-5 Pa. $3.50

THE HISTORY OF THE LEWIS AND CLARK EXPEDITION, Meriwether Lewis and William Clark. Edited by Eliott Coues. Great classic edition of Lewis and Clark's day-by-day journals. Complete 1893 edition, edited by Eliott Coues from Biddle's authorized 1814 history. 1508pp. 5⅜ × 8½.
21268-8, 21269-6, 21270-X Pa. Three-vol. set $22.50

ORLEY FARM, Anthony Trollope. Three-dimensional tale of great criminal case. Original Millais illustrations illuminate marvelous panorama of Victorian society. Plot was author's favorite. 736pp. 5⅜ × 8½. 24181-5 Pa. $8.95

THE CLAVERINGS, Anthony Trollope. Major novel, chronicling aspects of British Victorian society, personalities. 16 plates by M. Edwards; first reprint of full text. 412pp. 5⅜ × 8½. 23464-9 Pa. $6.00

EINSTEIN'S THEORY OF RELATIVITY, Max Born. Finest semi-technical account; much explanation of ideas and math not readily available elsewhere on this level. 376pp. 5⅜ × 8½. 60769-0 Pa. $5.00

COMPUTABILITY AND UNSOLVABILITY, Martin Davis. Classic graduate-level introduction th theory of computability, usually referred to as theory of recurrent functions. New preface and appendix. 288pp. 5⅜ × 8½. 61471-9 Pa. $6.50

THE GODS OF THE EGYPTIANS, E.A. Wallis Budge. Never excelled for richness, fullness: all gods, goddesses, demons, mythical figures of Ancient Egypt; their legends, rites, incarnations, etc. Over 225 illustrations, plus 6 color plates. 988pp. 6⅛ × 9¼. (EBE) 22055-9, 22056-7 Pa., Two-vol. set $20.00

THE I CHING (THE BOOK OF CHANGES), translated by James Legge. Most penetrating divination manual ever prepared. Indispensable to study of early Oriental civilizations, to modern inquiring reader. 448pp. 5⅜ × 8½.
21062-6 Pa. $6.50

THE CRAFTSMAN'S HANDBOOK, Cennino Cennini. 15th-century handbook, school of Giotto, explains applying gold, silver leaf; gesso; fresco painting, grinding pigments, etc. 142pp. 6⅛ × 9¼. 20054-X Pa. $3.50

AN ATLAS OF ANATOMY FOR ARTISTS, Fritz Schider. Finest text, working book. Full text, plus anatomical illustrations; plates by great artists showing anatomy. 593 illustrations. 192pp. 7⅞ × 10¾. 20241-0 Pa. $6.00

EASY-TO-MAKE STAINED GLASS LIGHTCATCHERS, Ed Sibbett, Jr. 67 designs for most enjoyable ornaments: fruits, birds, teddy bears, trumpet, etc. Full size templates. 64pp. 8¼ × 11. 24081-9 Pa. $3.95

TRIAD OPTICAL ILLUSIONS AND HOW TO DESIGN THEM, Harry Turner. Triad explained in 32 pages of text, with 32 pages of Escher-like patterns on coloring stock. 92 figures. 32 plates. 64pp. 8¼ × 11. 23549-1 Pa. $2.50

CATALOG OF DOVER BOOKS

SOURCE BOOK OF MEDICAL HISTORY, edited by Logan Clendening, M.D. Original accounts ranging from Ancient Egypt and Greece to discovery of X-rays: Galen, Pasteur, Lavoisier, Harvey, Parkinson, others. 685pp. 5⅜ × 8½.
20621-1 Pa. $10.95

THE ROSE AND THE KEY, J.S. Lefanu. Superb mystery novel from Irish master. Dark doings among an ancient and aristocratic English family. Well-drawn characters; capital suspense. Introduction by N. Donaldson. 448pp. 5⅜ × 8½.
24377-X Pa. $6.95

SOUTH WIND, Norman Douglas. Witty, elegant novel of ideas set on languorous Mediterranean island of Nepenthe. Elegant prose, glittering epigrams, mordant satire. 1917 masterpiece. 416pp. 5⅜ × 8½. (Available in U.S. only)
24361-3 Pa. $5.95

RUSSELL'S CIVIL WAR PHOTOGRAPHS, Capt. A.J. Russell. 116 rare Civil War Photos: Bull Run, Virginia campaigns, bridges, railroads, Richmond, Lincoln's funeral car. Many never seen before. Captions. 128pp. 9⅜ × 12¼.
24283-8 Pa. $6.95

PHOTOGRAPHS BY MAN RAY: 105 Works, 1920-1934. Nudes, still lifes, landscapes, women's faces, celebrity portraits (Dali, Matisse, Picasso, others), rayographs. Reprinted from rare gravure edition. 128pp. 9⅜ × 12¼. (Available in U.S. only)
23842-3 Pa. $6.95

STAR NAMES: THEIR LORE AND MEANING, Richard H. Allen. Star names, the zodiac, constellations: folklore and literature associated with heavens. The basic book of its field, fascinating reading. 563pp. 5⅜ × 8½.
21079-0 Pa. $7.95

BURNHAM'S CELESTIAL HANDBOOK, Robert Burnham, Jr. Thorough guide to the stars beyond our solar system. Exhaustive treatment. Alphabetical by constellation: Andromeda to Cetus in Vol. 1; Chamaeleon to Orion in Vol. 2; and Pavo to Vulpecula in Vol. 3. Hundreds of illustrations. Index in Vol. 3. 2000pp. 6⅛ × 9¼.
23567-X, 23568-8, 23673-0 Pa. Three-vol. set $32.85

THE ART NOUVEAU STYLE BOOK OF ALPHONSE MUCHA, Alphonse Mucha. All 72 plates from *Documents Decoratifs* in original color. Stunning, essential work of Art Nouveau. 80pp. 9⅜ × 12¼.
24044-4 Pa. $7.95

DESIGNS BY ERTE; FASHION DRAWINGS AND ILLUSTRATIONS FROM "HARPER'S BAZAR," Erte. 310 fabulous line drawings and 14 *Harper's Bazar* covers, 8 in full color. Erte's exotic temptresses with tassels, fur muffs, long trains, coifs, more. 129pp. 9⅜ × 12¼.
23397-9 Pa. $6.95

HISTORY OF STRENGTH OF MATERIALS, Stephen P. Timoshenko. Excellent historical survey of the strength of materials with many references to the theories of elasticity and structure. 245 figures. 452pp. 5⅜ × 8½. 61187-6 Pa. $8.95

Prices subject to change without notice.
Available at your book dealer or write for free catalog to Dept. GI, Dover Publications, Inc., 31 East 2nd St. Mineola, N.Y. 11501. Dover publishes more than 175 books each year on science, elementary and advanced mathematics, biology, music, art, literary history, social sciences and other areas.

Kirriley Library
Columbia College
Columbia, Missouri 65216

For reference

Not To Be Taken From the Room